The Multiverse

" The Theories of Multiple Universes "

Edited by Paul F. Kisak

Contents

1 Multiverse — 1
   1.1 Explanation — 1
   1.2 Multiverse hypotheses in physics — 1
      1.2.1 Categories — 1
      1.2.2 Cyclic theories — 3
      1.2.3 M-theory — 3
      1.2.4 Black-hole cosmology — 3
      1.2.5 Anthropic principle — 4
      1.2.6 Search for evidence — 4
      1.2.7 Criticism — 4
   1.3 Multiverse hypotheses in philosophy and logic — 6
      1.3.1 Modal realism — 6
      1.3.2 Trans-world identity — 6
   1.4 See also — 6
   1.5 References — 6
      1.5.1 Notes — 6
      1.5.2 Bibliography — 8
   1.6 External links — 9

2 Universe — 10
   2.1 Definition — 10
   2.2 Etymology — 10
      2.2.1 Synonyms — 11
   2.3 Chronology and the Big Bang — 11
   2.4 Properties — 11
      2.4.1 Shape — 12
      2.4.2 Size and regions — 12
      2.4.3 Age and expansion — 13
      2.4.4 Spacetime — 13
   2.5 Contents — 14

		2.5.1	Dark energy	15
		2.5.2	Dark matter	15
		2.5.3	Ordinary Matter	15
		2.5.4	Particles	16
	2.6	Cosmological models		17
		2.6.1	Model of the Universe based on general relativity	17
		2.6.2	Multiverse hypothesis	18
		2.6.3	Fine-tuned Universe	19
	2.7	Historical development		19
		2.7.1	Mythologies	19
		2.7.2	Philosophical models	20
		2.7.3	Astronomical concepts	20
	2.8	See also		22
	2.9	References		23
3	**Many-worlds interpretation**			**29**
	3.1	Outline		29
	3.2	Interpreting wavefunction collapse		30
	3.3	Probability		31
		3.3.1	Everett, Gleason and Hartle	31
		3.3.2	DeWitt and Graham	31
		3.3.3	Deutsch *et al.*	31
	3.4	Brief overview		32
	3.5	Relative state		32
	3.6	Properties of the theory		33
	3.7	Comparative properties and possible experimental tests		34
		3.7.1	Copenhagen interpretation	34
		3.7.2	The universe decaying to a new vacuum state	34
		3.7.3	Many-minds	34
	3.8	Common objections		34
	3.9	Reception		36
		3.9.1	Polls	37
	3.10	Speculative implications		38
		3.10.1	Quantum suicide thought experiment	38
		3.10.2	Weak coupling	38
		3.10.3	Similarity to modal realism	39
		3.10.4	Time travel	39
	3.11	Many-worlds in literature and science fiction		39
	3.12	See also		39

3.13	Notes	40
3.14	Further reading	43
3.15	External links	44

4 Eternal inflation

4.1	Inflation and the multiverse	45
4.2	History	45
4.3	Quantum fluctuations of the inflation field	46
4.4	Differential decay	46
4.5	False vacuum and true vacuum	46
	4.5.1 Evidence from the fluctuation level in our universe	46
4.6	See also	47
4.7	References	47
4.8	External links	48

5 Cosmological principle

5.1	Origin	49
5.2	Implications	50
5.3	Justification	50
5.4	Criticism	50
5.5	See also	51
5.6	References	51

6 Inflation (cosmology)

6.1	Overview	52
	6.1.1 Space expands	53
	6.1.2 Few inhomogeneities remain	53
	6.1.3 Duration	54
	6.1.4 Reheating	54
6.2	Motivations	54
	6.2.1 Horizon problem	54
	6.2.2 Flatness problem	54
	6.2.3 Magnetic-monopole problem	55
6.3	History	55
	6.3.1 Precursors	55
	6.3.2 Early inflationary models	56
	6.3.3 Slow-roll inflation	56
	6.3.4 Effects of asymmetries	56
6.4	Observational status	56

- 6.5 Theoretical status ... 57
  - 6.5.1 Fine-tuning problem ... 58
  - 6.5.2 Eternal inflation ... 58
  - 6.5.3 Initial conditions ... 59
  - 6.5.4 Hybrid inflation ... 59
  - 6.5.5 Inflation and string cosmology ... 59
  - 6.5.6 Inflation and loop quantum gravity ... 60
- 6.6 Alternatives ... 60
  - 6.6.1 Big bounce ... 60
  - 6.6.2 String theory ... 60
  - 6.6.3 Ekpyrotic and cyclic models ... 60
  - 6.6.4 Varying C ... 60
- 6.7 Criticisms ... 61
- 6.8 See also ... 61
- 6.9 Notes ... 61
- 6.10 References ... 66
- 6.11 External links ... 67

7 Spontaneous symmetry breaking — 68

- 7.1 Spontaneous symmetry breaking in physics ... 68
  - 7.1.1 Particle physics ... 68
  - 7.1.2 Condensed matter physics ... 69
  - 7.1.3 Dynamical symmetry breaking ... 70
- 7.2 Generalisation and technical usage ... 70
- 7.3 A pedagogical example: the Mexican hat potential ... 70
- 7.4 Other examples ... 71
- 7.5 Nobel Prize ... 72
- 7.6 See also ... 72
- 7.7 Notes ... 72
- 7.8 References ... 72
- 7.9 External links ... 73

8 Cyclic model — 74

- 8.1 Overview ... 74
- 8.2 The Steinhardt–Turok model ... 74
- 8.3 The Baum–Frampton model ... 75
- 8.4 Other cyclic models ... 75
- 8.5 See also ... 75
- 8.6 Notes ... 75

8.7	Further reading	76
8.8	External links	76

9 Lee Smolin — 77

- 9.1 Early life — 77
- 9.2 Education and career — 77
- 9.3 Theories and work — 77
  - 9.3.1 Loop quantum gravity — 77
  - 9.3.2 Background independent approaches to string theory — 77
  - 9.3.3 Experimental tests of quantum gravity — 77
  - 9.3.4 Foundations of quantum mechanics — 78
  - 9.3.5 Cosmological natural selection — 78
  - 9.3.6 Contributions to philosophy of physics — 79
- 9.4 The Trouble with Physics — 79
- 9.5 Views — 79
- 9.6 Publications — 80
- 9.7 Awards and honors — 80
- 9.8 Personal life — 80
- 9.9 References — 80
- 9.10 External links — 81

10 Multiple histories — 82

- 10.1 See also — 82

11 Many-minds interpretation — 83

- 11.1 The central problems — 83
  - 11.1.1 The many-worlds interpretation — 83
  - 11.1.2 Continuous infinity of minds — 84
- 11.2 Objections — 84
- 11.3 See also — 85
- 11.4 External links — 85

12 Mathematical universe hypothesis — 86

- 12.1 Description — 86
- 12.2 Criticisms and responses — 86
  - 12.2.1 Definition of the Ensemble — 86
  - 12.2.2 Consistency with Gödel's theorem — 86
  - 12.2.3 Observability — 87
  - 12.2.4 Plausibility of Radical Platonism — 87
  - 12.2.5 Coexistence of all mathematical structures — 87

- 12.2.6 Consistency with our "simple universe" 88
- 12.2.7 Occam's razor 88
- 12.3 Major books 88
  - 12.3.1 *Our Mathematical Universe* 88
- 12.4 See also 88
- 12.5 References 88
- 12.6 Further reading 89
- 12.7 External links 89

13 Brane cosmology — 90

- 13.1 Brane and bulk 90
- 13.2 Why gravity is weak and the cosmological constant is small 90
- 13.3 Models of brane cosmology 90
- 13.4 Empirical tests 91
- 13.5 See also 91
- 13.6 References 91
- 13.7 External links 91

14 Brane — 92

- 14.1 D-branes 92
- 14.2 Mathematical viewpoint 93
- 14.3 See also 93
- 14.4 Notes 93
- 14.5 References 94

15 D-brane — 95

- 15.1 Theoretical background 95
- 15.2 Braneworld cosmology 95
- 15.3 D-brane scattering 96
- 15.4 Gauge theories 96
- 15.5 Black holes 97
- 15.6 History 97
- 15.7 See also 98
- 15.8 References 98

16 M-theory — 99

- 16.1 Background 99
  - 16.1.1 Quantum gravity and strings 99
  - 16.1.2 Number of dimensions 100
  - 16.1.3 Dualities 100

- 16.1.4 Supersymmetry 101
- 16.1.5 Branes 101
- 16.2 History and development 102
  - 16.2.1 Kaluza–Klein theory 102
  - 16.2.2 Early work on supergravity 102
  - 16.2.3 Relationships between string theories 103
  - 16.2.4 Membranes and fivebranes 103
  - 16.2.5 Second superstring revolution 103
  - 16.2.6 Origin of the term 104
- 16.3 Matrix theory 104
  - 16.3.1 BFSS matrix model 104
  - 16.3.2 Noncommutative geometry 104
- 16.4 AdS/CFT correspondence 105
  - 16.4.1 Overview 105
  - 16.4.2 6D (2,0) superconformal field theory 106
  - 16.4.3 ABJM superconformal field theory 106
- 16.5 Phenomenology 107
  - 16.5.1 Overview 107
  - 16.5.2 Compactification on G_2 manifolds 107
  - 16.5.3 Heterotic M-theory 107
- 16.6 References 108
  - 16.6.1 Notes 108
  - 16.6.2 Citations 108
  - 16.6.3 Bibliography 109
- 16.7 External links 111

17 Ekpyrotic universe — 112
- 17.1 See also 113
- 17.2 Notes and references 113
- 17.3 Further reading 113

18 String theory landscape — 114
- 18.1 Anthropic principle 114
- 18.2 Bayesian probability 114
- 18.3 Simplified approaches 115
- 18.4 Criticism 115
- 18.5 See also 115
- 18.6 References 115
- 18.7 External links 116

19 Holographic principle — 117
- 19.1 Black hole entropy . . . 117
- 19.2 Black hole information paradox . . . 118
- 19.3 Limit on information density . . . 119
- 19.4 High-level summary . . . 119
    - 19.4.1 Unexpected connection . . . 119
    - 19.4.2 Energy, matter, and information equivalence . . . 119
- 19.5 Experimental tests . . . 120
- 19.6 Tests of Maldacena's conjecture . . . 120
- 19.7 See also . . . 120
- 19.8 Notes . . . 120
- 19.9 References . . . 120
- 19.10 External links . . . 121

20 Simulated reality — 122
- 20.1 Types of simulation . . . 122
    - 20.1.1 Brain-computer interface . . . 122
    - 20.1.2 Virtual people . . . 122
- 20.2 Arguments . . . 122
    - 20.2.1 Simulation argument . . . 122
    - 20.2.2 Relativity of reality . . . 123
    - 20.2.3 Computationalism . . . 123
    - 20.2.4 Dreaming . . . 124
    - 20.2.5 Computability of physics . . . 124
    - 20.2.6 CantGoTu environments . . . 125
    - 20.2.7 Nested simulations . . . 125
    - 20.2.8 Peer-to-Peer Explanation of Quantum Phenomena . . . 125
- 20.3 In fiction . . . 126
- 20.4 See also . . . 126
    - 20.4.1 Major contributing thinkers . . . 126
- 20.5 Bibliography . . . 127
- 20.6 References . . . 127
- 20.7 External links . . . 128

21 Black-hole cosmology — 130
- 21.1 References . . . 130

22 Anthropic principle — 131
- 22.1 Definition and basis . . . 131

- 22.2 Anthropic coincidences ... 131
- 22.3 Origin ... 132
- 22.4 Variants ... 133
- 22.5 Character of anthropic reasoning ... 134
- 22.6 Observational evidence ... 135
- 22.7 Applications of the principle ... 136
  - 22.7.1 The nucleosynthesis of carbon-12 ... 136
  - 22.7.2 Cosmic inflation ... 136
  - 22.7.3 String theory ... 136
  - 22.7.4 Spacetime ... 137
- 22.8 *The Anthropic Cosmological Principle* ... 137
- 22.9 Criticisms ... 138
- 22.10 See also ... 139
- 22.11 Footnotes ... 139
- 22.12 References ... 141
- 22.13 External links ... 142

23 Wilkinson Microwave Anisotropy Probe — 143
- 23.1 Objectives ... 144
- 23.2 Development ... 144
- 23.3 Spacecraft ... 144
- 23.4 Launch, trajectory, and orbit ... 145
- 23.5 Foreground radiation subtraction ... 146
- 23.6 Measurements and discoveries ... 146
  - 23.6.1 One-year data release ... 146
  - 23.6.2 Three-year data release ... 147
  - 23.6.3 Five-year data release ... 147
  - 23.6.4 Seven-year data release ... 148
  - 23.6.5 Nine-year data release ... 148
- 23.7 Main result ... 148
- 23.8 Follow-on missions and future measurements ... 149
- 23.9 See also ... 149
- 23.10 Further reading ... 149
- 23.11 References ... 149
  - 23.11.1 Footnotes ... 149
  - 23.11.2 Primary sources ... 150
- 23.12 External links ... 151

24 Kolmogorov complexity — 152

- 24.1 Definition 152
- 24.2 Invariance theorem 152
  - 24.2.1 Informal treatment 153
  - 24.2.2 A more formal treatment 153
- 24.3 History and context 153
- 24.4 Basic results 154
  - 24.4.1 Uncomputability of Kolmogorov complexity 154
  - 24.4.2 Chain rule for Kolmogorov complexity 154
- 24.5 Compression 154
- 24.6 Chaitin's incompleteness theorem 155
- 24.7 Minimum message length 156
- 24.8 Kolmogorov randomness 156
- 24.9 Relation to entropy 156
- 24.10 Conditional versions 156
- 24.11 See also 157
- 24.12 Notes 157
- 24.13 References 157
- 24.14 External links 158

25 Possible world — 159

- 25.1 Possibility, necessity, and contingency 159
- 25.2 Formal semantics of modal logics 160
- 25.3 From modal logic to philosophical tool 160
- 25.4 Possible-world theory in literary studies 160
- 25.5 See also 161
- 25.6 Notes 161
- 25.7 Further reading 161
- 25.8 External links 161

26 Modal realism — 162

- 26.1 The term *possible world* 162
- 26.2 Main tenets of modal realism 162
- 26.3 Reasons given by Lewis 162
- 26.4 Details and alternatives 163
- 26.5 Criticisms 163
  - 26.5.1 Lewis's own critique 163
  - 26.5.2 Stalnaker's response 164
  - 26.5.3 Kripke's response 164
- 26.6 See also 164

- 26.7 References ... 164
- 26.8 Bibliography ... 164

27 Counterpart theory — 166

- 27.1 Differences from the Kripkean View ... 166
  - 27.1.1 The basics ... 166
- 27.2 The counterpart relation ... 166
  - 27.2.1 Parthood relation ... 167
- 27.3 The formal theory ... 167
  - 27.3.1 Comments on the axioms ... 167
  - 27.3.2 Principles that are not accepted in normal CT ... 167
- 27.4 Motivations for Counterpart theory ... 168
  - 27.4.1 In possible worlds ... 168
  - 27.4.2 Temporal parts ... 169
- 27.5 Counterpart theory and the necessity of identity ... 169
- 27.6 Counterpart theory and rigid designators ... 170
  - 27.6.1 Arguments for inconstancy ... 170
  - 27.6.2 Counterpart theory compared to qua-theory and individual concepts ... 170
  - 27.6.3 Counterpart theory and epistemic possibility ... 170
- 27.7 Arguments against Counterpart theory ... 171
- 27.8 See also ... 171
- 27.9 References ... 172
- 27.10 External links ... 172

28 The Fabric of Reality — 173

- 28.1 Overview ... 173
  - 28.1.1 The four strands ... 173
  - 28.1.2 Deutsch's TOE ... 173
- 28.2 Reception ... 173
- 28.3 See also ... 173
- 28.4 References ... 173
- 28.5 Text and image sources, contributors, and licenses ... 175
  - 28.5.1 Text ... 175
  - 28.5.2 Images ... 183
  - 28.5.3 Content license ... 187

Chapter 1

Multiverse

For other uses, see Multiverse (disambiguation).
See also: Many-worlds interpretation

The **multiverse** (or **meta-universe**) is the hypothetical set of infinite or finite possible universes (including the Universe we consistently experience) that together comprise everything that exists: the entirety of space, time, matter, and energy as well as the physical laws and constants that describe them. The various universes within the multiverse are also called "**parallel universes**" or "alternate universes".

1.1 Explanation

The structure of the multiverse, the nature of each universe within it and the relationships among the various constituent universes, depend on the specific multiverse hypothesis considered. Multiple universes have been hypothesized in cosmology, physics, astronomy, religion, philosophy, transpersonal psychology, and fiction, particularly in science fiction and fantasy. In these contexts, parallel universes are also called "alternate universes", "quantum universes", "interpenetrating dimensions", "parallel dimensions", "parallel worlds", "alternate realities", "alternate timelines", and "dimensional planes", among other names. The American philosopher and psychologist William James coined the term *multiverse* in 1895, but in a different context.[1]

The physics community continues to fiercely debate the multiverse hypothesis. Prominent physicists disagree about whether the multiverse may exist, and whether it is even a legitimate topic of scientific inquiry.[2] Serious concerns have been raised about whether attempts to exempt the multiverse from experimental verification may erode public confidence in science and ultimately damage the nature of fundamental physics.[3] Some have argued that the multiverse question is philosophical rather than scientific because it lacks falsifiability; the ability to disprove a theory by means of scientific experiment has always been part of the accepted scientific method.[4] Paul Steinhardt has famously argued that no experiment can rule out a theory if it provides for all possible outcomes.[5]

Supporters of one of the multiverse hypotheses include Stephen Hawking,[6] Brian Greene,[7][8] Max Tegmark,[9] Alan Guth,[10] Andrei Linde,[11] Michio Kaku,[12] David Deutsch,[13] Leonard Susskind,[14] Raj Pathria,[15] Alexander Vilenkin,[16] Laura Mersini-Houghton,[17][18] Neil deGrasse Tyson[19] and Sean Carroll.[20]

Scientists who are not proponents of the multiverse include: Nobel laureate Steven Weinberg,[21] Nobel laureate David Gross,[22] Paul Steinhardt,[23] Neil Turok,[24] Viatcheslav Mukhanov,[25] Michael S. Turner,[26] Roger Penrose,[27] George Ellis,[28][29] Joe Silk,[30] Adam Frank,[31] Marcelo Gleiser,[31] Jim Baggott,[32] and Paul Davies.[33]

In 2007, Steven Weinberg suggested that if the multiverse were true, "the hope of finding a rational explanation for the precise values of quark masses and other constants of the standard model that we observe in our big bang is doomed, for their values would be an accident of the particular part of the multiverse in which we live."[34]

1.2 Multiverse hypotheses in physics

1.2.1 Categories

Max Tegmark and Brian Greene have devised classification schemes that categorize the various theoretical types of multiverse, or types of universe that might theoretically comprise a multiverse ensemble.

Max Tegmark's four levels

Cosmologist Max Tegmark has provided a taxonomy of universes beyond the familiar observable universe. The levels according to Tegmark's classification are arranged such that subsequent levels can be understood to encompass and expand upon previous levels, and they are briefly described below.[35][36]

Level I: Beyond our cosmological horizon A generic prediction of chaotic inflation is an infinite ergodic universe, which, being infinite, must contain Hubble volumes realizing all initial conditions.

Accordingly, an infinite universe will contain an infinite number of Hubble volumes, all having the same physical laws and physical constants. In regard to configurations such as the distribution of matter, almost all will differ from our Hubble volume. However, because there are infinitely many, far beyond the cosmological horizon, there will eventually be Hubble volumes with similar, and even identical, configurations. Tegmark estimates that an identical volume to ours should be about $10^{10^{115}}$ meters away from us.[9] Given infinite space, there would, in fact, be an infinite number of Hubble volumes identical to ours in the Universe.[37] This follows directly from the cosmological principle, wherein it is assumed our Hubble volume is not special or unique.

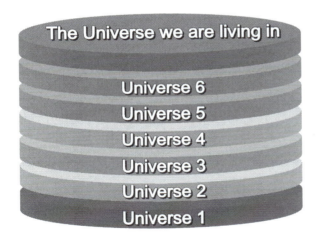

"Bubble universes": every disk is a bubble universe (Universe 1 to Universe 6 are different bubbles; they have physical constants that are different from our universe); our universe is just one of the bubbles.

Level II: Universes with different physical constants In the chaotic inflation theory, a variant of the cosmic inflation theory, the multiverse as a whole is stretching and will continue doing so forever,[38] but some regions of space stop stretching and form distinct bubbles, like gas pockets in a loaf of rising bread. Such bubbles are embryonic level I multiverses. Linde and Vanchurin calculated the number of these universes to be on the scale of $10^{10^{10,000,000}}$.[39]

Different bubbles may experience different spontaneous symmetry breaking resulting in different properties such as different physical constants.[37]

This level also includes John Archibald Wheeler's oscillatory universe theory and Lee Smolin's fecund universes theory.

Level III: Many-worlds interpretation of quantum mechanics Hugh Everett's many-worlds interpretation (MWI) is one of several mainstream interpretations of quantum mechanics. In brief, one aspect of quantum mechanics is that certain observations cannot be predicted absolutely. Instead, there is a range of possible observations, each with a different probability. According to the MWI, each of these possible observations corresponds to a different universe. Suppose a six-sided die is thrown and that the result of the throw corresponds to a quantum mechanics observable. All six possible ways the die can fall correspond to six different universes.

Tegmark argues that a level III multiverse does not contain more possibilities in the Hubble volume than a level I-II multiverse. In effect, all the different "worlds" created by "splits" in a level III multiverse with the same physical constants can be found in some Hubble volume in a level I multiverse. Tegmark writes that "The only difference between Level I and Level III is where your doppelgängers reside. In Level I they live elsewhere in good old three-dimensional space. In Level III they live on another quantum branch in infinite-dimensional Hilbert space." Similarly, all level II bubble universes with different physical constants can in effect be found as "worlds" created by "splits" at the moment of spontaneous symmetry breaking in a level III multiverse.[37] According to Yasunori Nomura[40] and Raphael Bousso and Leonard Susskind,[14] this is because global spacetime appearing in the (eternally) inflating multiverse is a redundant concept. This implies that the multiverses of Level I, II, and III are, in fact, the same thing. This hypothesis is referred to as "Multiverse = Quantum Many Worlds".

Related to the *many-worlds* idea are Richard Feynman's *multiple histories* interpretation and H. Dieter Zeh's *many-minds* interpretation.

Level IV: Ultimate ensemble The ultimate ensemble or mathematical universe hypothesis is the hypothesis of Tegmark himself.[41] This level considers equally real all universes that can be described by different mathematical structures. Tegmark writes that "abstract mathematics is so general that any Theory Of Everything (TOE) that is definable in purely formal terms (independent of vague human terminology) is also a mathematical structure. For instance, a TOE involving a set of different types of entities (denoted by words, say) and relations between them (denoted by additional words) is nothing but what mathematicians call a set-theoretical model, and one can generally find a formal system that it is a model of." He argues this "implies that any conceivable parallel universe theory can be described at Level IV" and "subsumes all other ensembles, therefore brings closure to the hierarchy of multiverses, and there cannot be say a Level V."[9]

Jürgen Schmidhuber, however, says the "set of mathematical structures" is not even well-defined, and admits only universe representations describable by constructive mathematics, that is, computer programs. He explicitly includes universe representations describable by non-halting programs whose output bits converge after finite time, although the convergence time itself may not be predictable by a halting program, due to Kurt Gödel's limitations.[42][43][44] He also explicitly discusses the more restricted ensemble of quickly computable universes.[45]

Brian Greene's nine types

American theoretical physicist and string theorist Brian Greene discussed nine types of parallel universes:[46]

Quilted The quilted multiverse works only in an infinite universe. With an infinite amount of space, every possible event will occur an infinite number of times. However, the speed of light prevents us from being aware of these other identical areas.

Inflationary The inflationary multiverse is composed of various pockets where inflation fields collapse and form new universes.

Brane The brane multiverse follows from M-theory and states that our universe is a 3-dimensional brane that exists with many others on a higher-dimensional brane or "bulk". Particles are bound to their respective branes except for gravity.

Cyclic The cyclic multiverse (via the ekpyrotic scenario) has multiple branes (each a universe) that collided, causing Big Bangs. The universes bounce back and pass through time, until they are pulled back together and again collide, destroying the old contents and creating them anew.

Landscape The landscape multiverse relies on string theory's Calabi–Yau shapes. Quantum fluctuations drop the shapes to a lower energy level, creating a pocket with a different set of laws from the surrounding space.

Quantum The quantum multiverse creates a new universe when a diversion in events occurs, as in the many-worlds interpretation of quantum mechanics.

Holographic The holographic multiverse is derived from the theory that the surface area of a space can simulate the volume of the region.

Simulated The simulated multiverse exists on complex computer systems that simulate entire universes.

Ultimate The ultimate multiverse contains every mathematically possible universe under different laws of physics.

1.2.2 Cyclic theories

Main article: Cyclic model

In several theories there is a series of infinite, self-sustaining cycles (for example: an eternity of Big Bang-Big crunches and/or Big Freezes.

1.2.3 M-theory

See also: Introduction to M-theory, M-theory, Brane cosmology and String theory landscape

A multiverse of a somewhat different kind has been envisaged within string theory and its higher-dimensional extension, M-theory.[47] These theories require the presence of 10 or 11 spacetime dimensions respectively. The extra 6 or 7 dimensions may either be compactified on a very small scale, or our universe may simply be localized on a dynamical (3+1)-dimensional object, a D-brane. This opens up the possibility that there are other branes which could support "other universes".[48][49] This is unlike the universes in the "quantum multiverse", but both concepts can operate at the same time.

Some scenarios postulate that our big bang was created, along with our universe, by the collision of two branes.[48][49]

1.2.4 Black-hole cosmology

Main article: Black-hole cosmology

A black-hole cosmology is a cosmological model in which the observable universe is the interior of a black hole existing as one of possibly many inside a larger universe. This includes the theory of white holes of which are on the opposite side of space time. While a black hole sucks everything in including light, a white hole releases matter and light, hence the name "white hole".

1.2.5 Anthropic principle

Main article: Anthropic principle

The concept of other universes has been proposed to explain how our own universe appears to be fine-tuned for conscious life as we experience it. If there were a large (possibly infinite) number of universes, each with possibly different physical laws (or different fundamental physical constants), some of these universes, even if very few, would have the combination of laws and fundamental parameters that are suitable for the development of matter, astronomical structures, elemental diversity, stars, and planets that can exist long enough for life to emerge and evolve. The weak anthropic principle could then be applied to conclude that we (as conscious beings) would only exist in one of those few universes that happened to be finely tuned, permitting the existence of life with developed consciousness. Thus, while the probability might be extremely small that any particular universe would have the requisite conditions for life (as we understand life) to emerge and evolve, this does not require intelligent design as an explanation for the conditions in the Universe that promote our existence in it.

1.2.6 Search for evidence

Around 2010, scientists such as Stephen M. Feeney analyzed Wilkinson Microwave Anisotropy Probe (WMAP) data and claimed to find preliminary evidence suggesting that our universe collided with other (parallel) universes in the distant past.[50][51][52][53] However, a more thorough analysis of data from the WMAP and from the Planck satellite, which has a resolution 3 times higher than WMAP, failed to find any statistically significant evidence of such a bubble universe collision.[54][55] In addition, there is no evidence of any gravitational pull of other universes on ours.[56][57]

1.2.7 Criticism

Non-scientific claims

In his 2003 NY Times opinion piece, *A Brief History of the Multiverse,* author and cosmologist, Paul Davies, offers a variety of arguments that multiverse theories are non-scientific :[58]

> For a start, how is the existence of the other universes to be tested? To be sure, all cosmologists accept that there are some regions of the universe that lie beyond the reach of our telescopes, but somewhere on the slippery slope between that and the idea that there are an infinite number of universes, credibility reaches a limit. As one slips down that slope, more and more must be accepted on faith, and less and less is open to scientific verification. Extreme multiverse explanations are therefore reminiscent of theological discussions. Indeed, invoking an infinity of unseen universes to explain the unusual features of the one we do see is just as ad hoc as invoking an unseen Creator. The multiverse theory may be dressed up in scientific language, but in essence it requires the same leap of faith.
> — Paul Davies, *A Brief History of the Multiverse*

Taking cosmic inflation as a popular case in point, George Ellis, writing in August 2011, provides a balanced criticism of not only the science, but as he suggests, the scientific philosophy, by which multiverse theories are generally substantiated. He, like most cosmologists, accepts Tegmark's level I "domains", even though they lie far beyond the cosmological horizon. Likewise, the multiverse of cosmic inflation is said to exist very far away. It would be so far away, however, that it's very unlikely any evidence of an early interaction will be found. He argues that for many theorists, the lack of empirical testability or falsifiability is not a major concern. "Many physicists who talk about the multiverse, especially advocates of the string landscape, do not care much about parallel universes per se. For them, objections to the multiverse as a concept are unimportant. Their theories live or die based on internal consistency and, one hopes, eventual laboratory testing." Although he believes there's little hope that will ever be possible, he grants that the theories on which the speculation is based, are not without scientific merit. He concludes that multiverse theory is a "productive research program":[59]

> As skeptical as I am, I think the contemplation of the multiverse is an excellent opportunity to reflect on the nature of science and on the ultimate nature of existence: why we are here... In looking at this concept, we need an open mind, though not too open. It is a delicate path to tread. Parallel universes may or may not exist; the case is unproved. We are going to have to live with that uncertainty. Nothing is wrong with scientifically based philosophical speculation, which is what multiverse proposals are. But we should name it for what it is.
> — George Ellis, *Scientific American*, Does the Multiverse Really Exist?

1.2. MULTIVERSE HYPOTHESES IN PHYSICS

Occam's razor

Proponents and critics disagree about how to apply Occam's razor. Critics argue that to postulate a practically infinite number of unobservable universes just to explain our own seems contrary to Occam's razor.[60] In contrast, proponents argue that, in terms of Kolmogorov complexity, the proposed multiverse is simpler than a single idiosyncratic universe.[37]

For example, multiverse proponent Max Tegmark argues:

> [A]n entire ensemble is often much simpler than one of its members. This principle can be stated more formally using the notion of algorithmic information content. The algorithmic information content in a number is, roughly speaking, the length of the shortest computer program that will produce that number as output. For example, consider the set of all integers. Which is simpler, the whole set or just one number? Naively, you might think that a single number is simpler, but the entire set can be generated by quite a trivial computer program, whereas a single number can be hugely long. Therefore, the whole set is actually simpler... (Similarly), the higher-level multiverses are simpler. Going from our universe to the Level I multiverse eliminates the need to specify initial conditions, upgrading to Level II eliminates the need to specify physical constants, and the Level IV multiverse eliminates the need to specify anything at all.... A common feature of all four multiverse levels is that the simplest and arguably most elegant theory involves parallel universes by default. To deny the existence of those universes, one needs to complicate the theory by adding experimentally unsupported processes and ad hoc postulates: finite space, wave function collapse and ontological asymmetry. Our judgment therefore comes down to which we find more wasteful and inelegant: many worlds or many words. Perhaps we will gradually get used to the weird ways of our cosmos and find its strangeness to be part of its charm.[37]
>
> — Max Tegmark, *"Parallel universes. Not just a staple of science fiction, other universes are a direct implication of cosmological observations."* Scientific American 2003 May;288(5):40–51

Princeton cosmologist Paul Steinhardt used the 2014 Annual Edge Question to voice his opposition to multiverse theorizing:

> A pervasive idea in fundamental physics and cosmology that should be retired: the notion that we live in a multiverse in which the laws of physics and the properties of the cosmos vary randomly from one patch of space to another. According to this view, the laws and properties within our observable universe cannot be explained or predicted because they are set by chance. Different regions of space too distant to ever be observed have different laws and properties, according to this picture. Over the entire multiverse, there are infinitely many distinct patches. Among these patches, in the words of Alan Guth, "anything that can happen will happen—and it will happen infinitely many times". Hence, I refer to this concept as a Theory of Anything. Any observation or combination of observations is consistent with a Theory of Anything. No observation or combination of observations can disprove it. Proponents seem to revel in the fact that the Theory cannot be falsified. The rest of the scientific community should be up in arms since an unfalsifiable idea lies beyond the bounds of normal science. Yet, except for a few voices, there has been surprising complacency and, in some cases, grudging acceptance of a Theory of Anything as a logical possibility. The scientific journals are full of papers treating the Theory of Anything seriously. What is going on?[23]
>
> — Paul Steinhardt, *"Theories of Anything"* edge.com'

Steinhardt claims that multiverse theories have gained currency mostly because too much has been invested in theories that have failed, e.g. inflation or string theory. He tends to see in them an attempt to redefine the values of science to which he objects even more strongly:

> A Theory of Anything is useless because it does not rule out any possibility and worthless because it submits to no do-or-die tests. (Many papers discuss potential observable consequences, but these are only possibilities, not certainties, so the Theory is never really put at risk.)[23]
>
> — Paul Steinhardt, *"Theories of Anything"* edge.com'

1.3 Multiverse hypotheses in philosophy and logic

1.3.1 Modal realism

Possible worlds are a way of explaining probability, hypothetical statements and the like, and some philosophers such as David Lewis believe that all possible worlds exist, and are just as real as the actual world (a position known as modal realism).[61]

1.3.2 Trans-world identity

A metaphysical issue that crops up in multiverse schema that posit infinite identical copies of any given universe is that of the notion that there can be identical objects in different possible worlds. According to the counterpart theory of David Lewis, the objects should be regarded as similar rather than identical.[62][63]

1.4 See also

- Holographic principle
- Hugh Everett
- Impossible world
- Laura Mersini-Houghton
- Martin Rees, Astronomer Royal
- Modal realism
- Multiverse (religion)
- Parallel universe (fiction)
- Philosophy of physics
- Philosophy of space and time
- Reductionism
- Roger Penrose
- Simulated reality
- *The Fabric of Reality*

1.5 References

1.5.1 Notes

[1] James, William, *The Will to Believe*, 1895; and earlier in 1895, as cited in OED's new 2003 entry for "multiverse": James, William (October 1895), " "Is Life Worth Living?", *Internat. Jrnl. Ethics* **6**: 10, Visible nature is all plasticity and indifference, a multiverse, as one might call it, and not a universe.

[2] Kragh, H. (2009). "Contemporary History of Cosmology and the Controversy over the Multiverse". *Annals of Science* **66** (4): 529. doi:10.1080/00033790903047725.

[3] Ellis, George; Silk, Joe (December 16, 2014). "Scientific Method: Defend the Integrity of Physics". *Nature*.

[4] "Feynman on Scientific Method". *YouTube*. Retrieved July 28, 2012.

[5] Steinhardt, Paul (June 3, 2014). "Big Bang blunder bursts the Multiverse bubble". *Nature*.

[6] *Universe or Multiverse*. p. 19. ISBN 9780521848411. Some physicists would prefer to believe that string theory, or M-theory, will answer these questions and uniquely predict the features of the Universe. Others adopt the view that the initial state of the Universe is prescribed by an outside agency, code-named God, or that there are many universes, with ours being picked out by the anthropic principle. Hawking argues that string theory is unlikely to predict the distinctive features of the Universe. But neither is he is an advocate of God. He therefore opts for the last approach, favouring the type of multiverse which arises naturally within the context of his own work in quantum cosmology.

[7] Greene, Brian (January 24, 2011). *A Physicist Explains Why Parallel Universes May Exist*. npr.org. Interview with Terry Gross. Archived from the original on September 12, 2014. Retrieved September 12, 2014.

[8] Greene, Brian (January 24, 2011). *Transcript:A Physicist Explains Why Parallel Universes May Exist*. npr.org. Interview with Terry Gross. Archived from the original on September 12, 2014. Retrieved September 12, 2014.

[9] Tegmark, Max (2003). "Parallel Universes". *In "Science and Ultimate Reality: from Quantum to Cosmos", honoring John Wheeler's th birthday. J. D. Barrow, P.C.W. Davies, & C.L. Harper eds. Cambridge University Press ().* v1 **90** (2003). arXiv:astro-ph/0302131. Bibcode:2003SciAm.288e..40T. doi:10.1038/scientificamerican0503-40.

[10] "Alan Guth: Inflationary Cosmology: Is Our Universe Part of a Multiverse?". *YouTube*. Retrieved 6 October 2014.

[11] Linde, Andrei (January 27, 2012). "Inflation in Supergravity and String Theory: Brief History of the Multiverse" (PDF). ctc.cam.ac.uk. Archived (PDF) from the original on September 13, 2014. Retrieved September 13, 2014.

1.5. REFERENCES

[12] Parallel Worlds: A Journey Through Creation, Higher Dimensions, and the Future of the Cosmos

[13] David Deutsch (1997). "The Ends of the Universe". The Fabric of Reality: The Science of Parallel Universes—and Its Implications. London: Penguin Press. ISBN 0-7139-9061-9.

[14] Bousso, R.; Susskind, L. (2012). "Multiverse interpretation of quantum mechanics". *Physical Review D* **85** (4). arXiv:1105.3796. doi:10.1103/PhysRevD.85.045007.

[15] Pathria, R. K. (1972). "The Universe as a Black Hole". *Nature* **240** (5379): 298. Bibcode:1972Natur.240..298P. doi:10.1038/240298a0.

[16] Vilenkin, Alex (2007). *Many Worlds in One: The Search for Other Universes*. ISBN 9780374707149.

[17] Catchpole, Heather (November 24, 2009). "Weird data suggests something big beyond the edge of the universe". *Cosmos (magazine)*. Retrieved July 27, 2014.

[18] Moon, Timur (May 19, 2013). "Planck Space Data Yields Evidence of Universes Beyond Our Own". *International Business Times*. Retrieved July 27, 2014.

[19] Freeman, David (March 4, 2014). "Why Revive 'Cosmos?' Neil DeGrasse Tyson Says Just About Everything We Know Has Changed". *huffingtonpost.com*. Archived from the original on September 12, 2014. Retrieved September 12, 2014.

[20] Sean Carroll (October 18, 2011). "Welcome to the Multiverse". *Discover (magazine)*. Retrieved May 5, 2015.

[21] Falk, Dan (March 17, 2015). "Science's Path from Myth to Multiverse". *Quanta Magazine* (New York: Simons Foundation).

[22] Davies, Paul (2008). "Many Scientists Hate the Multiverse Idea". *The Goldilocks Enigma: Why Is the Universe Just Right for Life?*. Houghton Mifflin Harcourt. p. 207. ISBN 9780547348469.

[23] Steinhardt, Paul (March 9, 2014). "Theories of Anything". *edge.org*. 2014 : WHAT SCIENTIFIC IDEA IS READY FOR RETIREMENT?. Archived from the original on March 9, 2014. Retrieved March 9, 2014.

[24] Gibbons, G.W.; Turok, Neil (2008). "The Measure Problem in Cosmology". *Phys.Rev.D* **77** (6): 063516. arXiv:hep-th/0609095. Bibcode:2008PhRvD..77f3516G. doi:10.1103/PhysRevD.77.063516.

[25] Mukhanov, Viatcheslav (2014). "Inflation without Selfreproduction". *Fortschritte der Physik* **63** (1): 36–41. doi:10.1002/prop.201400074.

[26] Woit, Peter (June 9, 2015). "A Crisis at the (Western) Edge of Physics". *Not Even Wrong*.

[27] Woit, Peter (June 14, 2015). "CMB @ 50". *Not Even Wrong*.

[28] Ellis, George F. R. (August 1, 2011). "Does the Multiverse Really Exist?". *Scientific American* (New York: Nature Publishing Group) **305** (2): 38–43. doi:10.1038/scientificamerican0811-38. ISSN 0036-8733. LCCN 04017574. OCLC 828582568. Retrieved September 12, 2014. (subscription required (help)).

[29] Ellis, George (2012). "The Multiverse: Conjecture, Proof, and Science" (PDF). *Slides for a talk at Nicolai Fest Golm 2012*. Archived (PDF) from the original on September 12, 2014. Retrieved September 12, 2014.

[30] Ellis, George; Silk, Joe (December 16, 2014), "Scientific Method: Defend the Integrity of Physics", *Nature*

[31] Frank, Adam; Gleiser, Marcelo (June 5, 2015). "A Crisis at the Edge of Physics". *New York Times*.

[32] Baggott, Jim (August 1, 2013). *Farewell to Reality: How Modern Physics Has Betrayed the Search for Scientific Truth*. Pegasus. ISBN 978-1-60598-472-8. ISBN 978-1-60598-574-9.

[33] Davies, Paul (April 12, 2003). "A Brief History of the Multiverse". *New York Times*.

[34] Weinberg, Steven (November 20, 2007). "Physics: What we do and don't know". *The New York Review of Books*.

[35] Tegmark, Max (May 2003). "Parallel Universes". *Scientific American*.

[36] Tegmark, Max (23 January 2003). *Parallel Universes* (PDF). Retrieved 7 February 2006.

[37] "Parallel universes. Not just a staple of science fiction, other universes are a direct implication of cosmological observations.", Tegmark M., Sci Am. 2003 May;288(5):40–51.

[38] "First Second of the Big Bang". *How The Universe Works 3*. 2014. Discovery Science.

[39] Zyga, Lisa "Physicists Calculate Number of Parallel Universes", *PhysOrg*, 16 October 2009.

[40] Nomura, Y. (2011). "Physical theories, eternal inflation, and the quantum universe". *Journal of High Energy Physics* **2011** (11). arXiv:1104.2324. doi:10.1007/JHEP11(2011)063.

[41] Tegmark, Max (2014). *Our Mathematical Universe: My Quest for the Ultimate Nature of Reality*. Knopf Doubleday Publishing Group. ISBN 9780307599803.

[42] J. Schmidhuber (1997): A Computer Scientist's View of Life, the Universe, and Everything. Lecture Notes in Computer Science, pp. 201–208, Springer: IDSIA – Dalle Molle Institute for Artificial Intelligence

[43] Schmidhuber, Juergen (2000). "Algorithmic Theories of Everything". *Sections in: Hierarchies of generalized Kolmogorov complexities and nonenumerable universal measures computable in the limit. International Journal of Foundations of Computer Science* ():587-612 (2002). Section 6

in: the Speed Prior: A New Simplicity Measure Yielding Near-Optimal Computable Predictions. in J. Kivinen and R. H. Sloan, editors, Proceedings of the 15th Annual Conference on Computational Learning Theory (COLT 2002), Sydney, Australia, Lecture Notes in Artificial Intelligence, pages 216-–228. Springer, 2002 **13** (4): 1–5. arXiv:quant-ph/0011122. Bibcode:2000quant.ph.11122S.

[44] J. Schmidhuber (2002): Hierarchies of generalized Kolmogorov complexities and nonenumerable universal measures computable in the limit. International Journal of Foundations of Computer Science 13(4):587–612 IDSIA – Dalle Molle Institute for Artificial Intelligence

[45] J. Schmidhuber (2002): The Speed Prior: A New Simplicity Measure Yielding Near-Optimal Computable Predictions. Proc. 15th Annual Conference on Computational Learning Theory (COLT 2002), Sydney, Australia, Lecture Notes in Artificial Intelligence, pp. 216–228. Springer: IDSIA – Dalle Molle Institute for Artificial Intelligence

[46] In The Hidden Reality: Parallel Universes and the Deep Laws of the Cosmos, 2011

[47] Weinberg, Steven (2005). "Living in the Multiverse". arXiv:hep-th/0511037v1.

[48] Richard J Szabo, *An introduction to string theory and D-brane dynamics* (2004)

[49] Maurizio Gasperini, *Elements of String Cosmology* (2007)

[50] Lisa Zyga (December 17, 2010). "Scientists find first evidence that many universes exist". *PhysOrg.com.* phys.org. Retrieved 12 October 2013.

[51] "Astronomers Find First Evidence Of Other Universe". technologyreview.com. December 13, 2010. Retrieved 12 October 2013.

[52] Max Tegmark; Alexander Vilenkin (July 19, 2011). "The Case for Parallel Universes". Retrieved 12 October 2013.

[53] "Is Our Universe Inside a Bubble? First Observational Test of the 'Multiverse'". *Science Daily.* sciencedaily.com. Aug 3, 2011. Retrieved 12 October 2013.

[54] Feeney, Stephen M.; et al. (2011). "First observational tests of eternal inflation: Analysis methods and WMAP 7-year results". *Physical Review D* **84** (4): 43507. arXiv:1012.3667. Bibcode:2011PhRvD..84d3507F. doi:10.1103/PhysRevD.84.043507.

[55] Feeney; et al. (2011). "First observational tests of eternal inflation". *Physical review letters* **107** (7). arXiv:1012.1995. Bibcode:2011PhRvL.107g1301F. doi:10.1103/PhysRevLett.107.071301.. Bousso, Raphael; Harlow, Daniel; Senatore, Leonardo (2013). "Inflation after False Vacuum Decay: Observational Prospects after Planck". *Physical Review D* **91** (8). arXiv:1309.4060. Bibcode:2015PhRvD..91h3527B. doi:10.1103/PhysRevD.91.083527.

[56] Collaboration, Planck; Ade, P. A. R.; Aghanim, N.; Arnaud, M.; Ashdown, M.; Aumont, J.; Baccigalupi, C.; Balbi, A.; Banday, A. J.; Barreiro, R. B.; Battaner, E.; Benabed, K.; Benoit-Levy, A.; Bernard, J. -P.; Bersanelli, M.; Bielewicz, P.; Bikmaev, I.; Bobin, J.; Bock, J. J.; Bonaldi, A.; Bond, J. R.; Borrill, J.; Bouchet, F. R.; Burigana, C.; Butler, R. C.; Cabella, P.; Cardoso, J. -F.; Catalano, A.; Chamballu, A.; et al. (2013-03-20). "[1303.5090] Planck intermediate results. XIII. Constraints on peculiar velocities". arXiv:[//arxiv.org/abs/1303.5090 1303.5090] [astro-ph.CO].

[57] "Blow for 'dark flow' in Planck's new view of the cosmos". *New Scientist.* 3 April 2013. Retrieved 10 March 2014.

[58] Davies, Paul (12 April 2003). "A Brief History of the Multiverse". *New York Times.* Retrieved 16 August 2011.

[59] Ellis, George F. R. (August 1, 2011). "Does the Multiverse Really Exist?". *Scientific American* (New York: Nature Publishing Group) **305** (2): 38–43. doi:10.1038/scientificamerican0811-38. ISSN 0036-8733. LCCN 04017574. OCLC 828582568. Retrieved August 16, 2011. (subscription required (help)).

[60] Trinh, Xuan Thuan (2006). Staune, Jean, ed. *Science & the Search for Meaning: Perspectives from International Scientists.* West Conshohocken, PA: Templeton Foundation. p. 186. ISBN 1-59947-102-7.

[61] Lewis, David (1986). *On the Plurality of Worlds.* Basil Blackwell. ISBN 0-631-22426-2.

[62] Deutsch, Harry (Summer 2002). Edward N. Zalta, ed. "Relative Identity". *The Stanford Encyclopedia of Philosophy.* Retrieved 6 October 2014.

[63] "Paul B. Kantor "The Interpretation of Cultures and Possible Worlds", 1 October 2002". Retrieved 6 October 2014.

1.5.2 Bibliography

- Bernard Carr, ed. (2007) *Universe or Multiverse?* Cambridge Univ. Press.

- Deutsch, David (1985). "Quantum theory, the Church–Turing principle and the universal quantum computer" (pdf). *Proceedings of the Royal Society of London A* (400): 97–117.

- Ellis, George F.R.; William R. Stoeger; Stoeger, W. R. (2004). "Multiverses and physical cosmology". *Monthly Notices of the Royal Astronomical Society* **347** (3): 921–936. arXiv:astro-ph/0305292. Bibcode:2004MNRAS.347..921E. doi:10.1111/j.1365-2966.2004.07261.x.

- Surya-Siddhanta: A Text Book of Hindu Astronomy by Ebenezer Burgess, ed. Phanindralal Gangooly (1989/1997) with a 45-page commentary by P. C. Sengupta (1935).

1.6 External links

- Interview with Tufts cosmologist Alex Vilenkin on his new book, "Many Worlds in One: The Search for Other Universes" on the podcast and public radio interview program ThoughtCast.

- Joseph Pine II about Multiverse, Presentation at Mobile Monday Amsterdam, 2008

- Multiverse – Radio-discussion on BBC Four with Melvyn Bragg

Chapter 2

Universe

For other uses, see Universe (disambiguation).

The **Universe is all of time and space and its contents**.[8][9][10][11] The Universe includes planets, stars, galaxies, the contents of intergalactic space, the smallest subatomic particles, and all matter and energy. The *observable universe is about 28 billion parsecs (91 billion light-years) in diameter at the present time*.[2] The size of the whole Universe is not known and may be infinite.[12] Observations and the development of physical theories have led to inferences about the composition and evolution of the Universe.

Throughout recorded history, cosmologies and cosmogonies, including scientific models, have been proposed to explain observations of the Universe. The earliest quantitative geocentric models were developed by ancient Greek philosophers and Indian philosophers.[13][14] Over the centuries, more precise astronomical observations led to Nicolaus Copernicus's heliocentric model of the Solar System and Johannes Kepler's improvement on that model with elliptical orbits, which was eventually explained by Isaac Newton's theory of gravity. Further observational improvements led to the realization that the Solar System is located in a galaxy composed of billions of stars, the Milky Way. It was subsequently discovered that our galaxy is one of many. On the largest scales, it is assumed that the distribution of galaxies is uniform and the same in all directions, meaning that the Universe has neither an edge nor a center. Observations of the distribution of these galaxies and their spectral lines have led to many of the theories of modern physical cosmology. The discovery in the early 20th century that galaxies are systematically redshifted suggested that the Universe is expanding, and the discovery of the cosmic microwave background radiation suggested that the Universe had a beginning.[15] Finally, observations in the late 1990s indicated the rate of the expansion of the Universe is increasing[16] indicating that the majority of energy is most likely in an unknown form called dark energy. The majority of mass in the universe also appears to exist in an unknown form, called dark matter.

The Big Bang theory is the prevailing cosmological model describing the development of the Universe. Space and time were created in the Big Bang, and these were imbued with a fixed amount of energy and matter; as space expands, the density of that matter and energy decreases. After the initial expansion, the Universe cooled sufficiently to allow the formation first of subatomic particles and later of simple atoms. Giant clouds of these primordial elements later coalesced through gravity to form stars. Assuming that the prevailing model is correct, the age of the Universe is measured to be 13.799 ± 0.021 billion years.[1]

There are many competing hypotheses about the ultimate fate of the Universe. Physicists and philosophers remain unsure about what, if anything, preceded the Big Bang. Many refuse to speculate, doubting that any information from any such prior state could ever be accessible. There are various multiverse hypotheses, in which some physicists have suggested that the Universe might be one among many universes that likewise exist.[17][18]

2.1 Definition

The Universe is customarily defined as everything that exists, everything that has existed, and everything that will exist.[19][20][21] According to our current understanding, the Universe consists of three constituents: spacetime, forms of energy (including electromagnetic radiation and matter), and the physical laws that relate them. The Universe also encompasses all of life, all of history, and some philosophers and scientists even suggest that it encompasses ideas such as mathematics and logic.[22][23][24]

2.2 Etymology

The word *universe* derives from the Old French word *univers*, which in turn derives from the Latin word *univer-*

sum.[25] The Latin word was used by Cicero and later Latin authors in many of the same senses as the modern English word is used.[26]

2.2.1 Synonyms

A term for "universe" among the ancient Greek philosophers from Pythagoras onwards was τὸ πᾶν *tò pân* ("the all"), defined as all matter and all space, and τὸ ὅλον *tò hólon* ("all things"), which did not necessarily include the void.[27][28] Another synonym was ὁ κόσμος *ho kósmos* (meaning the world, the cosmos).[29] Synonyms are also found in Latin authors (*totum, mundus, natura*)[30] and survive in modern languages, e.g., the German words *Das All*, *Weltall*, and *Natur* for *Universe*. The same synonyms are found in English, such as everything (as in the theory of everything), the cosmos (as in cosmology), the world (as in the many-worlds interpretation), and nature (as in natural laws or natural philosophy).[31]

2.3 Chronology and the Big Bang

Main articles: Big Bang and Chronology of the Universe

The prevailing model for the evolution of the Universe is the Big Bang theory.[32][33] The Big Bang model states that the earliest state of the Universe was extremely hot and dense and that it subsequently expanded. The model is based on general relativity and on simplifying assumptions such as homogeneity and isotropy of space. A version of the model with a cosmological constant (Lambda) and cold dark matter, known as the Lambda-CDM model, is the simplest model that provides a reasonably good account of various observations about the Universe. The Big Bang model accounts for observations such as the correlation of distance and redshift of galaxies, the ratio of the number of hydrogen to helium atoms, and the microwave radiation background.

The initial hot, dense state is called the Planck epoch, a brief period extending from time zero to one Planck time unit of approximately 10^{-43} seconds. During the Planck epoch, all types of matter and all types of energy were concentrated into a dense state, where gravitation is believed to have been as strong as the other fundamental forces, and all the forces may have been unified. Since the Planck epoch, the Universe has been expanding to its present form, possibly with a very brief period of cosmic inflation which caused the Universe to reach a much larger size in less than 10^{-32} seconds.[34]

After the Planck epoch and inflation came the quark, hadron, and lepton epochs. Together, these epochs encompassed less than 10 seconds of time following the Big Bang. The observed abundance of the elements can be explained by combining the overall expansion of space with nuclear and atomic physics. As the Universe expands, the energy density of electromagnetic radiation decreases more quickly than does that of matter because the energy of a photon decreases with its wavelength. As the Universe expanded and cooled, elementary particles associated stably into ever larger combinations. Thus, in the early part of the matter-dominated era, stable protons and neutrons formed, which then formed atomic nuclei through nuclear reactions. This process, known as Big Bang nucleosynthesis, led to the present abundances of lighter nuclei, particularly hydrogen, deuterium, and helium. Big Bang nucleosynthesis ended about 20 minutes after the Big Bang, when the Universe had cooled enough so that nuclear fusion could no longer occur. At this stage, matter in the Universe was mainly a hot, dense plasma of negatively charged electrons, neutral neutrinos and positive nuclei. This era, called the photon epoch, lasted about 380 thousand years.

Eventually, at a time known as recombination, electrons and nuclei formed stable atoms, which are transparent to most wavelengths of radiation. With photons decoupled from matter, the Universe entered the matter-dominated era. Light from this era could now travel freely, and it can still be seen in the Universe as the cosmic microwave background (CMB). After around 100 million years, the first stars formed; these were likely very massive, luminous, and responsible for the reionization of the Universe. Having no elements heavier than lithium, these stars also produced the first heavy elements through stellar nucleosynthesis.[35] The Universe also contains a mysterious energy called dark energy; the energy density of dark energy does not change over time. After about 9.8 billion years, the Universe had expanded sufficiently so that the density of matter was less than the density of dark energy, marking the beginning of the present dark-energy-dominated era.[36] In this era, the expansion of the Universe is accelerating due to dark energy.

2.4 Properties

Main articles: Observable universe, Age of the Universe and Metric expansion of space

The spacetime of the Universe is usually interpreted from a Euclidean perspective, with space as consisting of three dimensions, and time as consisting of one dimension, the "fourth dimension".[37] By combining space and time into a single manifold called Minkowski space, physicists have simplified a large number of physical theories, as well as described in a more uniform way the workings of the Universe at both the supergalactic and subatomic levels.

Spacetime events are not absolutely defined spatially and temporally but rather are known relative to the motion of an observer. Minkowski space approximates the Universe without gravity; the pseudo-Riemannian manifolds of general relativity describe spacetime with matter and gravity. String theory postulates the existence of additional dimensions.

Of the four fundamental interactions, gravitation is dominant at cosmological length scales, including galaxies and larger-scale structures. Gravity's effects are cumulative; by contrast, the effects of positive and negative charges tend to cancel one another, making electromagnetism relatively insignificant on cosmological length scales. The remaining two interactions, the weak and strong nuclear forces, decline very rapidly with distance; their effects are confined mainly to sub-atomic length scales.

The Universe appears to have much more matter than antimatter, an asymmetry possibly related to the observations of CP violation.[38] The Universe also appears to have neither net momentum nor angular momentum. The absence of net charge and momentum would follow from accepted physical laws (Gauss's law and the non-divergence of the stress-energy-momentum pseudotensor, respectively) if the Universe were finite.[39]

2.4.1 Shape

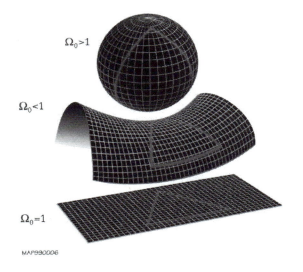

The three possible options of the shape of the Universe.

Main article: Shape of the Universe

General relativity describes how spacetime is curved and bent by mass and energy. The topology or geometry of the Universe includes both local geometry in the observable universe and global geometry. Cosmologists often work with a given space-like slice of spacetime called the comoving coordinates. The section of spacetime which can be observed is the backward light cone, which delimits the cosmological horizon. The cosmological horizon (also called the particle horizon or the light horizon) is the maximum distance from which particles can have traveled to the observer in the age of the Universe. This horizon represents the boundary between the observable and the unobservable regions of the Universe.[40][41] The existence, properties, and significance of a cosmological horizon depend on the particular cosmological model.

An important parameter determining the future evolution of the Universe theory is the density parameter, Omega (Ω), defined as the average matter density of the universe divided by a critical value of that density. This selects one of three possible geometries depending on whether Ω is equal to, less than, or greater than 1. These are called, respectively, the flat, open and closed universes.[42]

Observations, including the Cosmic Background Explorer (COBE), Wilkinson Microwave Anisotropy Probe (WMAP), and Planck maps of the CMB, suggest that the Universe is infinite in extent with a finite age, as described by the Friedmann–Lemaître–Robertson–Walker (FLRW) models.[43][44][45][46] These FLRW models thus support inflationary models and the standard model of cosmology, describing a flat, homogeneous universe presently dominated by dark matter and dark energy.[47][48]

2.4.2 Size and regions

See also: Observable universe and Observational cosmology

The size of the Universe is somewhat difficult to define. According to a restrictive definition, the Universe is everything within our connected spacetime that could have a chance to interact with us and vice versa.[49] According to the general theory of relativity, some regions of space may never interact with ours even in the lifetime of the Universe due to the finite speed of light and the ongoing expansion of space. For example, radio messages sent from Earth may never reach some regions of space, even if the Universe were to exist forever: space may expand faster than light can traverse it.[50]

Distant regions of space are assumed to exist and be part of reality as much as we are even though we can never interact with them. The spatial region that we can affect and be affected by is the observable universe. The observable universe depends on the location of the observer. By traveling, an observer can come into contact with a greater region of spacetime than an observer who remains still. Nevertheless, even the most rapid traveler will not be able to interact with all of space. Typically, the observable universe is taken to mean the portion of the Universe that is observable from our vantage point in the Milky Way.

The proper distance—the distance as would be measured at a specific time, including the present—between Earth and the edge of the observable universe is 46 billion light-years (14 billion parsecs), making the diameter of the observable universe about 91 billion light-years (28×10^9 pc). The distance the light from the edge of the observable universe has travelled is very close to the age of the Universe times the speed of light, 13.8 billion light-years (4.2×10^9 pc), but this does not represent the distance at any given time because the edge of the Universe and the Earth have moved since further apart.[51] For comparison, the diameter of a typical galaxy is 30,000 light-years, and the typical distance between two neighboring galaxies is 3 million light-years.[52] As an example, the Milky Way is roughly 100,000 light years in diameter,[53] and the nearest sister galaxy to the Milky Way, the Andromeda Galaxy, is located roughly 2.5 million light years away.[54] Because we cannot observe space beyond the edge of the observable universe, it is unknown whether the size of the Universe is finite or infinite.[12][55][56]

2.4.3 Age and expansion

Main articles: Age of the universe and Metric expansion of space

Astronomers calculate the age of the Universe by assuming that the Lambda-CDM model accurately describes the evolution of the Universe from a very uniform, hot, dense primordial state to its present state and measuring the cosmological parameters which constitute the model. This model is well understood theoretically and supported by recent high-precision astronomical observations such as WMAP and Planck. Commonly, the set of observations fitted includes the cosmic microwave background anisotropy, the brightness/redshift relation for Type Ia supernovae, and large-scale galaxy clustering including the baryon acoustic oscillation feature. Other observations, such as the Hubble constant, the abundance of galaxy clusters, weak gravitational lensing and globular cluster ages, are generally consistent with these, providing a check of the model, but are less accurately measured at present. With the prior that the Lambda-CDM model is correct, the measurements of the parameters using a variety of techniques by numerous experiments yield a best value of the age of the Universe as of 2015 of 13.799 ± 0.021 billion years.[1]

Over time, the Universe and its contents have evolved; for example, the relative population of quasars and galaxies has changed[57] and space itself has expanded. Due to this expansion, scientists on Earth can observe the light from a galaxy 30 billion light years away even though that light has traveled for only 13 billion years; the very space between them has expanded. This expansion is consistent with the observation that the light from distant galaxies has been redshifted; the photons emitted have been stretched to longer wavelengths and lower frequency during their journey. Analyses of Type Ia supernovae indicate that the spatial expansion is accelerating.[58][59]

The more matter there is in the Universe, the stronger the mutual gravitational pull of the matter. If the Universe were *too* dense then it would re-collapse into a gravitational singularity. However, if the Universe contained too *little* matter then the expansion would accelerate too rapidly for planets and planetary systems to form. Since the Big Bang, the universe has expanded monotonically. Surprisingly, our universe has just the right mass density of about 5 protons per cubic meter which has allowed it to expand for the last 13.8 billion years, giving time to form the universe as observed today.[60]

There are dynamical forces acting on the particles in the Universe which affect the expansion rate. Before 1998, it was expected that the rate of increase of the Hubble Constant would be decreasing as time went on due to the influence of gravitational interactions in the Universe, and thus there is an additional observable quantity in the Universe called the deceleration parameter which cosmologists expected to be directly related to the matter density of the Universe. In 1998, the deceleration parameter was measured by two different groups to be consistent with −1 but not zero, which implied that the present-day rate of increase of the Hubble Constant is increasing over time.[61][16]

2.4.4 Spacetime

Main articles: Spacetime and World line
See also: Lorentz transformation

Spacetimes are the arenas in which all physical events take place—an event is a point in spacetime specified by its time and place. The basic elements of spacetime are events. In any given spacetime, an event is a unique position at a unique time. Because events are spacetime points, an example of an event in classical relativistic physics is (x, y, z, t), the location of an elementary (point-like) particle at a particular time. A spacetime is the union of all events in the same way that a line is the union of all of its points, formally organized into a manifold.

The Universe appears to be a smooth spacetime continuum consisting of three spatial dimensions and one temporal (time) dimension. On the average, space is observed to be very nearly flat (close to zero curvature), meaning that Euclidean geometry is empirically true with high accuracy throughout most of the Universe.[62] Spacetime also appears to have a simply connected topology, in analogy with a sphere, at least on the length-scale of the observable Universe. However, present observations cannot exclude the possibilities that the Universe has more dimensions and that its spacetime may have a multiply connected global topology, in analogy with the cylindrical or toroidal topologies of two-dimensional spaces.[44][63]

2.5 Contents

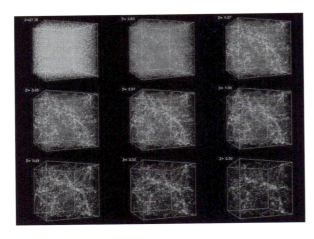

The formation of clusters and large-scale filaments in the Cold Dark Matter model with dark energy. The frames show the evolution of structures in a 43 million parsecs (or 140 million light years) box from redshift of 30 to the present epoch (upper left z=30 to lower right z=0).

See also: Galaxy formation and evolution, Galaxy cluster, Illustris project and Nebula

The Universe is composed almost completely of dark energy, dark matter, and ordinary matter. Other contents are electromagnetic radiation (estimated to be from 0.005% to close to 0.01%) and antimatter.[64][65][66] The total amount of electromagnetic radiation generated within the universe has decreased by 1/2 in the past 2 billion years.[67][68]

Ordinary matter, which includes atoms, stars, galaxies, and life, accounts for only 4.9% of the contents of the Universe.[6] The present overall density of this type of matter is very low, roughly 4.5×10^{-31} grams per cubic centimetre, corresponding to a density of the order of only one proton for every four cubic meters of volume.[4] The nature of both dark energy and dark matter is unknown. Dark matter, a mysterious form of matter that has not yet been identified, accounts for 26.8% of the contents. Dark energy, which is the energy of empty space and that is causing the expansion of the Universe to accelerate, accounts for the remaining 68.3% of the contents.[69][70][6]

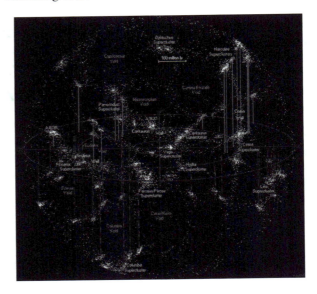

A map of the Superclusters and voids nearest to Earth

Matter, dark matter, and dark energy are distributed homogeneously throughout the Universe over length scales longer than 300 million light-years or so.[71] However, over shorter length-scales, matter tends to clump hierarchically; many atoms are condensed into stars, most stars into galaxies, most galaxies into clusters, superclusters and, finally, large-scale galactic filaments. The observable Universe contains approximately 300 sextillion (3×10^{23}) stars[72] and more than 100 billion (10^{11}) galaxies.[73] Typical galaxies range from dwarfs with as few as ten million[74] (10^7) stars up to giants with one trillion[75] (10^{12}) stars. Between the structures are voids, which are typically 10–150 Mpc (33 million–490 million ly) in diameter. The Milky Way is in the Local Group of galaxies, which in turn is in the Laniakea Supercluster.[76] This supercluster spans over 500 million light years, while the Local Group spans over 10 million light years.[77] The Universe also has vast regions of relative emptiness; the largest known void measures 1.8 billion ly (550 Mpc) across.[78]

The observable Universe is isotropic on scales significantly larger than superclusters, meaning that the statistical properties of the Universe are the same in all directions as observed from Earth. The Universe is bathed in highly isotropic microwave radiation that corresponds to a thermal equilibrium blackbody spectrum of roughly 2.72548 kelvin.[5] The hypothesis that the large-scale Universe is homogeneous and isotropic is known as the cosmological principle.[80] A Universe that is both homo-

2.5. CONTENTS

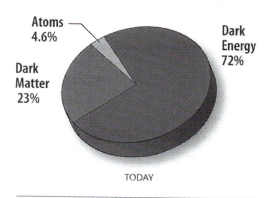

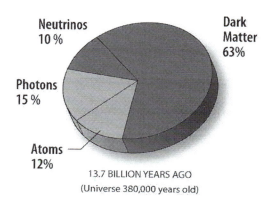

Comparison of the contents of the Universe today to 380,000 years after the Big Bang as measured with 5 year WMAP data (from 2008).[79] (Due to rounding errors, the sum of these numbers is not 100%). This reflects the 2008 limits of WMAP's ability to define Dark Matter and Dark Energy.

geneous and isotropic looks the same from all vantage points[81] and has no center.[82]

2.5.1 Dark energy

Main article: Dark energy

An explanation for why the expansion of the Universe is accelerating remains elusive. It is often attributed to "dark energy", an unknown form of energy that is hypothesized to permeate space.[83] On a mass–energy equivalence basis, the density of dark energy (6.91×10^{-27} kg/m^3) is much less than the density of ordinary matter or dark matter within galaxies. However, in the present dark-energy era, it dominates the mass–energy of the universe because it is uniform across space.[84]

Two proposed forms for dark energy are the cosmological constant, a *constant* energy density filling space homogeneously,[85] and scalar fields such as quintessence

or moduli, *dynamic* quantities whose energy density can vary in time and space. Contributions from scalar fields that are constant in space are usually also included in the cosmological constant. The cosmological constant can be formulated to be equivalent to vacuum energy. Scalar fields having only a slight amont of spatial inhomogeneity would be difficult to distinguish from a cosmological constant.

2.5.2 Dark matter

Main article: Dark matter

Dark matter is a hypothetical kind of matter that cannot be seen with telescopes, but which accounts for most of the matter in the Universe. The existence and properties of dark matter are inferred from its gravitational effects on visible matter, radiation, and the large-scale structure of the Universe. Other than neutrinos, a form of hot dark matter, dark matter has not been detected directly, making it one of the greatest mysteries in modern astrophysics. Dark matter neither emits nor absorbs light or any other electromagnetic radiation at any significant level. Dark matter is estimated to constitute 26.8% of the total mass–energy and 84.5% of the total matter in the Universe.[69][86]

2.5.3 Ordinary Matter

Main article: Matter

The remaining 4.9% of the mass–energy of the Universe is ordinary matter, that is, atoms, ions, electrons and the objects they form. This matter includes stars, which produce nearly all of the light we see from galaxies, as well as interstellar gas in the interstellar and intergalactic media, planets, and all the objects from everyday life that we can bump into, touch or squeeze.[87] Ordinary matter commonly exists in four states (or phases): solid, liquid, gas, and plasma. However, advances in experimental techniques have revealed other previously theoretical phases, such as Bose–Einstein condensates and fermionic condensates.

Ordinary matter is composed of two types of elementary particles: quarks and leptons.[88] For example, the proton is formed of two up quarks and one down quark; the neutron is formed of two down quarks and one up quark; and the electron is a kind of lepton. An atom consists of an atomic nucleus, made up of protons and neutrons, and electrons that orbit the nucleus. Because most of the mass of an atom is concentrated in its nucleus, which is made up of baryons, astronomers often use the term *baryonic matter* to describe ordinary matter, although a small fraction of this "baryonic matter" is electrons.

Soon after the Big Bang, primordial protons and neutrons formed from the quark–gluon plasma of the early Universe as it cooled below two trillion degrees. A few minutes later, in a process known as Big Bang nucleosynthesis, nuclei formed from the primordial protons and neutrons. This nucleosynthesis formed lighter elements, those with small atomic numbers up to lithium and beryllium, but the abundance of heavier elements dropped off sharply with increasing atomic number. Some boron may have been formed at this time, but the next heavier element, carbon, was not be formed in significant amounts. Big Bang nucleosynthesis shut down after about 20 minutes due to the rapid drop in temperature and density of the expanding Universe. Subsequent formation of heavier elements resulted from stellar nucleosynthesis and supernova nucleosynthesis.[89]

2.5.4 Particles

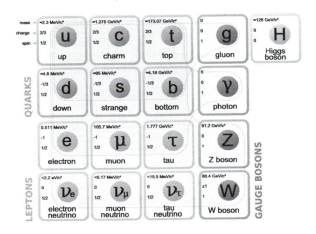

Standard model of elementary particles: the 12 fundamental fermions and 4 fundamental bosons. Brown loops indicate which bosons (red) couple to which fermions (purple and green). Columns are three generations of matter (fermions) and one of forces (bosons). In the first three columns, two rows contain quarks and two leptons. The top two rows' columns contain up (u) and down (d) quarks, charm (c) and strange (s) quarks, top (t) and bottom (b) quarks, and photon (γ) and gluon (g), respectively. The bottom two rows' columns contain electron neutrino (νe) and electron (e), muon neutrino ($\nu \mu$) and muon (μ), tau neutrino ($\nu \tau$) and tau (τ), and the Z^0 and $W^{\pm}$ carriers of the weak force. Mass, charge, and spin are listed for each particle.

Main article: Particle physics

Ordinary matter and the forces that act on matter can be described in terms of elementary particles.[90] These particles are sometimes described as being fundamental, since they have an unknown substructure, and it is unknown whether or not they are composed of smaller and even more fundamental particles.[91][92] Of central importance is the Standard Model, a theory that is concerned with electromagnetic interactions and the weak and strong nuclear interactions.[93] The Standard Model is supported by the experimental confirmation of the existence of particles that compose matter: quarks and leptons, and their corresponding "antimatter" duals, as well as the force particles that mediate interactions: the photon, the W and Z bosons, and the gluon.[91] The Standard Model predicted the existence of the recently discovered Higgs boson, a particle that is a manifestation of a field within the Universe that can endow particles with mass.[94][95] Because of its success in explaining a wide variety of experimental results, the Standard Model is sometimes regarded as a "theory of almost everything".[93] The Standard Model does not, however, accommodate gravity. A true force-particle "theory of everything" has not been attained.[96]

Hadrons

Main article: Hadron

A hadron is a composite particle made of quarks held together by the strong force. Hadrons are categorized into two families: baryons (such as protons and neutrons) made of three quarks, and mesons (such as pions) made of one quark and one antiquark. Of the hadrons, protons are stable, and neutrons bound within atomic nuclei are stable. Other hadrons are unstable under ordinary conditions and are thus insignificant constituents of the modern Universe. From approximately 10^{-6} seconds after the Big Bang, during a period is known as the hadron epoch, the temperature of the universe had fallen sufficiently to allow quarks to bind together into hadrons, and the mass of the Universe was dominated by hadrons. Initially the temperature was high enough to allow the formation of hadron/anti-hadron pairs, which kept matter and antimatter in thermal equilibrium. However, as the temperature of the Universe continued to fall, hadron/anti-hadron pairs were no longer produced. Most of the hadrons and anti-hadrons were then eliminated in particle-antiparticle annihilation reactions, leaving a small residual of hadrons by the time the Universe was about one second old.[97]:244–266

Leptons

Main article: Lepton

A lepton is an elementary, half-integer spin particle that does not undergo strong interactions but is subject to the Pauli exclusion principle; no two leptons of the same species can be in exactly the same state at the same time.[98] Two main classes of leptons exist: charged leptons (also known as the *electron-like* leptons), and neutral leptons (better known as neutrinos). Electrons are stable and the most common charged lepton in the Universe, whereas muons and taus are unstable particle that quickly decay after being produced in high energy collisions, such as those involving cosmic rays or carried out in particle accelerators.[99][100] Charged leptons can combine with other particles to form various composite particles such as atoms and positronium. The electron governs nearly all of chemistry, as it is found in atoms and is directly tied to all chemical properties. Neutrinos rarely interact with anything, and are consequently rarely observed. Neutrinos stream throughout the Universe but rarely interact with normal matter.[101]

The lepton epoch was the period in the evolution of the early Universe in which the leptons dominated the mass of the Universe. It started roughly 1 second after the Big Bang, after the majority of hadrons and anti-hadrons annihilated each other at the end of the hadron epoch. During the lepton epoch the temperature of the Universe was still high enough to create lepton/anti-lepton pairs, so leptons and anti-leptons were in thermal equilibrium. Approximately 10 seconds after the Big Bang, the temperature of the Universe had fallen to the point where lepton/anti-lepton pairs were no longer created.[102] Most leptons and anti-leptons were then eliminated in annihilation reactions, leaving a small residue of leptons. The mass of the Universe was then dominated by photons as it entered the following photon epoch.

Photons

Main article: Photon epoch
See also: Photino

A photon is the quantum of light and all other forms of electromagnetic radiation. It is the force carrier for the electromagnetic force, even when static via virtual photons. The effects of this force are easily observable at the microscopic and at the macroscopic level because the photon has zero rest mass; this allows long distance interactions. Like all elementary particles, photons are currently best explained by quantum mechanics and exhibit wave–particle duality, exhibiting properties of waves and of particles.

The photon epoch started after most leptons and anti-leptons were annihilated at the end of the lepton epoch, about 10 seconds after the Big Bang. Atomic nuclei were created in the process of nucleosynthesis which occurred during the first few minutes of the photon epoch. For the remainder of the photon epoch the Universe contained a hot dense plasma of nuclei, electrons and photons. About 380,000 years after the Big Bang, the temperature of the Universe fell to the point where nuclei could combine with electrons to create neutral atoms. As a result, photons no longer interacted frequently with matter and the Universe became transparent. The highly redshifted photons from this period form the cosmic microwave background. Tiny variations in temperature and density detectable in the CMB were the early "seeds" from which all subsequent structure formation took place.[97]:244–266

2.6 Cosmological models

2.6.1 Model of the Universe based on general relativity

Main article: Solutions of the Einstein field equations
See also: Big Bang and Ultimate fate of the Universe

General relativity is the geometric theory of gravitation published by Albert Einstein in 1915 and the current description of gravitation in modern physics. It is the basis of current cosmological models of the Universe. General relativity generalizes special relativity and Newton's law of universal gravitation, providing a unified description of gravity as a geometric property of space and time, or spacetime. In particular, the curvature of spacetime is directly related to the energy and momentum of whatever matter and radiation are present. The relation is specified by the Einstein field equations, a system of partial differential equations. In general relativity, the distribution of matter and energy determines the geometry of spacetime, which in turn describes the acceleration of matter. Therefore, solutions of the Einstein field equations describe the evolution of the Universe. Combined with measurements of the amount, type, and distribution of matter in the Universe, the equations of general relativity describe the evolution of the Universe over time.[103]

With the assumption of the cosmological principle that the Universe is homogeneous and isotropic everywhere, a specific solution of the field equations that describes the Universe is the metric tensor called the Friedmann–Lemaître–Robertson–Walker metric,

$$ds^2 = -c^2dt^2 + R(t)^2\left(\frac{dr^2}{1-kr^2} + r^2d\theta^2 + r^2\sin^2\theta\, d\phi^2\right)$$

where (r, θ, φ) correspond to a spherical coordinate system. This metric has only two undetermined parameters. An overall dimensionless length scale factor R describes the size scale of the Universe as a function of time; an increase in R is the expansion of the Universe.[104] A curvature index k describes the geometry. The index k is defined so that it can be only 0, corresponding to flat Euclidean geometry,

1, corresponding to a space of positive curvature, or −1, a space of positive or negative curvature.[105] The value of R as a function of time t depends upon k and the cosmological constant Λ.[103] The cosmological constant represents the energy density of the vacuum of space and could be related to dark energy.[70] The equation describing how R varies with time is known as the Friedmann equation after its inventor, Alexander Friedmann.[106]

The solutions for $R(t)$ depend on k and Λ, but some qualitative features of such solutions are general. First and most importantly, the length scale R of the Universe can remain constant *only* if the Universe is perfectly isotropic with positive curvature ($k=1$) and has one precise value of density everywhere, as first noted by Albert Einstein.[103] However, this equilibrium is unstable: because the Universe is known to be inhomogeneous on smaller scales, R must change over time. When R changes, all the spatial distances in the Universe change in tandem; there is an overall expansion or contraction of space itself. This accounts for the observation that galaxies appear to be flying apart; the space between them is stretching. The stretching of space also accounts for the apparent paradox that two galaxies can be 40 billion light years apart, although they started from the same point 13.8 billion years ago[107] and never moved faster than the speed of light.

Second, all solutions suggest that there was a gravitational singularity in the past, when R went to zero and matter and energy were infinitely dense. It may seem that this conclusion is uncertain because it is based on the questionable assumptions of perfect homogeneity and isotropy (the cosmological principle) and that only the gravitational interaction is significant. However, the Penrose–Hawking singularity theorems show that a singularity should exist for very general conditions. Hence, according to Einstein's field equations, R grew rapidly from an unimaginably hot, dense state that existed immediately following this singularity (when R had a small, finite value); this is the essence of the Big Bang model of the Universe. Understanding the singularity of the Big Bang likely requires a quantum theory of gravity, which does not yet exist.[108]

Third, the curvature index k determines the sign of the mean spatial curvature of spacetime[105] averaged over sufficiently large length scales (greater than about a billion light years). If $k=1$, the curvature is positive and the Universe has a finite volume.[109] Such universes are often visualized as a three-dimensional sphere embedded in a four-dimensional space. Conversely, if k is zero or negative, the Universe has infinite volume.[109] It may seem counter-intuitive that an infinite and yet infinitely dense Universe could be created in a single instant at the Big Bang when $R=0$, but exactly that is predicted mathematically when k does not equal 1. By analogy, an infinite plane has zero curvature but infinite area, whereas an infinite cylinder is finite in one direction and a torus is finite in both. A toroidal Universe could behave like a normal Universe with periodic boundary conditions.

The ultimate fate of the Universe is still unknown, because it depends critically on the curvature index k and the cosmological constant Λ. If the Universe were sufficiently dense, k would equal +1, meaning that its average curvature throughout is positive and the Universe will eventually recollapse in a Big Crunch,[110] possibly starting a new Universe in a Big Bounce. Conversely, if the Universe were insufficiently dense, k would equal 0 or −1 and the Universe would expand forever, cooling off and eventually reaching the Big Freeze and the heat death of the Universe.[103] Modern data suggests that the expansion speed of the Universe is not decreasing as originally expected, but increasing; if this continues indefinitely, the Universe may eventually reach a Big Rip. Observationally, the Universe appears to be flat ($k = 0$), with an overall density that is very close to the critical value between recollapse and eternal expansion.[111]

2.6.2 Multiverse hypothesis

Main articles: Multiverse, Many-worlds interpretation, Bubble universe theory and Parallel universe (fiction)
See also: Eternal inflation

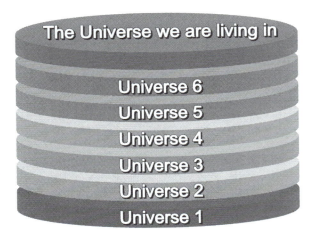

Depiction of a multiverse of seven "bubble" universes, which are separate spacetime continua, each having different physical laws, physical constants, and perhaps even different numbers of dimensions or topologies.

Some speculative theories have proposed that our Universe is but one of a set of disconnected universes, collectively denoted as the multiverse, challenging or enhancing more limited definitions of the Universe.[17][112] Scientific multiverse models are distinct from concepts such as alternate planes of consciousness and simulated reality.

Max Tegmark developed a four-part classification scheme for the different types of multiverses that scientists have suggested in various problem domains. An example of such a model is the chaotic inflation model of the early universe.[113] Another is the many-worlds interpretation of quantum mechanics. Parallel worlds are generated in a manner similar to quantum superposition and decoherence, with all states of the wave function being realized in separate worlds. Effectively, the multiverse evolves as a universal wavefunction. If the Big Bang that created our multiverse created an ensemble of multiverses, the wave function of the ensemble would be entangled in this sense.[114]

The least controversial category of multiverse in Tegmark's scheme is Level I, which describes distant spacetime events "in our own universe", but suggests that statistical analysis exploiting the anthropic principle provides an opportunity to test multiverse theories in some cases. If space is infinite, or sufficiently large and uniform, identical instances of the history of Earth's entire Hubble volume occur every so often, simply by chance. Tegmark calculated our nearest so-called doppelgänger, is $10^{10^{115}}$ meters away from us (a double exponential function larger than a googolplex).[115][116] In principle, it would be impossible to scientifically verify an identical Hubble volume. However, it does follow as a fairly straightforward consequence from otherwise unrelated scientific observations and theories.

It is possible to conceive of disconnected spacetimes, each existing but unable to interact with one another.[115][117] An easily visualized metaphor is a group of separate soap bubbles, in which observers living on one soap bubble cannot interact with those on other soap bubbles, even in principle.[118] According to one common terminology, each "soap bubble" of spacetime is denoted as a *universe*, whereas our particular spacetime is denoted as *the Universe*,[17] just as we call our moon *the Moon*. The entire collection of these separate spacetimes is denoted as the multiverse.[17] With this terminology, different *Universes* are not causally connected to each other.[17] In principle, the other unconnected *Universes* may have different dimensionalities and topologies of spacetime, different forms of matter and energy, and different physical laws and physical constants, although such possibilities are purely speculative.[17] Others consider each of several bubbles created as part of chaotic inflation to be separate *Universes*, though in this model these universes all share a causal origin.[17]

2.6.3 Fine-tuned Universe

Main article: Fine-tuned Universe

The fine-tuned Universe is the proposition that the conditions that allow life in the Universe can only occur when certain universal fundamental physical constants lie within a very narrow range, so that if any of several fundamental constants were only slightly different, the Universe would be unlikely to be conducive to the establishment and development of matter, astronomical structures, elemental diversity, or life as it is understood.[119] The proposition is discussed among philosophers, scientists, theologians, and proponents and detractors of creationism.

2.7 Historical development

See also: Cosmology, Timeline of cosmology, Nicolaus Copernicus § Copernican system and Philosophiæ Naturalis Principia Mathematica § Beginnings of the Scientific Revolution

Historically, there have been many ideas of the cosmos (cosmologies) and its origin (cosmogonies). Theories of an impersonal Universe governed by physical laws were first proposed by the Greeks and Indians.[14] Ancient Chinese philosophy encompassed the notion of the Universe including both all of space and all of time.[120][121] Over the centuries, improvements in astronomical observations and theories of motion and gravitation led to ever more accurate descriptions of the Universe. The modern era of cosmology began with Albert Einstein's 1915 general theory of relativity, which made it possible to quantitatively predict the origin, evolution, and conclusion of the Universe as a whole. Most modern, accepted theories of cosmology are based on general relativity and, more specifically, the predicted Big Bang.[122]

2.7.1 Mythologies

Main articles: Creation myth, Creator deity and Religious cosmology

Many cultures have stories describing the origin of the world and universe. Cultures generally regard these stories as having some truth. There are however many differing beliefs in how these stories apply amongst those believing in a supernatural origin, ranging from a god directly creating the Universe as it is now to a god just setting the "wheels in motion" (for example via mechanisms such as the big bang and evolution).[123]

Ethnologists and anthropologists who study myths have developed various classification schemes for the various themes that appear in creation stories.[124][125] For example, in one type of story, the world is born from a world egg; such stories include the Finnish epic poem *Kalevala*, the Chinese story of Pangu or the Indian Brahmanda Purana. In related stories, the Universe is created by a single entity emanating or producing something by him- or herself, as in the Tibetan Buddhism concept of Adi-Buddha, the ancient Greek story of Gaia (Mother Earth), the Aztec goddess Coatlicue myth, the ancient Egyptian god Atum story, and the Judeo-Christian Genesis creation narrative in which the Abrahamic God created the Universe. In another type of story, the Universe is created from the union of male and female deities, as in the Maori story of Rangi and Papa. In other stories, the Universe is created by crafting it from pre-existing materials, such as the corpse of a dead god — as from Tiamat in the Babylonian epic *Enuma Elish* or from the giant Ymir in Norse mythology – or from chaotic materials, as in Izanagi and Izanami in Japanese mythology. In other stories, the Universe emanates from fundamental principles, such as Brahman and Prakrti, the creation myth of the Serers,[126] or the yin and yang of the Tao.

2.7.2 Philosophical models

Further information: Cosmology
See also: Pre-Socratic philosophy, Physics (Aristotle), Hindu cosmology, Islamic cosmology and Philosophy of space and time

The pre-Socratic Greek philosophers and Indian philosophers developed some of the earliest philosophical concepts of the Universe.[14][127] The earliest Greek philosophers noted that appearances can be deceiving, and sought to understand the underlying reality behind the appearances. In particular, they noted the ability of matter to change forms (e.g., ice to water to steam) and several philosophers proposed that all the physical materials in the world are different forms of a single primordial material, or *arche*. The first to do so was Thales, who proposed this material to be water. Thales' student, Anaximander, proposed that everything came from the limitless *apeiron*. Anaximenes proposed the primordial material to be air on account of its perceived attractive and repulsive qualities that cause the *arche* to condense or dissociate into different forms. Anaxagoras proposed the principle of *Nous* (Mind), while Heraclitus proposed fire (and spoke of *logos*). Empedocles proposed the elements to be earth, water, air and fire. His four-element model became very popular. Like Pythagoras, Plato believed that all things were composed of number, with Empedocles' elements taking the form of the Platonic solids. Democritus, and later philosophers—most notably Leucippus—proposed that the Universe is composed of indivisible atoms moving through void (vacuum), although Aristotle did not believe that to be feasible because air, like water, offers resistance to motion. Air will immediately rush in to fill a void, and moreover, without resistance, it would do so indefinitely fast.[14]

Although Heraclitus argued for eternal change, his contemporary Parmenides made the radical suggestion that all change is an illusion, that the true underlying reality is eternally unchanging and of a single nature. Parmenides denoted this reality as τὸ ἕν (The One). Parmenides' idea seemed implausible to many Greeks, but his student Zeno of Elea challenged them with several famous paradoxes. Aristotle responded to these paradoxes by developing the notion of a potential countable infinity, as well as the infinitely divisible continuum. Unlike the eternal and unchanging cycles of time, he believed that the world is bounded by the celestial spheres and that cumulative stellar magnitude is only finitely multiplicative.

The Indian philosopher Kanada, founder of the Vaisheshika school, developed a notion of atomism and proposed that light and heat were varieties of the same substance.[128] In the 5th century AD, the Buddhist atomist philosopher Dignāga proposed atoms to be point-sized, durationless, and made of energy. They denied the existence of substantial matter and proposed that movement consisted of momentary flashes of a stream of energy.[129]

The notion of temporal finitism was inspired by the doctrine of creation shared by the three Abrahamic religions: Judaism, Christianity and Islam. The Christian philosopher, John Philoponus, presented the philosophical arguments against the ancient Greek notion of an infinite past and future. Philoponus' arguments against an infinite past were used by the early Muslim philosopher, Al-Kindi (Alkindus); the Jewish philosopher, Saadia Gaon (Saadia ben Joseph); and the Muslim theologian, Al-Ghazali (Algazel).[130]

2.7.3 Astronomical concepts

Main articles: History of astronomy and Timeline of astronomy

Astronomical models of the Universe were proposed soon after astronomy began with the Babylonian astronomers, who viewed the Universe as a flat disk floating in the ocean, and this forms the premise for early Greek maps like those of Anaximander and Hecataeus of Miletus.

Later Greek philosophers, observing the motions of the heavenly bodies, were concerned with developing models of the Universe-based more profoundly on empirical evidence. The first coherent model was proposed by Eudoxus of Cnidos. According to Aristotle's physical interpretation of the model, celestial spheres eternally rotate with uniform motion around a stationary Earth. Normal matter is entirely

2.7. HISTORICAL DEVELOPMENT

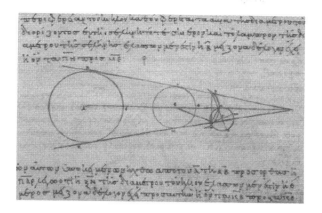

Aristarchus's 3rd century BCE calculations on the relative sizes of from left the Sun, Earth and Moon, from a 10th-century AD Greek copy

contained within the terrestrial sphere.

De Mundo (composed before 250 BC or between 350 and 200 BC), stated, *Five elements, situated in spheres in five regions, the less being in each case surrounded by the greater — namely, earth surrounded by water, water by air, air by fire, and fire by ether — make up the whole Universe.*[131]

This model was also refined by Callippus and after concentric spheres were abandoned, it was brought into nearly perfect agreement with astronomical observations by Ptolemy. The success of such a model is largely due to the mathematical fact that any function (such as the position of a planet) can be decomposed into a set of circular functions (the Fourier modes). Other Greek scientists, such as the Pythagorean philosopher Philolaus, postulated (according to Stobaeus account) that at the center of the Universe was a "central fire" around which the Earth, Sun, Moon and Planets revolved in uniform circular motion.[132]

The Greek astronomer Aristarchus of Samos was the first known individual to propose a heliocentric model of the Universe. Though the original text has been lost, a reference in Archimedes' book *The Sand Reckoner* describes Aristarchus's heliocentric model. Archimedes wrote: (translated into English):

> "You, King Gelon, are aware the Universe is the name given by most astronomers to the sphere the center of which is the center of the Earth, while its radius is equal to the straight line between the center of the Sun and the center of the Earth. This is the common account as you have heard from astronomers. But Aristarchus has brought out a book consisting of certain hypotheses, wherein it appears, as a consequence of the assumptions made, that the Universe is many times greater than the Universe just mentioned. His hypotheses are that the fixed stars and the Sun remain unmoved, that the Earth revolves about the Sun on the circumference of a circle, the Sun lying in the middle of the orbit, and that the sphere of fixed stars, situated about the same center as the Sun, is so great that the circle in which he supposes the Earth to revolve bears such a proportion to the distance of the fixed stars as the center of the sphere bears to its surface"

Aristarchus thus believed the stars to be very far away, and saw this as the reason why stellar parallax had not been observed, that is, the stars had not been observed to move relative each other as the Earth moved around the Sun. The stars are in fact much farther away than the distance that was generally assumed in ancient times, which is why stellar parallax is only detectable with precision instruments. The geocentric model, consistent with planetary parallax, was assumed to be an explanation for the unobservability of the parallel phenomenon, stellar parallax. The rejection of the heliocentric view was apparently quite strong, as the following passage from Plutarch suggests (*On the Apparent Face in the Orb of the Moon*):

> "Cleanthes [a contemporary of Aristarchus and head of the Stoics] thought it was the duty of the Greeks to indict Aristarchus of Samos on the charge of impiety for putting in motion the Hearth of the Universe [i.e. the Earth], ... supposing the heaven to remain at rest and the Earth to revolve in an oblique circle, while it rotates, at the same time, about its own axis"

Flammarion engraving, Paris 1888

The only other astronomer from antiquity known by name who supported Aristarchus's heliocentric model was Seleucus of Seleucia, a Hellenistic astronomer who lived a century after Aristarchus.[133][134][135] According to Plutarch, Seleucus was the first to prove the heliocentric system through reasoning, but it is not known what arguments he used. Seleucus' arguments for a heliocentric cosmology were probably related to the phenomenon of tides.[136] According to Strabo (1.1.9), Seleucus was the first to state that the tides are due to the attraction of the Moon, and that the height of the tides depends on the Moon's position relative to the Sun.[137] Alternatively, he may have proved heliocentricity by determining the constants of a geometric model for it, and by developing methods to compute planetary positions using this model, like what Nicolaus Copernicus later did in the 16th century.[138] During the Middle Ages, heliocentric models were also proposed by the Indian astronomer Aryabhata,[139] and by the Persian astronomers Albumasar[140] and Al-Sijzi.[141]

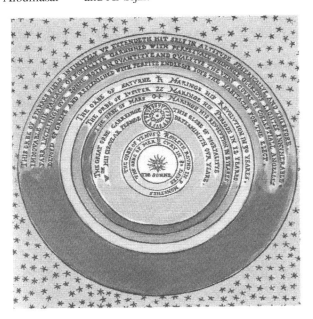

Model of the Copernican Universe by Thomas Digges in 1576, with the amendment that the stars are no longer confined to a sphere, but spread uniformly throughout the space surrounding the planets.

The Aristotelian model was accepted in the Western world for roughly two millennia, until Copernicus revived Aristarchus's perspective that the astronomical data could be explained more plausibly if the earth rotated on its axis and if the sun were placed at the center of the Universe.

> In the center rests the Sun. For who would place this lamp of a very beautiful temple in another or better place than this wherefrom it can illuminate everything at the same time?
> — Nicolaus Copernicus, in Chapter 10, Book 1 of *De Revolutionibus Orbium Coelestrum* (1543)

As noted by Copernicus himself, the notion that the Earth rotates is very old, dating at least to Philolaus (c. 450 BC), Heraclides Ponticus (c. 350 BC) and Ecphantus the Pythagorean. Roughly a century before Copernicus, the Christian scholar Nicholas of Cusa also proposed that the Earth rotates on its axis in his book, *On Learned Ignorance* (1440).[142] Aryabhata (476–550 AD/CE)[143] and Al-Sijzi[144] also proposed that the Earth rotates on its axis. Empirical evidence for the Earth's rotation on its axis, using the phenomenon of comets, was given by Tusi (1201–1274) and Ali Qushji (1403–1474).[145]

This cosmology was accepted by Isaac Newton, Christiaan Huygens and later scientists.[146] Edmund Halley (1720)[147] and Jean-Philippe de Chéseaux (1744)[148] noted independently that the assumption of an infinite space filled uniformly with stars would lead to the prediction that the nighttime sky would be as bright as the Sun itself; this became known as Olbers' paradox in the 19th century.[149] Newton believed that an infinite space uniformly filled with matter would cause infinite forces and instabilities causing the matter to be crushed inwards under its own gravity.[146] This instability was clarified in 1902 by the Jeans instability criterion.[150] One solution to these paradoxes is the Charlier Universe, in which the matter is arranged hierarchically (systems of orbiting bodies that are themselves orbiting in a larger system, *ad infinitum*) in a fractal way such that the Universe has a negligibly small overall density; such a cosmological model had also been proposed earlier in 1761 by Johann Heinrich Lambert.[52][151] A significant astronomical advance of the 18th century was the realization by Thomas Wright, Immanuel Kant and others of nebulae.[147]

The modern era of physical cosmology began in 1917, when Albert Einstein first applied his general theory of relativity to model the structure and dynamics of the Universe.[152]

2.8 See also

- Cosmic Calendar (scaled down timeline)
- Cosmic latte
- Esoteric cosmology
- False vacuum
- Illustris project
- Galaxy And Mass Assembly survey
- History of the Center of the Universe
- Nucleocosmochronology
- Non-standard cosmology

- Rare Earth hypothesis
- Religious cosmology
- Vacuum genesis
- World view
- Zero-energy Universe

2.9 References

[1] Planck Collaboration (2015). "Planck 2015 results. XIII. Cosmological parameters (See Table 4 on page 31 of pfd).". arXiv:1502.01589.

[2] Itzhak Bars; John Terning (2009). *Extra Dimensions in Space and Time*. Springer. pp. 27ff. ISBN 978-0-387-77637-8. Retrieved 2011-05-01.

[3] Paul Davies (2006). *The Goldilocks Enigma*. First Mariner Books. p. 43ff. ISBN 978-0-618-59226-5. Retrieved 2013-07-01.

[4] NASA/WMAP Science Team (24 January 2014). "Universe 101: What is the Universe Made Of?". NASA. Retrieved 2015-02-17.

[5] Fixsen, D. J. (2009). "The Temperature of the Cosmic Microwave Background". *The Astrophysical Journal* **707** (2): 916–920. arXiv:0911.1955. Bibcode:2009ApJ...707..916F. doi:10.1088/0004-637X/707/2/916.

[6] "First Planck results: the Universe is still weird and interesting". *Matthew Francis*. Ars technica. 2013-03-21. Retrieved 2015-08-21.

[7] NASA/WMAP Science Team (24 January 2014). "Universe 101: Will the Universe expand forever?". NASA. Retrieved 16 April 2015.

[8] *Universe*. Webster's New World College Dictionary, Wiley Publishing, Inc. 2010.

[9] "Universe". *Dictionary.com*. Retrieved 2012-09-21.

[10] "Universe". *Merriam-Webster Dictionary*. Retrieved 2012-09-21.

[11] Zeilik, Michael; Gregory, Stephen A. (1998). *Introductory Astronomy & Astrophysics* (4th ed.). Saunders College Publishing. ISBN 0030062284. The totality of all space and time; all that is, has been, and will be.

[12] Brian Greene (2011). *The Hidden Reality*. Alfred A. Knopf.

[13] Dold-Samplonius, Yvonne (2002). *From China to Paris: 2000 Years Transmission of Mathematical Ideas*. Franz Steiner Verlag.

[14] Thomas F. Glick; Steven Livesey; Faith Wallis. *Medieval Science Technology and Medicine: An Encyclopedia*. Routledge.

[15] Hawking, Stephen (1988). *A Brief History of Time*. Bantam Books. p. 125. ISBN 0-553-05340-X.

[16] "The Nobel Prize in Physics 2011". Retrieved 16 April 2015.

[17] Ellis, George F.R.; U. Kirchner; W.R. Stoeger (2004). "Multiverses and physical cosmology". *Monthly Notices of the Royal Astronomical Society* **347** (3): 921–936. arXiv:astro-ph/0305292. Bibcode:2004MNRAS.347..921E. doi:10.1111/j.1365-2966.2004.07261.x.

[18] Palmer, Jason. (2011-08-03) BBC News – 'Multiverse' theory suggested by microwave background. Retrieved 2011-11-28.

[19] Paul Copan; William Lane Craig (2004). *Creation Out of Nothing: A Biblical, Philosophical, and Scientific Exploration*. Baker Academic. p. 220. ISBN 9780801027338.

[20] Alexander Bolonkin (November 2011). *Universe, Human Immortality and Future Human Evaluation*. Elsevier. pp. 3–. ISBN 978-0-12-415801-6.

[21] Duco A. Schreuder (3 December 2014). *Vision and Visual Perception*. Archway Publishing. pp. 135–. ISBN 978-1-4808-1294-9.

[22] Tegmark, Max. "The Mathematical Universe". *Foundations of Physics* **38** (2): 101–150. arXiv:0704.0646. Bibcode:2008FoPh...38..101T. doi:10.1007/s10701-007-9186-9. a short version of which is available at *Shut up and calculate*. (in reference to David Mermin's famous quote "shut up and calculate")

[23] Jim Holt (2012). *Why Does the World Exist?*. Liveright Publishing. p. 308.

[24] Timothy Ferris (1997). *The Whole Shebang: A State-of-the-Universe(s) Report*. Simon & Schuster. p. 400.

[25] *The Compact Edition of the Oxford English Dictionary*, volume II, Oxford: Oxford University Press, 1971, p. 3518.

[26] Lewis, C. T. and Short, S (1879) *A Latin Dictionary*, Oxford University Press, ISBN 0-19-864201-6, pp. 1933, 1977–1978.

[27] Liddell; Scott. "A Greek-English Lexicon". πᾶς

[28] Liddell; Scott. "A Greek-English Lexicon". ὅλος

[29] Liddell; Scott. "A Greek–English Lexicon". κόσμος

[30] Lewis, C. T.; Short, S (1879). *A Latin Dictionary*. Oxford University Press. pp. 1881–1882, 1175, 1189–1190. ISBN 0-19-864201-6.

[31] *The Compact Edition of the Oxford English Dictionary* **II**. Oxford: Oxford University Press. 1971. pp. 909, 569, 3821–3822, 1900. ISBN 978-0198611172.

[32] Joseph Silk (2009). *Horizons of Cosmology*. Templeton Pressr. p. 208.

[33] Simon Singh (2005). *Big Bang: The Origin of the Universe*. Harper Perennial. p. 560.

[34] C. Sivaram (1986). "Evolution of the Universe through the Planck epoch". *Astrophysics & Space Science* **125**: 189. Bibcode:1986Ap&SS.125..189S. doi:10.1007/BF00643984.

[35] Richard B. Larson and Volker Bromm (March 2002). "The First Stars in the Universe". *Scientific American*.

[36] Ryden, Barbara, "Introduction to Cosmology", 2006, eqn. 6.33

[37] Brill, Dieter; Jacobsen, Ted (2006). "Spacetime and Euclidean geometry". *General Relativity and Gravitation* **38**: 643. arXiv:gr-qc/0407022. Bibcode:2006GReGr..38..643B. doi:10.1007/s10714-006-0254-9.

[38] "Antimatter". Particle Physics and Astronomy Research Council. October 28, 2003. Retrieved 2006-08-10.

[39] Landau & Lifshitz (1975, p. 361): "It is interesting to note that in a closed space the total electric charge must be zero. Namely, every closed surface in a finite space encloses on each side of itself a finite region of space. Therefore the flux of the electric field through this surface is equal, on th eone hand, to the total charge located in the interior of the surface, and on the other hand to the total charge outside of it, with opposite sign. Consequently, the sum of the charges on the two sides of the surface is zero."

[40] Edward Robert Harrison (2000). *Cosmology: the science of the universe*. Cambridge University Press. pp. 447–. ISBN 978-0-521-66148-5. Retrieved 1 May 2011.

[41] Andrew R. Liddle; David Hilary Lyth (13 April 2000). *Cosmological inflation and large-scale structure*. Cambridge University Press. pp. 24–. ISBN 978-0-521-57598-0. Retrieved 1 May 2011.

[42] "What is the Ultimate Fate of the Universe?". *National Aeronautics and Space Administration*. NASA. Retrieved 23 August 2015.

[43] Will the Universe expand forever?, WMAP website at NASA.

[44] Luminet, Jean-Pierre; Weeks, Jeffrey R.; Riazuelo, Alain; Lehoucq, Roland; Uzan, Jean-Philippe (2003-10-09). "Dodecahedral space topology as an explanation for weak wide-angle temperature correlations in the cosmic microwave background". *Nature* **425** (6958): 593–5. arXiv:astro-ph/0310253. Bibcode:2003Natur.425..593L. doi:10.1038/nature01944. PMID 14534579.

[45] Roukema, Boudewijn; Zbigniew Buliński; Agnieszka Szaniewska; Nicolas E. Gaudin (2008). "A test of the Poincare dodecahedral space topology hypothesis with the WMAP CMB data". *Astronomy and Astrophysics* **482** (3): 747. arXiv:0801.0006. Bibcode:2008A&A...482..747L. doi:10.1051/0004-6361:20078777.

[46] Aurich, Ralf; Lustig, S.; Steiner, F.; Then, H. (2004). "Hyperbolic Universes with a Horned Topology and the CMB Anisotropy". *Classical and Quantum Gravity* **21** (21): 4901–4926. arXiv:astro-ph/0403597. Bibcode:2004CQGra..21.4901A. doi:10.1088/0264-9381/21/21/010.

[47] Planck collaboration (2014). "Planck 2013 results. XVI. Cosmological parameters". *Astronomy & Astrophysics*. arXiv:1303.5076. Bibcode:2014A&A...571A..16P. doi:10.1051/0004-6361/201321591.

[48] "Planck reveals 'almost perfect' universe". *Michael Banks*. Physics World. 2013-03-21. Retrieved 2013-03-21.

[49] McCall, Storrs. *A Model of the Universe: Space-time, Probability, and Decision*. Oxford University. p. 23.

[50] Michio Kaku (11 March 2008). *Physics of the Impossible: A Scientific Exploration into the World of Phasers, Force Fields, Teleportation, and Time Travel*. Knopf Doubleday Publishing Group. pp. 202–. ISBN 978-0-385-52544-2.

[51] Christopher Crockett (February 20, 2013). "What is a light-year?". *EarthSky*.

[52] Rindler, p. 196.

[53] Christian, Eric; Samar, Safi-Harb. "How large is the Milky Way?". Retrieved 2007-11-28.

[54] I. Ribas; C. Jordi; F. Vilardell; E.L. Fitzpatrick; R.W. Hilditch; F. Edward Guinan (2005). "First Determination of the Distance and Fundamental Properties of an Eclipsing Binary in the Andromeda Galaxy". *Astrophysical Journal* **635** (1): L37–L40. arXiv:astro-ph/0511045. Bibcode:2005ApJ...635L..37R. doi:10.1086/499161. McConnachie, A. W.; Irwin, M. J.; Ferguson, A. M. N.; Ibata, R. A.; Lewis, G. F.; Tanvir, N. (2005). "Distances and metallicities for 17 Local Group galaxies". *Monthly Notices of the Royal Astronomical Society* **356** (4): 979–997. arXiv:astro-ph/0410489. Bibcode:2005MNRAS.356..979M. doi:10.1111/j.1365-2966.2004.08514.x.

[55] "How can space travel faster than the speed of light?". *Vannesa Janek*. Universe Today. 20 February 2015. Retrieved 6 June 2015.

[56] "Is faster-than-light travel or communication possible? Section: Expansion of the Universe". *Philip Gibbs*. 1997. Retrieved 6 June 2015.

[57] Phil Berardelli (March 25, 2010). "Galaxy Collisions Give Birth to Quasars". *Science News*.

2.9. REFERENCES

[58] Riess, Adam G.; Filippenko; Challis; Clocchiatti; Diercks; Garnavich; Gilliland; Hogan; Jha; Kirshner; Leibundgut; Phillips; Reiss; Schmidt; Schommer; Smith; Spyromilio; Stubbs; Suntzeff; Tonry (1998). "Observational evidence from supernovae for an accelerating universe and a cosmological constant". *Astronomical J.* **116** (3): 1009–38. arXiv:astro-ph/9805201. Bibcode:1998AJ....116.1009R. doi:10.1086/300499.

[59] Perlmutter, S.; Aldering; Goldhaber; Knop; Nugent; Castro; Deustua; Fabbro; Goobar; Groom; Hook; Kim; Kim; Lee; Nunes; Pain; Pennypacker; Quimby; Lidman; Ellis; Irwin; McMahon; Ruiz-Lapuente; Walton; Schaefer; Boyle; Filippenko; Matheson; Fruchter; et al. (1999). "Measurements of Omega and Lambda from 42 high redshift supernovae". *Astrophysical Journal* **517** (2): 565–86. arXiv:astro-ph/9812133. Bibcode:1999ApJ...517..565P. doi:10.1086/307221.

[60] Sean Carroll and Michio Kaku (2014). *How the Universe Works 3*. End of the Universe. Discovery Channel.

[61] Overbye, Dennis (October 11, 2003). "A 'Cosmic Jerk' That Reversed the Universe". *New York Times*.

[62] WMAP Mission: Results – Age of the Universe. Map.gsfc.nasa.gov. Retrieved 2011-11-28.

[63] Luminet, Jean-Pierre; Boudewijn F. Roukema (1999). "Topology of the Universe: Theory and Observations". *Proceedings of Cosmology School held at Cargese, Corsica, August 1998*. arXiv:astro-ph/9901364.

[64] Fritzsche, Hellmut. "electromagnetic radiation | physics". Encyclopedia Britannica. p. 1. Retrieved 2015-07-26.

[65] "Physics 7:Relativity, SpaceTime and Cosmology" (PDF). *Physics 7:Relativity, SpaceTime and Cosmology*. University of California Riverside. Retrieved 2015-07-26.

[66] "Physics - for the 21st Century". *www.learner.org*. Harvard-Smithsonian Center for Astrophysics Annenberg Learner. Retrieved 2015-07-27.

[67] Redd,SPACE.com, Nola Taylor. "It's Official: The Universe Is Dying Slowly". Retrieved 2015-08-11.

[68] "RIP Universe - Your Time Is Coming... Slowly | Video". *Will Parr, et al.* Space.com. Retrieved 2015-08-20.

[69] Sean Carroll, Ph.D., Cal Tech, 2007, The Teaching Company, *Dark Matter, Dark Energy: The Dark Side of the Universe*, Guidebook Part 2 page 46, Accessed Oct. 7, 2013, "...dark matter: An invisible, essentially collisionless component of matter that makes up about 25 percent of the energy density of the universe... it's a different kind of particle... something not yet observed in the laboratory..."

[70] Peebles, P. J. E. & Ratra, Bharat (2003). "The cosmological constant and dark energy". *Reviews of Modern Physics* **75** (2): 559–606. arXiv:astro-ph/0207347. Bibcode:2003RvMP...75..559P. doi:10.1103/RevModPhys.75.559.

[71] Mandolesi, N.; Calzolari, P.; Cortiglioni, S.; Delpino, F.; Sironi, G.; Inzani, P.; Deamici, G.; Solheim, J. -E.; Berger, L.; Partridge, R. B.; Martenis, P. L.; Sangree, C. H.; Harvey, R. C. (1986). "Large-scale homogeneity of the Universe measured by the microwave background". *Nature* **319** (6056): 751. doi:10.1038/319751a0.

[72] Vergano, Dan (1 December 2010). "Universe holds billions more stars than previously thought". *USA Today*. Retrieved 2010-12-14.

[73] Mackie, Glen (February 1, 2002). "To see the Universe in a Grain of Taranaki Sand". Swinburne University. Retrieved 2006-12-20.

[74] "Unveiling the Secret of a Virgo Dwarf Galaxy". ESO. 2000-05-03. Retrieved 2007-01-03.

[75] "Hubble's Largest Galaxy Portrait Offers a New High-Definition View". NASA. 2006-02-28. Retrieved 2007-01-03.

[76] "Earth's new address: 'Solar System, Milky Way, Laniakea'". *Elizabeth Gibney*. Nature. 3 September 2014. Retrieved 21 August 2015.

[77] "Local Group". *Fraser Cain*. Universe Today. 4 May 2009. Retrieved 21 August 2015.

[78] "Astronomers discover largest known structure in the universe is ... a big hole". The Guardian. 20 April 2015.

[79] "Content of the Universe - WMAP 9yr Pie Chart". *wmap.gsfc.nasa.gov*. Retrieved 2015-07-26.

[80] Rindler, p. 202.

[81] Andrew Liddle (2003). *An Introduction to Modern Cosmology (2nd ed.)*. John Wiley & Sons. ISBN 978-0-470-84835-7.. p. 2.

[82] Livio, Mario (2001). *The Accelerating Universe: Infinite Expansion, the Cosmological Constant, and the Beauty of the Cosmos*. John Wiley and Sons. p. 53. Retrieved 31 March 2012.

[83] Peebles, P. J. E. and Ratra, Bharat (2003). "The cosmological constant and dark energy". *Reviews of Modern Physics* **75** (2): 559–606. arXiv:astro-ph/0207347. Bibcode:2003RvMP...75..559P. doi:10.1103/RevModPhys.75.559.

[84] "Dark Energy". *Hyperphysics*. Retrieved January 4, 2014.

[85] Carroll, Sean (2001). "The cosmological constant". *Living Reviews in Relativity* **4**. Retrieved 2006-09-28.

[86] "Planck captures portrait of the young Universe, revealing earliest light". University of Cambridge. 21 March 2013. Retrieved 21 March 2013.

[87] P. Davies (1992). *The New Physics: A Synthesis*. Cambridge University Press. p. 1. ISBN 0-521-43831-4.

[88] G. 't Hooft (1997). *In search of the ultimate building blocks*. Cambridge University Press. p. 6. ISBN 0-521-57883-3.

[89] Clayton, Donald D. (1983). *Principles of Stellar Evolution and Nucleosynthesis*. The University of Chicago Press. pp. 362–435. ISBN 0-226-10953-4.

[90] Veltman, Martinus (2003). *Facts and Mysteries in Elementary Particle Physics*. World Scientific. ISBN 981-238-149-X.

[91] Sylvie Braibant; Giorgio Giacomelli; Maurizio Spurio (2012). *Particles and Fundamental Interactions: An Introduction to Particle Physics* (2nd ed.). Springer. pp. 1–3. ISBN 978-94-007-2463-1.

[92] Close, Frank (2012). *Particle Physics: A Very Short Introduction*. Oxford University Press. ISBN 978-0192804341.

[93] R. Oerter (2006). *The Theory of Almost Everything: The Standard Model, the Unsung Triumph of Modern Physics* (Kindle ed.). Penguin Group. p. 2. ISBN 0-13-236678-9.

[94] Onyisi, P. (23 October 2012). "Higgs boson FAQ". University of Texas ATLAS group. Retrieved 2013-01-08.

[95] Strassler, M. (12 October 2012). "The Higgs FAQ 2.0". *ProfMattStrassler.com*. Retrieved 2013-01-08. [Q] Why do particle physicists care so much about the Higgs particle? [A] Well, actually, they don't. What they really care about is the Higgs *field*, because it is *so* important. [emphasis in original]

[96] Steven Weinberg. *Dreams of a Final Theory: The Scientist's Search for the Ultimate Laws of Nature*. Knopf Doubleday Publishing Group. ISBN 978-0-307-78786-6.

[97] Allday, Jonathan (2002). *Quarks, Leptons and the Big Bang* (Second ed.). IOP Publishing. ISBN 0-7503-0806-0.

[98] "Lepton (physics)". *Encyclopædia Britannica*. Retrieved 2010-09-29.

[99] Harari, H. (1977). "Beyond charm". In Balian, R.; Llewellyn-Smith, C.H. *Weak and Electromagnetic Interactions at High Energy, Les Houches, France, Jul 5- Aug 14, 1976*. Les Houches Summer School Proceedings **29**. North-Holland. p. 613.

[100] Harari H. (1977). "Three generations of quarks and leptons" (PDF). In E. van Goeler, Weinstein R. (eds.). *Proceedings of the XII Rencontre de Moriond*. p. 170. SLAC-PUB-1974.

[101] "Experiment confirms famous physics model" (Press release). MIT News Office. 18 April 2007.

[102] The Timescale of Creation

[103] Zeilik, Michael; Gregory, Stephen A. (1998). "25-2". *Introductory Astronomy & Astrophysics* (4th ed.). Saunders College Publishing. ISBN 0030062284.

[104] Raine & Thomas (2001, p. 12)

[105] Raine & Thomas (2001, p. 66)

[106] Friedmann A. (1922). "Über die Krümmung des Raumes" (PDF). *Zeitschrift für Physik* **10** (1): 377–386. Bibcode:1922ZPhy...10..377F. doi:10.1007/BF01332580.

[107] "Cosmic Detectives". The European Space Agency (ESA). 2013-04-02. Retrieved 2013-04-15.

[108] Raine & Thomas (2001, p. 122–123)

[109] Raine & Thomas (2001, p. 70)

[110] Raine & Thomas (2001, p. 84)

[111] Raine & Thomas (2001, p. 88, 110–113)

[112] Munitz MK (1959). "One Universe or Many?". *Journal of the History of Ideas* **12** (2): 231–255. doi:10.2307/2707516. JSTOR 2707516.

[113] Linde A. (1986). "Eternal chaotic inflation". *Mod. Phys. Lett.* **A1** (2): 81–85. Bibcode:1986MPLA....1...81L. doi:10.1142/S0217732386000129.
Linde A. (1986). "Eternally existing self-reproducing chaotic inflationary Universe" (PDF). *Phys. Lett.* **B175** (4): 395–400. Bibcode:1986PhLB..175..395L. doi:10.1016/0370-2693(86)90611-8. Retrieved 2011-03-17.

[114] Everett, Hugh (1957). "Relative State Formulation of Quantum Mechanics". *Reviews of Modern Physics* **29**: 454–462. Bibcode:1957RvMP...29..454E. doi:10.1103/RevModPhys.29.454.

[115] Tegmark M. (2003). "Parallel universes. Not just a staple of science fiction, other universes are a direct implication of cosmological observations". *Scientific American* **288** (5): 40–51. doi:10.1038/scientificamerican0503-40. PMID 12701329.

[116] Tegmark, Max (2003). J. D. Barrow; P.C.W. Davies; C.L. Harper, eds. "Parallel Universes". *Scientific American: "Science and Ultimate Reality: from Quantum to Cosmos", honoring John Wheeler's 90th birthday* (Cambridge University Press): 2131. arXiv:astro-ph/0302131. Bibcode:2003astro.ph..2131T. doi:10.1038/scientificamerican0503-40.

[117] Ellis G. F (2011). "Does the Multiverse Really Exist?". *Scientific American* **305** (2): 38–43. doi:10.1038/scientificamerican0811-38.

[118] Clara Moskowitz (August 12, 2011). "Weird! Our Universe May Be a 'Multiverse,' Scientists Say". *livescience*.

[119] Mark Isaak (ed.) (2005). "CI301: The Anthropic Principle". *Index to Creationist Claims*. TalkOrigins Archive. Retrieved 2007-10-31.

[120] Gernet, J. (1993–1994). "Space and time: Science and religion in the encounter between China and Europe". *Chinese Science* **11**. pp. 93–102.

2.9. REFERENCES

[121] Ng, Tai (2007). "III.3". *Chinese Culture, Western Culture: Why Must We Learn from Each Other?*. iUniverse, Inc.

[122] Blandford R. D. "A century of general relativity: Astrophysics and cosmology". *Science* **347** (6226): 103–108. Bibcode:2015Sci...347.1103B. doi:10.1126/science.aaa4033.

[123] Leeming, David A. (2010). *Creation Myths of the World*. ABC-CLIO. p. xvii. ISBN 978-1-59884-174-9. In common usage the word 'myth' refers to narratives or beliefs that are untrue or merely fanciful; the stories that make up national or ethnic mythologies describe characters and events that common sense and experience tell us are impossible. Nevertheless, all cultures celebrate such myths and attribute to them various degrees of literal or symbolic *truth*.

[124] Eliade, Mircea (1964). *Myth and Reality (Religious Traditions of the World)*. Allen & Unwin. ISBN 978-0-04-291001-7.

[125] Leonard, Scott A.; McClure, Michael (2004). *Myth and Knowing: An Introduction to World Mythology* (1st ed.). McGraw-Hill. ISBN 978-0-7674-1957-4.

[126] (Henry Gravrand, "La civilisation Sereer -Pangool") [in] Universität Frankfurt am Main, Frobenius-Institut, Deutsche Gesellschaft für Kulturmorphologie, Frobenius Gesellschaft, "Paideuma: Mitteilungen zur Kulturkunde, Volumes 43–44", F. Steiner (1997), pp. 144–5, ISBN 3515028420

[127] B. Young, Louise. *The Unfinished Universe*. Oxford University Press. p. 21.

[128] Will Durant, *Our Oriental Heritage*:

"Two systems of Hindu thought propound physical theories suggestively similar to those of Greece. Kanada, founder of the Vaisheshika philosophy, held that the world is composed of atoms as many in kind as the various elements. The Jains more nearly approximated to Democritus by teaching that all atoms were of the same kind, producing different effects by diverse modes of combinations. Kanada believed light and heat to be varieties of the same substance; Udayana taught that all heat comes from the Sun; and Vachaspati, like Newton, interpreted light as composed of minute particles emitted by substances and striking the eye."

[129] Stcherbatsky, F. Th. (1930, 1962), *Buddhist Logic*, Volume 1, p. 19, Dover, New York:

"The Buddhists denied the existence of substantial matter altogether. Movement consists for them of moments, it is a staccato movement, momentary flashes of a stream of energy... "Everything is evanescent",... says the Buddhist, because there is no stuff... Both systems [Sānkhya, and later Indian Buddhism] share in common a tendency to push the analysis of existence up to its minutest, last elements which are imagined as absolute qualities, or things possessing only one unique quality. They are called "qualities" (*guna-dharma*) in both systems in the sense of absolute qualities, a kind of atomic, or intra-atomic, energies of which the empirical things are composed. Both systems, therefore, agree in denying the objective reality of the categories of Substance and Quality,... and of the relation of Inference uniting them. There is in Sānkhya philosophy no separate existence of qualities. What we call quality is but a particular manifestation of a subtle entity. To every new unit of quality corresponds a subtle quantum of matter which is called *guna*, "quality", but represents a subtle substantive entity. The same applies to early Buddhism where all qualities are substantive... or, more precisely, dynamic entities, although they are also called *dharmas* ('qualities')."

[130] Donald Wayne Viney (1985). "The Cosmological Argument". *Charles Hartshorne and the Existence of God*. SUNY Press. pp. 65–68. ISBN 0-87395-907-8.

[131] Aristotle; Forster, E. S. (Edward Seymour), 1879-1950; Dobson, J. F. (John Frederic), 1875-1947 (1914). *De Mundo*. p. 2.

[132] Boyer, C. (1968) *A History of Mathematics*. Wiley, p. 54.

[133] Neugebauer, Otto E. (1945). "The History of Ancient Astronomy Problems and Methods". *Journal of Near Eastern Studies* **4** (1): 1–38. doi:10.1086/370729. JSTOR 595168. the Chaldaean Seleucus from Seleucia

[134] Sarton, George (1955). "Chaldaean Astronomy of the Last Three Centuries B. C". *Journal of the American Oriental Society* **75** (3): 166–173 (169). doi:10.2307/595168. JSTOR 595168. the heliocentrical astronomy invented by Aristarchos of Samos and still defended a century later by Seleucos the Babylonian

[135] William P. D. Wightman (1951, 1953), *The Growth of Scientific Ideas*, Yale University Press p. 38, where Wightman calls him Seleukos the Chaldean.

[136] Lucio Russo, *Flussi e riflussi*, Feltrinelli, Milano, 2003, ISBN 88-07-10349-4.

[137] Bartel (1987, p. 527)

[138] Bartel (1987, pp. 527–9)

[139] Bartel (1987, pp. 529–34)

[140] Bartel (1987, pp. 534–7)

[141] Nasr, Seyyed H. (1993) [1964]. *An Introduction to Islamic Cosmological Doctrines* (2nd ed.). 1st edition by Harvard University Press, 2nd edition by State University of New York Press. pp. 135–6. ISBN 0-7914-1515-5.

[142] Misner, Thorne and Wheeler, p. 754.

[143] Abhishek Parakh (2006). "A Note on Aryabhata's Principle of Relativity". arXiv:physics/0610095.

[144] Ālī, Ema Ākabara. *Science in the Quran* **1**. Malik Library. p. 218.

[145] Ragep, F. Jamil (2001), "Tusi and Copernicus: The Earth's Motion in Context", *Science in Context* (Cambridge University Press) **14** (1–2): 145–163, doi:10.1017/s0269889701000060

[146] Misner, Thorne and Wheeler, p. 755–756.

[147] Misner, Thorne and Wheeler, p. 756.

[148] de Cheseaux JPL (1744). *Traité de la Comète*. Lausanne. pp. 223ff.. Reprinted as Appendix II in Dickson FP (1969). *The Bowl of Night: The Physical Universe and Scientific Thought*. Cambridge, MA: M.I.T. Press. ISBN 978-0-262-54003-2.

[149] Olbers HWM (1826). "Unknown title". *Bode's Jahrbuch* **111**.. Reprinted as Appendix I in Dickson FP (1969). *The Bowl of Night: The Physical Universe and Scientific Thought*. Cambridge, MA: M.I.T. Press. ISBN 978-0-262-54003-2.

[150] Jeans, J. H. (1902). "The Stability of a Spherical Nebula" (PDF). *Philosophical Transactions of the Royal Society A* **199** (312–320): 1–53. Bibcode:1902RSPTA.199....1J. doi:10.1098/rsta.1902.0012. JSTOR 90845. Retrieved 2011-03-17.

[151] Misner, Thorne and Wheeler, p. 757.

[152] Einstein, A (1917). "Kosmologische Betrachtungen zur allgemeinen Relativitätstheorie". *Preussische Akademie der Wissenschaften, Sitzungsberichte*. 1917. (part 1): 142–152.

Sources

- Bartel, Leendert van der Waerden (1987). "The Heliocentric System in Greek, Persian and Hindu Astronomy". *Annals of the New York Academy of Sciences* **500** (1): 525–545. Bibcode:1987NYASA.500..525V. doi:10.1111/j.1749-6632.1987.tb37224.x.

- Landau L, Lifshitz EM (1975). *The Classical Theory of Fields (Course of Theoretical Physics)* **2** (revised 4th English ed.). New York: Pergamon Press. pp. 358–397. ISBN 978-0-08-018176-9.

- Liddell, H. G. & Scott, R. (1968). *A Greek-English Lexicon*. Oxford University Press. ISBN 0-19-864214-8.

- Misner, C.W., Thorne, Kip, Wheeler, J.A. (1973). *Gravitation*. San Francisco: W. H. Freeman. pp. 703–816. ISBN 978-0-7167-0344-0.

- Raine, D. J.; Thomas, E. G. (2001). *An Introduction to the Science of Cosmology*. Institute of Physics Publishing.

- Rindler, W. (1977). *Essential Relativity: Special, General, and Cosmological*. New York: Springer Verlag. pp. 193–244. ISBN 0-387-10090-3.

Chapter 3

Many-worlds interpretation

The quantum-mechanical "Schrödinger's cat" paradox according to the many-worlds interpretation. In this interpretation, every event is a branch point; the cat is both alive and dead, even before the box is opened, but the "alive" and "dead" cats are in different branches of the universe, both of which are equally real, but which do not interact with each other.[1]

The **many-worlds interpretation** is an interpretation of quantum mechanics that asserts the objective reality of the universal wavefunction and denies the actuality of wavefunction collapse. Many-worlds implies that all possible alternate histories and futures are real, each representing an actual "world" (or "universe"). In lay terms, the hypothesis states there is a very large—perhaps infinite[2]—number of universes, and everything that could possibly have happened in our past, but did not, has occurred in the past of some other universe or universes. The theory is also referred to as **MWI**, the **relative state formulation**, the **Everett interpretation**, the **theory of the universal wavefunction**, **many-universes interpretation**, or just **many-worlds**.

The original relative state formulation is due to Hugh Everett in 1957.[3][4] Later, this formulation was popularized and renamed *many-worlds* by Bryce Seligman DeWitt in the 1960s and 1970s.[1][5][6][7] The decoherence approaches to interpreting quantum theory have been further explored and developed,[8][9][10] becoming quite popular. MWI is one of many multiverse hypotheses in physics and philosophy. It is currently considered a mainstream interpretation along with the other decoherence interpretations, collapse theories (including the historical Copenhagen interpretation),[11] and hidden variable theories such as the Bohmian mechanics.

Before many-worlds, reality had always been viewed as a single unfolding history. Many-worlds, however, views reality as a many-branched tree, wherein every possible quantum outcome is realised.[12] Many-worlds reconciles the observation of non-deterministic events, such as random radioactive decay, with the fully deterministic equations of quantum physics.

In many-worlds, the subjective appearance of wavefunction collapse is explained by the mechanism of quantum decoherence, and this is supposed to resolve all of the correlation paradoxes of quantum theory, such as the EPR paradox[13][14] and Schrödinger's cat,[1] since every possible outcome of every event defines or exists in its own "history" or "world".

3.1 Outline

Although several versions of many-worlds have been proposed since Hugh Everett's original work,[4] they all contain one key idea: the equations of physics that model the time evolution of systems *without* embedded observers are sufficient for modelling systems which *do* contain observers; in particular there is no observation-triggered wave function collapse which the Copenhagen interpretation proposes. Provided the theory is linear with respect to the wavefunction, the exact form of the quantum dynamics modelled, be it the non-relativistic Schrödinger equation, relativistic quantum field theory or some form of quantum gravity or string theory, does not alter the validity of MWI since MWI is a metatheory applicable to all linear quantum theories, and there is no experimental evidence for any non-linearity of the wavefunction in physics.[15][16] MWI's main conclusion is that the universe (or multiverse in this context) is

Hugh Everett III (1930–1982) was the first physicist who proposed the many-worlds interpretation (MWI) of quantum physics, which he termed his "relative state" formulation.

composed of a quantum superposition of very many, possibly even non-denumerably infinitely[2] many, increasingly divergent, non-communicating parallel universes or quantum worlds.[7]

The idea of MWI originated in Everett's Princeton Ph.D. thesis "The Theory of the Universal Wavefunction",[7] developed under his thesis advisor John Archibald Wheeler, a shorter summary of which was published in 1957 entitled "Relative State Formulation of Quantum Mechanics" (Wheeler contributed the title "relative state";[17] Everett originally called his approach the "Correlation Interpretation", where "correlation" refers to quantum entanglement). The phrase "many-worlds" is due to Bryce DeWitt,[7] who was responsible for the wider popularisation of Everett's theory, which had been largely ignored for the first decade after publication. DeWitt's phrase "many-worlds" has become so much more popular than Everett's "Universal Wavefunction" or Everett–Wheeler's "Relative State Formulation" that many forget that this is only a difference of terminology; the content of both of Everett's papers and DeWitt's popular article is the same.

The many-worlds interpretation shares many similarities with later, other "post-Everett" interpretations of quantum mechanics which also use decoherence to explain the process of measurement or wavefunction collapse. MWI treats the other histories or worlds as real since it regards the universal wavefunction as the "basic physical entity"[18] or "the fundamental entity, obeying at all times a deterministic wave equation".[19] The other decoherent interpretations, such as consistent histories, the Existential Interpretation etc., either regard the extra quantum worlds as metaphorical in some sense, or are agnostic about their reality; it is sometimes hard to distinguish between the different varieties. MWI is distinguished by two qualities: it assumes realism,[18][19] which it assigns to the wavefunction, and it has the minimal formal structure possible, rejecting any hidden variables, quantum potential, any form of a collapse postulate (i.e., Copenhagenism) or mental postulates (such as the many-minds interpretation makes).

Decoherent interpretations of many-worlds using einselection to explain how a small number of classical pointer states can emerge from the enormous Hilbert space of superpositions have been proposed by Wojciech H. Zurek. "Under scrutiny of the environment, only pointer states remain unchanged. Other states decohere into mixtures of stable pointer states that can persist, and, in this sense, exist: They are einselected."[20] These ideas complement MWI and bring the interpretation in line with our perception of reality.

Many-worlds is often referred to as a theory, rather than just an interpretation, by those who propose that many-worlds can make testable predictions (such as David Deutsch) or is falsifiable (such as Everett) or by those who propose that all the other, non-MW interpretations, are inconsistent, illogical or unscientific in their handling of measurements; Hugh Everett argued that his formulation was a metatheory, since it made statements about other interpretations of quantum theory; that it was the "only completely coherent approach to explaining both the contents of quantum mechanics and the appearance of the world."[21] Deutsch is dismissive that many-worlds is an "interpretation", saying that calling it an interpretation "is like talking about dinosaurs as an 'interpretation' of fossil records."[22]

3.2 Interpreting wavefunction collapse

As with the other interpretations of quantum mechanics, the many-worlds interpretation is motivated by behavior that can be illustrated by the double-slit experiment. When particles of light (or anything else) are passed through the double slit, a calculation assuming wave-like behavior of light can be used to identify where the particles are likely to be observed. Yet when the particles are observed in this experiment, they appear as particles (i.e., at definite places) and not as non-localized waves.

Some versions of the Copenhagen interpretation of quantum mechanics proposed a process of "collapse" in which an indeterminate quantum system would probabilistically collapse down onto, or select, just one determinate outcome to "explain" this phenomenon of observation. Wavefunction collapse was widely regarded as artificial and *ad hoc*, so an alternative interpretation in which the behavior of measurement could be understood from more fundamental physical principles was considered desirable.

Everett's Ph.D. work provided such an alternative interpretation. Everett stated that for a composite system – for example a subject (the "observer" or measuring apparatus) observing an object (the "observed" system, such as a particle) – the statement that either the observer or the observed has a well-defined state is meaningless; in modern parlance, the observer and the observed have become entangled; we can only specify the state of one *relative* to the other, i.e., the state of the observer and the observed are correlated *after* the observation is made. This led Everett to derive from the unitary, deterministic dynamics alone (i.e., without assuming wavefunction collapse) the notion of a *relativity of states*.

Everett noticed that the unitary, deterministic dynamics alone decreed that after an observation is made each element of the quantum superposition of the combined subject–object wavefunction contains two "relative states": a "collapsed" object state and an associated observer who has observed the same collapsed outcome; what the observer sees and the state of the object have become correlated by the act of measurement or observation. The subsequent evolution of each pair of relative subject–object states proceeds with complete indifference as to the presence or absence of the other elements, *as if* wavefunction collapse has occurred, which has the consequence that later observations are always consistent with the earlier observations. Thus the *appearance* of the object's wavefunction's collapse has emerged from the unitary, deterministic theory itself. (This answered Einstein's early criticism of quantum theory, that the theory should define what is observed, not for the observables to define the theory).[23] Since the wavefunction merely appears to have collapsed then, Everett reasoned, there was no need to actually assume that it had collapsed. And so, invoking Occam's razor, he removed the postulate of wavefunction collapse from the theory.

3.3 Probability

A consequence of removing wavefunction collapse from the quantum formalism is that the Born rule requires derivation, since many-worlds derives its interpretation from the formalism. Attempts have been made, by many-world advocates and others, over the years to *derive* the Born rule, rather than just conventionally *assume* it, so as to reproduce all the required statistical behaviour associated with quantum mechanics. There is no consensus on whether this has been successful.[24][25][26]

3.3.1 Everett, Gleason and Hartle

Everett (1957) briefly derived the Born rule by showing that the Born rule was the only possible rule, and that its derivation was as justified as the procedure for defining probability in classical mechanics. Everett stopped doing research in theoretical physics shortly after obtaining his Ph.D., but his work on probability has been extended by a number of people. Andrew Gleason (1957) and James Hartle (1965) independently reproduced Everett's work, known as Gleason's theorem[27][28] which was later extended.[29][30]

3.3.2 DeWitt and Graham

Bryce DeWitt and his doctoral student R. Neill Graham later provided alternative (and longer) derivations to Everett's derivation of the Born rule. They demonstrated that the norm of the worlds where the usual statistical rules of quantum theory broke down vanished, in the limit where the number of measurements went to infinity.

3.3.3 Deutsch *et al.*

An information-theoretic derivation of the Born rule from Everettarian assumptions, was produced by David Deutsch (1999)[31] and refined by Wallace (2002–2009)[32][33][34][35] and Saunders (2004).[36][37] Deutsch's derivation is a two-stage proof: first he shows that the number of orthonormal Everett-worlds after a branching is proportional to the conventional probability density. Then he uses game theory to show that these are all equally likely to be observed. The last step in particular has been criticised for circularity.[38][39] Some other reviews have been positive, although the status of these arguments remains highly controversial; some theoretical physicists have taken them as supporting the case for parallel universes.[40][41] In the *New Scientist* article, reviewing their presentation at a September 2007 conference,[42][43] Andy Albrecht, a physicist at the University of California at Davis, is quoted as saying "This work will go down as one of the most important developments in the history of science."[40]

Wojciech H. Zurek (2005)[44] has produced a derivation of the Born rule, where decoherence has replaced Deutsch's informatic assumptions.[45] Lutz Polley (2000) has produced Born rule derivations where the informatic assumptions are replaced by symmetry arguments.[46][47]

The Born rule and the collapse of the wave function have been obtained in the framework of the relative-state formulation of quantum mechanics by Armando V.D.B. Assis. He has proved that the Born rule and the collapse of the wave function follow from a game-theoretical strategy, namely the Nash equilibrium within a von Neumann zero-sum game between nature and observer.[48]

3.4 Brief overview

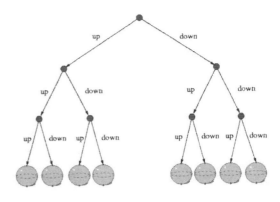

Schematic illustration of splitting as a result of a repeated measurement.

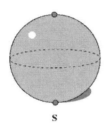 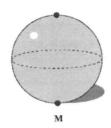

Schematic representation of pair of "smallest possible" quantum mechanical systems prior to interaction: Measured system S and measurement apparatus M. Systems such as S are referred to as 1-qubit systems.

In Everett's formulation, a measuring apparatus **M** and an object system **S** form a composite system, each of which prior to measurement exists in well-defined (but time-dependent) states. Measurement is regarded as causing **M** and **S** to interact. After **S** interacts with **M**, it is no longer possible to describe either system by an independent state. According to Everett, the only meaningful descriptions of each system are relative states: for example the relative state of **S** given the state of **M** or the relative state of **M** given the state of **S**. In DeWitt's formulation, the state of **S** after a sequence of measurements is given by a quantum superposition of states, each one corresponding to an alternative measurement history of **S**.

For example, consider the smallest possible truly quantum system **S**, as shown in the illustration. This describes for instance, the spin-state of an electron. Considering a specific axis (say the z-axis) the north pole represents spin "up" and the south pole, spin "down". The superposition states of the system are described by (the surface of) a sphere called the Bloch sphere. To perform a measurement on **S**, it is made to interact with another similar system **M**. After the interaction, the combined system is described by a state that ranges over a six-dimensional space (the reason for the number six is explained in the article on the Bloch sphere). This six-dimensional object can also be regarded as a quantum superposition of two "alternative histories" of the original system **S**, one in which "up" was observed and the other in which "down" was observed. Each subsequent binary measurement (that is interaction with a system **M**) causes a similar split in the history tree. Thus after three measurements, the system can be regarded as a quantum superposition of $8 = 2 \times 2 \times 2$ copies of the original system **S**.

The accepted terminology is somewhat misleading because it is incorrect to regard the universe as splitting at certain times; at any given instant there is one state in one universe.

3.5 Relative state

In his 1957 doctoral dissertation, Everett proposed that rather than modeling an isolated quantum system subject to external observation, one could mathematically model an object as well as its observers as purely physical systems within the mathematical framework developed by Paul Dirac, von Neumann and others, discarding altogether the *ad hoc* mechanism of wave function collapse. Since Everett's original work, there have appeared a number of similar formalisms in the literature. One such idea is discussed in the next section.

The relative state formulation makes two assumptions. The first is that the wavefunction is not simply a description of the object's state, but that it actually is entirely equivalent to the object, a claim it has in common with some other interpretations. The second is that observation or measurement has no special laws or mechanics, unlike in the Copenhagen interpretation which considers the wavefunction collapse as a special kind of event which occurs as a result of observation. Instead, measurement in the relative state formulation is the consequence of a configuration change in the memory of an observer described by the same basic wave physics as the object being modeled.

The many-worlds interpretation is DeWitt's popularisation of Everett's work, who had referred to the combined observer–object system as being split by an observation, each split corresponding to the different or multiple possible outcomes of an observation. These splits generate a possible tree as shown in the graphic below. Subsequently DeWitt introduced the term "world" to describe a complete measurement history of an observer, which corresponds roughly to a single branch of that tree. Note that "splitting" in this sense, is hardly new or even quantum mechanical. The idea of a space of complete alternative histories had already been used in the theory of probability since the mid-1930s for instance to model Brownian motion.

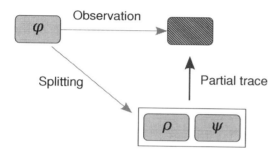

Partial trace as relative state. Light blue rectangle on upper left denotes system in pure state. Trellis shaded rectangle in upper right denotes a (possibly) mixed state. Mixed state from observation is partial trace of a linear superposition of states as shown in lower right-hand corner.

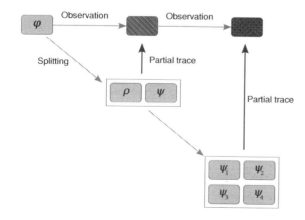

Successive measurements with successive splittings

Under the many-worlds interpretation, the Schrödinger equation, or relativistic analog, holds all the time everywhere. An observation or measurement of an object by an observer is modeled by applying the wave equation to the entire system comprising the observer *and* the object. One consequence is that every observation can be thought of as causing the combined observer–object's wavefunction to change into a quantum superposition of two or more non-interacting branches, or split into many "worlds". Since many observation-like events have happened, and are constantly happening, there are an enormous and growing number of simultaneously existing states.

If a system is composed of two or more subsystems, the system's state will be a superposition of products of the subsystems' states. Once the subsystems interact, their states are no longer independent. Each product of subsystem states in the overall superposition evolves over time independently of other products. The subsystems states have become correlated or entangled and it is no longer possible to consider them independent of one another. In Everett's terminology each subsystem state was now *correlated* with its *relative state*, since each subsystem must now be considered relative to the other subsystems with which it has interacted.

3.6 Properties of the theory

- MWI removes the observer-dependent role in the quantum measurement process by replacing wavefunction collapse with quantum decoherence. Since the role of the observer lies at the heart of most if not all "quantum paradoxes," this automatically resolves a number of problems; see for example Schrödinger's cat thought experiment, the EPR paradox, von Neumann's "boundary problem" and even wave-particle duality. Quantum cosmology also becomes intelligible, since there is no need anymore for an observer outside of the universe.

- MWI is a realist, deterministic, local theory, akin to classical physics (including the theory of relativity), at the expense of losing counterfactual definiteness. MWI achieves this by removing wavefunction collapse, which is indeterministic and non-local, from the deterministic and local equations of quantum theory.[49]

- MWI (or other, broader multiverse considerations) provides a context for the anthropic principle which may provide an explanation for the fine-tuned universe.[50][51]

- MWI, being a decoherent formulation, is axiomatically more streamlined than the Copenhagen and other collapse interpretations; and thus favoured under certain interpretations of Occam's razor.[52] Of course there are other decoherent interpretations that also possess this advantage with respect to the collapse interpretations.

3.7 Comparative properties and possible experimental tests

One of the salient properties of the many-worlds interpretation is that it does not require an exceptional method of wave function collapse to explain it. "It seems that there is no experiment distinguishing the MWI from other no-collapse theories such as Bohmian mechanics or other variants of MWI... In most no-collapse interpretations, the evolution of the quantum state of the Universe is the same. Still, one might imagine that there is an experiment distinguishing the MWI from another no-collapse interpretation based on the difference in the correspondence between the formalism and the experience (the results of experiments)."[53]

However, in 1985, David Deutsch published three related thought experiments which could test the theory vs the Copenhagen interpretation.[54] The experiments require macroscopic quantum state preparation and quantum erasure by a hypothetical quantum computer which is currently outside experimental possibility. Since then Lockwood (1989), Vaidman and others have made similar proposals.[53] These proposals also require an advanced technology which is able to place a macroscopic object in a coherent superposition, another task for which it is uncertain whether it will ever be possible. Many other controversial ideas have been put forward though, such as a recent claim that cosmological observations could test the theory,[55] and another claim by Rainer Plaga (1997), published in *Foundations of Physics*, that communication might be possible between worlds.[56]

3.7.1 Copenhagen interpretation

In the Copenhagen interpretation, the mathematics of quantum mechanics allows one to predict probabilities for the occurrence of various events. When an event occurs, it becomes part of the definite reality, and alternative possibilities do not. There is no necessity to say anything definite about what is not observed.

3.7.2 The universe decaying to a new vacuum state

Any event that changes the number of observers in the universe may have experimental consequences.[57] Quantum tunnelling to a new vacuum state would reduce the number of observers to zero (i.e., kill all life). Some cosmologists argue that the universe is in a false vacuum state and that consequently the universe should have already experienced quantum tunnelling to a true vacuum state. This has not happened and is cited as evidence in favor of many-worlds. In some worlds, quantum tunnelling to a true vacuum state has happened but most other worlds escape this tunneling and remain viable. This can be thought of as a variation on quantum suicide.

3.7.3 Many-minds

Main article: Many-minds interpretation

The *many-minds* interpretation is a multi-world interpretation that defines the splitting of reality on the level of the observers' minds. In this, it differs from Everett's many-worlds interpretation, in which there is no special role for the observer's mind.[56]

3.8 Common objections

- The many-worlds interpretation is very vague about the ways to determine when splitting happens, and nowadays usually the criterion is that the two branches have decohered. However, present day understanding of decoherence does not allow a completely precise, self-contained way to say when the two branches have decohered/"do not interact", and hence many-worlds interpretation remains arbitrary. This objection is saying that it is not clear what is precisely meant by branching, and point to the lack of self-contained criteria specifying branching.

  > MWI response: the decoherence or "splitting" or "branching" is complete when the measurement is complete. In Dirac notation a measurement is complete when:
  >
  > $\langle O_i | O_j \rangle = \delta_{ij}$ [58]
  >
  > where O_i represents the observer having detected the object system in the *i*th state. Before the measurement has started the observer states are identical; after the measurement is complete the observer states are orthonormal.[4][7] Thus a measurement defines the branching process: the branching is as well- or ill-defined

3.8. COMMON OBJECTIONS

as the measurement is; the branching is as complete as the measurement is complete – which is to say that the delta function above represents an idealised measurement. Although true "for all practical purposes" in reality the measurement, and hence the branching, is never fully complete, since delta functions are unphysical.[59]

Since the role of the observer and measurement per se plays no special role in MWI (measurements are handled as all other interactions are) there is no need for a precise definition of what an observer or a measurement is — just as in Newtonian physics no precise definition of either an observer or a measurement was required or expected. In all circumstances the universal wavefunction is still available to give a complete description of reality.

Also, it is a common misconception to think that branches are completely separate. In Everett's formulation, they may in principle quantum interfere (i.e., "merge" instead of "splitting") with each other in the future,[60] although this requires all "memory" of the earlier branching event to be lost, so no observer ever sees two branches of reality.[61][62]

- MWI states that there is no special role nor need for precise definition of measurement in MWI, yet Everett uses the word "measurement" repeatedly throughout its exposition.

    MWI response: "measurements" are treated as a subclass of interactions, which induce subject–object correlations in the combined wavefunction. There is nothing special about measurements (such as the ability to trigger a wave function collapse), that cannot be dealt with by the usual unitary time development process.[3] This is why there is no precise definition of measurement in Everett's formulation, although some other formulations emphasize that measurements must be effectively irreversible or create classical information.

- The splitting of worlds forward in time, but not backwards in time (i.e., merging worlds), is time asymmetric and incompatible with the time symmetric nature of Schrödinger's equation, or CPT invariance in general.[63]

    MWI response: The splitting is time asymmetric; this observed temporal asymmetry is due to the boundary conditions imposed by the Big Bang[64]

- There is circularity in Everett's measurement theory. Under the assumptions made by Everett, there are no 'good observations' as defined by him, and since his analysis of the observational process depends on the latter, it is void of any meaning. The concept of a 'good observation' is the projection postulate in disguise and Everett's analysis simply derives this postulate by having assumed it, without any discussion.[65]

    MWI response: Everett's treatment of observations / measurements covers *both* idealised good measurements and the more general bad or approximate cases.[66] Thus it is legitimate to analyse probability in terms of measurement; no circularity is present.

- Talk of probability in Everett presumes the existence of a preferred basis to identify measurement outcomes for the probabilities to range over. But the existence of a preferred basis can only be established by the process of decoherence, which is itself probabilistic[38] or arbitrary.[67]

    MWI response: Everett analysed branching using what we now call the "measurement basis". It is fundamental theorem of quantum theory that nothing measurable or empirical is changed by adopting a different basis. Everett was therefore free to choose whatever basis he liked. The measurement basis was simply the simplest basis in which to analyse the measurement process.[68][69]

- We cannot be sure that the universe is a quantum multiverse until we have a theory of everything and, in particular, a successful theory of quantum gravity.[70] If the final theory of everything is non-linear with respect to wavefunctions then many-worlds would be invalid.[1][4][5][6][7]

MWI response: All accepted quantum theories of fundamental physics are linear with respect to the wavefunction. While quantum gravity or string theory may be non-linear in this respect there is no evidence to indicate this at the moment.[15][16]

- Conservation of energy is grossly violated if at every instant near-infinite amounts of new matter are generated to create the new universes.

  MWI response: There are two responses to this objection. First, the law of conservation of energy says that energy is conserved *within each universe*. Hence, even if "new matter" were being generated to create new universes, this would not violate conservation of energy. Second, conservation of energy is not violated since the energy of each branch has to be weighted by its probability, according to the standard formula for the conservation of energy in quantum theory. This results in the total energy of the multiverse being conserved.[71]

- Occam's Razor rules against a plethora of unobservable universes – Occam would prefer just one universe; i.e., any non-MWI.

  MWI response: Occam's razor actually is a constraint on the complexity of physical theory, not on the number of universes. MWI is a simpler theory since it has fewer postulates.[52] Occams's razor is often cited by MWI adherents as an advantage of MWI.

- Unphysical universes: If a state is a superposition of two states Ψ_A and Ψ_B, i.e., $\Psi = (a\Psi_A + b\Psi_B)$, i.e., weighted by coefficients a and b, then if $b \ll a$, what principle allows a universe with vanishingly small probability b to be instantiated on an equal footing with the much more probable one with probability a? This seems to throw away the information in the probability amplitudes.

  MWI response: The magnitude of the coefficients provides the weighting that makes the branches or universes "unequal", as Everett and others have shown, leading the emergence of the conventional probabilistic rules.[1][4][5][6][7][72]

- Violation of the principle of locality, which contradicts special relativity: MWI splitting is instant and total: this may conflict with relativity, since an alien in the Andromeda galaxy can't know I collapse an electron over here before she collapses hers there: the relativity of simultaneity says we can't say which electron collapsed first – so which one splits off another universe first? This leads to a hopeless muddle with everyone splitting differently. Note: EPR is not a get-out here, as the alien's and my electrons need never have been part of the same quantum, i.e., entangled.

  MWI response: the splitting can be regarded as causal, local and relativistic, spreading at, or below, the speed of light (e.g., we are not split by Schrödinger's cat until we look in the box).[73] For spacelike separated splitting you can't say which occurred first — but this is true of all spacelike separated events, simultaneity is not defined for them. Splitting is no exception; many-worlds is a local theory.[49]

3.9 Reception

There is a wide range of claims that are considered "many-worlds" interpretations. It was often claimed by those who do not believe in MWI[74] that Everett himself was not entirely clear[75] as to what he believed; however, MWI adherents (such as DeWitt, Tegmark, Deutsch and others) believe they fully understand Everett's meaning as implying the literal existence of the other worlds. Additionally, recent biographical sources make it clear that Everett believed in the literal reality of the other quantum worlds.[22] Everett's son reported that Hugh Everett "never wavered in his belief over his many-worlds theory".[76] Also Everett was reported to believe "his many-worlds theory guaranteed him immortality".[77]

One of MWI's strongest advocates is David Deutsch.[78] According to Deutsch, the single photon interference pattern observed in the double slit experiment can be explained by interference of photons in multiple universes. Viewed in this way, the single photon interference experiment is indistinguishable from the multiple photon interference experiment. In a more practical vein, in one of the earliest papers on quantum computing,[79] he suggested that parallelism that results from the validity of MWI could lead to *"a method by which certain probabilistic tasks can be performed faster by a universal quantum computer than by any classical restriction of it"*. Deutsch has also proposed that when reversible computers become conscious that MWI will be testable (at least against "naive" Copenhagenism) via the reversible observation of spin.[61]

Asher Peres was an outspoken critic of MWI; for example, a section in his 1993 textbook had the title *Everett's interpretation and other bizarre theories*. In fact, Peres not only questioned whether MWI is really an "interpretation", but rather, if *any* interpretations of quantum mechanics are needed at all. Indeed, an interpretation can be regarded as a purely formal transformation, which adds nothing to the rules of the quantum mechanics. Peres seems to suggest that positing the existence of an infinite number of non-communicating parallel universes is highly suspect per those who interpret it as a violation of Occam's razor, i.e., that it does not minimize the number of hypothesized entities. However, it is understood that the number of elementary particles are not a gross violation of Occam's Razor, one counts the types, not the tokens. Max Tegmark remarks that the alternative to many-worlds is "many words", an allusion to the complexity of von Neumann's collapse postulate. On the other hand, the same derogatory qualification "many words" is often applied to MWI by its critics who see it as a word game which obfuscates rather than clarifies by confounding the von Neumann branching of possible worlds with the Schrödinger parallelism of many worlds in superposition.

MWI is considered by some to be unfalsifiable and hence unscientific because the multiple parallel universes are non-communicating, in the sense that no information can be passed between them. Others[61] claim MWI is directly testable. Everett regarded MWI as falsifiable since any test that falsifies conventional quantum theory would also falsify MWI.[21]

According to Martin Gardner, the "other" worlds of MWI have two different interpretations: real or unreal; he claims that Stephen Hawking and Steve Weinberg both favour the unreal interpretation.[80] Gardner also claims that the non-real interpretation is favoured by the majority of physicists, whereas the "realist" view is only supported by MWI experts such as Deutsch and Bryce DeWitt. Hawking has said that "according to Feynman's idea", all the other histories are as "equally real" as our own,[81] and Martin Gardner reports Hawking saying that MWI is "trivially true".[82] In a 1983 interview, Hawking also said he regarded the MWI as "self-evidently correct" but was dismissive towards questions about the interpretation of quantum mechanics, saying, "When I hear of Schrödinger's cat, I reach for my gun." In the same interview, he also said, "But, look: All that one does, really, is to calculate conditional probabilities—in other words, the probability of A happening, given B. I think that's all the many worlds interpretation is. Some people overlay it with a lot of mysticism about the wave function splitting into different parts. But all that you're calculating is conditional probabilities."[83] Elsewhere Hawking contrasted his attitude towards the "reality" of physical theories with that of his colleague Roger Penrose, saying, "He's a Platonist and I'm a positivist. He's worried that Schrödinger's cat is in a quantum state, where it is half alive and half dead. He feels that can't correspond to reality. But that doesn't bother me. I don't demand that a theory correspond to reality because I don't know what it is. Reality is not a quality you can test with litmus paper. All I'm concerned with is that the theory should predict the results of measurements. Quantum theory does this very successfully."[84] For his own part, Penrose agrees with Hawking that QM applied to the universe implies MW, although he considers the current lack of a successful theory of quantum gravity negates the claimed universality of conventional QM.[70]

3.9.1 Polls

Advocates of MWI often cite a poll of 72 "leading cosmologists and other quantum field theorists"[85] conducted by the American political scientist David Raub in 1995 showing 58% agreement with "Yes, I think MWI is true".[86]

The poll is controversial: for example, Victor J. Stenger remarks that Murray Gell-Mann's published work explicitly rejects the existence of simultaneous parallel universes. Collaborating with James Hartle, Gell-Mann is working toward the development a more "palatable" *post-Everett quantum mechanics*. Stenger thinks it's fair to say that most physicists dismiss the many-world interpretation as too extreme, while noting it "has merit in finding a place for the observer inside the system being analyzed and doing away with the troublesome notion of wave function collapse".[87]

Max Tegmark also reports the result of a "highly unscientific" poll taken at a 1997 quantum mechanics workshop.[88] According to Tegmark, "The many worlds interpretation (MWI) scored second, comfortably ahead of the consistent histories and Bohm interpretations." Such polls have been taken at other conferences, for example, in response to Sean Carroll's observation, "As crazy as it sounds, most working physicists buy into the many-worlds theory"[89] Michael Nielsen counters: "at a quantum computing conference at Cambridge in 1998, a many-worlder surveyed the audience of approximately 200 people... Many-worlds did just fine, garnering support on a level comparable to, but somewhat below, Copenhagen and decoherence." However, Nielsen notes that it seemed most attendees found it to be a waste of time: Asher Peres "got a huge and sustained round of applause... when he got up at the end of the polling and asked 'And who here believes the laws of physics are decided by a democratic vote?'"[90]

A 2005 poll of fewer than 40 students and researchers taken after a course on the Interpretation of Quantum Mechanics at the Institute for Quantum Computing University of Waterloo found "Many Worlds (and decoherence)" to be the least favored.[91]

A 2011 poll of 33 participants at an Austrian conference found 6 endorsed MWI, 8 "Information-based/information-theoretical", and 14 Copenhagen;[92] the authors remark that the results are similar to Tegmark's 1998 poll.

3.10 Speculative implications

Speculative physics deals with questions which are also discussed in science fiction.

3.10.1 Quantum suicide thought experiment

Main article: Quantum suicide and immortality

Quantum suicide, as a thought experiment, was published independently by Hans Moravec in 1987[93][94] and Bruno Marchal in 1988[95][96] and was independently developed further by Max Tegmark in 1998.[97] It attempts to distinguish between the Copenhagen interpretation of quantum mechanics and the Everett many-worlds interpretation by means of a variation of the Schrödinger's cat thought experiment, from the cat's point of view. **Quantum immortality** refers to the subjective experience of surviving quantum suicide regardless of the odds.[98]

3.10.2 Weak coupling

Another speculation is that the separate worlds remain weakly coupled (e.g., by gravity) permitting "communication between parallel universes". A possible test of this using quantum-optical equipment is described in a 1997 *Foundations of Physics* article by Rainer Plaga.[56] It involves an isolated ion in an ion trap, a quantum measurement that would yield two parallel worlds (their difference just being in the detection of a single photon), and the excitation of the ion from only one of these worlds. If the excited ion can be detected from the other parallel universe, then this would constitute direct evidence in support of the many-worlds interpretation and would automatically exclude the orthodox, "logical", and "many-histories" interpretations. The reason the ion is isolated is to make it not participate immediately in the decoherence which insulates the parallel world branches, therefore allowing it to act as a gateway between the two worlds, and if the measure apparatus could perform the measurements quickly enough before the gateway ion is decoupled then the test would succeed (with electronic computers the necessary time window between the two worlds would be in a time scale of milliseconds or nanoseconds, and if the measurements are taken by humans then a few seconds would still be enough). R. Plaga shows that macroscopic decoherence timescales are a possibility. The proposed test is based on technical equipment described in a 1993 *Physical Review* article by Itano et al.[99] and R. Plaga says that this level of technology is enough to realize the proposed inter-world communication experiment. The necessary technology for precision measurements of single ions already exists since the 1970s, and the ion recommended for excitation is $^{199}Hg^+$. The excitation methodology is described by Itano et al. and the time needed for it is given by the Rabi flopping formula[100]

Such a test as described by R. Plaga would mean that energy transfer is possible between parallel worlds. This does not violate the fundamental principles of physics because these require energy conservation only for the whole universe and not for the single parallel branches.[56] Neither the excitation of the single ion (which is a degree of freedom of the proposed system) leads to decoherence, something which is proven by Welcher Weg detectors which can excite atoms without momentum transfer (which causes the loss of coherence).[101]

The proposed test would allow for low-bandwidth inter-world communication, the limiting factors of bandwidth and time being dependent on the technology of the equipment. Because of the time needed to determine the state of the partially decohered isolated excited ion based on Itano et al.'s methodology, the ion would decohere by the time its state is determined during the experiment, so Plaga's proposal would pass just enough information between the two worlds to confirm their parallel existence and nothing more. The author contemplates that with increased bandwidth, one could even transfer television imagery across the parallel worlds.[56] For example, Itano et al.'s methodology could be improved (by lowering the time needed for state determination of the excited ion) if a more efficient process were found for the detection of fluorescence radiation using 194 nm photons.[56]

A 1991 article by J.Polchinski also supports the view that inter-world communication is a theoretical possibility.[102] Other authors in a 1994 preprint article also contemplated similar ideas.[103]

The reason inter-world communication seems like a possibility is because decoherence which separates the parallel worlds is never fully complete,[104][105] therefore weak influences from one parallel world to another can still pass between them,[104][106] and these should be measurable with advanced technology. Deutsch proposed such an experiment in a 1985 *International Journal of Theoretical Physics* article,[107] but the technology it requires involves human-level artificial intelligence.[56]

3.10.3 Similarity to modal realism

The many-worlds interpretation has some similarity to modal realism in philosophy, which is the view that the possible worlds used to interpret modal claims exist and are of a kind with the actual world. Unlike the possible worlds of philosophy, however, in quantum mechanics counterfactual alternatives can influence the results of experiments, as in the Elitzur–Vaidman bomb-testing problem or the Quantum Zeno effect. Also, while the worlds of the many-worlds interpretation all share the same physical laws, modal realism postulates a world for every way things could conceivably have been.

3.10.4 Time travel

The many-worlds interpretation could be one possible way to resolve the paradoxes[78] that one would expect to arise *if* time travel turns out to be permitted by physics (permitting closed timelike curves and thus violating causality). Entering the past would itself be a quantum event causing branching, and therefore the timeline accessed by the time traveller simply would be another timeline of many. In that sense, it would make the Novikov self-consistency principle unnecessary.

3.11 Many-worlds in literature and science fiction

Main article: Parallel universe (fiction)
See also: Alternate History

The many-worlds interpretation (and the somewhat related concept of possible worlds) has been associated to numerous themes in literature, art and science fiction.

Some of these stories or films violate fundamental principles of causality and relativity, and are extremely misleading since the information-theoretic structure of the path space of multiple universes (that is information flow between different paths) is very likely extraordinarily complex. Also see Michael Clive Price's FAQ referenced in the external links section below where these issues (and other similar ones) are dealt with more decisively.

A map from Robert Sobel's novel For Want of a Nail, *an artistic illustration of how small events – in this example the branching or point of divergence from our timeline's history is in October 1777 – can profoundly alter the course of history. According to the many-worlds interpretation every event, even microscopic, is a branch point; all possible alternative histories actually exist.*[1]

Another kind of popular illustration of many-worlds splittings, which does not involve information flow between paths, or information flow backwards in time considers alternate outcomes of historical events. According to the many-worlds interpretation, all of the historical speculations entertained within the alternate history genre are realized in parallel universes.[1]

The many-worlds interpretation of reality was anticipated with remarkable fidelity in Olaf Stapledon's 1937 science fiction novel Star Maker, in a paragraph describing one of the many universes created by the Star Maker god of the title. "In one inconceivably complex cosmos, whenever a creature was faced with several possible courses of action, it took them all, thereby creating many distinct temporal dimensions and distinct histories of the cosmos. Since in every evolutionary sequence of the cosmos there were very many creatures, and each was constantly faced with many possible courses, and the combinations of all their courses were innumerable, an infinity of distinct universes exfoliated from every moment of every temporal sequence in this cosmos."

3.12 See also

- Consistent histories

- EPR paradox
- *Fabric of Reality*
- Garden of Forking Paths
- Interpretations of quantum mechanics
- Many-minds interpretation
- Multiverse
- Multiple histories
- *The Beginning of Infinity*
- Quantum immortality – a thought experiment.
- Wave function collapse

3.13 Notes

[1] Bryce Seligman DeWitt, Quantum Mechanics and Reality: Could the solution to the dilemma of indeterminism be a universe in which all possible outcomes of an experiment actually occur?, *Physics Today*, 23(9) pp 30–40 (September 1970) "every quantum transition taking place on every star, in every galaxy, in every remote corner of the universe is splitting our local world on earth into myriads of copies of itself." See also *Physics Today*, letters followup, 24(4), (April 1971), pp 38–44

[2] Osnaghi, Stefano; Freitas, Fabio; Olival Freire, Jr (2009). "The Origin of the Everettian Heresy" (PDF). *Studies in History and Philosophy of Modern Physics* **40**: 97–123. doi:10.1016/j.shpsb.2008.10.002.

[3] Hugh Everett Theory of the Universal Wavefunction, Thesis, Princeton University, (1956, 1973), pp 1–140

[4] Everett, Hugh (1957). "Relative State Formulation of Quantum Mechanics". *Reviews of Modern Physics* **29**: 454–462. Bibcode:1957RvMP...29..454E. doi:10.1103/RevModPhys.29.454.

[5] Cecile M. DeWitt, John A. Wheeler eds, The Everett–Wheeler Interpretation of Quantum Mechanics, *Battelle Rencontres: 1967 Lectures in Mathematics and Physics* (1968)

[6] Bryce Seligman DeWitt, The Many-Universes Interpretation of Quantum Mechanics, *Proceedings of the International School of Physics "Enrico Fermi" Course IL: Foundations of Quantum Mechanics*, Academic Press (1972)

[7] Bryce Seligman DeWitt, R. Neill Graham, eds, *The Many-Worlds Interpretation of Quantum Mechanics*, Princeton Series in Physics, Princeton University Press (1973), ISBN 0-691-08131-X Contains Everett's thesis: The Theory of the Universal Wavefunction, pp 3–140.

[8] H. Dieter Zeh, On the Interpretation of Measurement in Quantum Theory, *Foundation of Physics*, vol. 1, pp. 69–76, (1970).

[9] Wojciech Hubert Zurek, Decoherence and the transition from quantum to classical, *Physics Today*, vol. 44, issue 10, pp. 36–44, (1991).

[10] Wojciech Hubert Zurek, Decoherence, einselection, and the quantum origins of the classical, *Reviews of Modern Physics*, 75, pp 715–775, (2003)

[11] The Many Worlds Interpretation of Quantum Mechanics

[12] David Deutsch argues that a great deal of fiction is close to a fact somewhere in the so called multiverse, *Beginning of Infinity*, p. 294

[13] Bryce Seligman DeWitt, R. Neill Graham, eds, *The Many-Worlds Interpretation of Quantum Mechanics*, Princeton Series in Physics, Princeton University Press (1973), ISBN 0-691-08131-X Contains Everett's thesis: The Theory of the Universal Wavefunction, where the claim to resolves all paradoxes is made on pg 118, 149.

[14] Hugh Everett, Relative State Formulation of Quantum Mechanics, *Reviews of Modern Physics* vol 29, (July 1957) pp 454–462. The claim to resolve EPR is made on page 462

[15] Steven Weinberg, *Dreams of a Final Theory: The Search for the Fundamental Laws of Nature* (1993), ISBN 0-09-922391-0, pg 68–69

[16] Steven Weinberg *Testing Quantum Mechanics*, Annals of Physics Vol 194 #2 (1989), pg 336–386

[17] John Archibald Wheeler, *Geons, Black Holes & Quantum Foam*, ISBN 0-393-31991-1. pp 268–270

[18] Everett 1957, section 3, 2nd paragraph, 1st sentence

[19] Everett [1956]1973, "Theory of the Universal Wavefunction", chapter 6 (e)

[20] Zurek, Wojciech (March 2009). "Quantum Darwinism". *Nature Physics* **5** (3): 181–188. arXiv:0903.5082. Bibcode:2009NatPh...5..181Z. doi:10.1038/nphys1202.

[21] Everett

[22] Peter Byrne, The Many Worlds of Hugh Everett III: Multiple Universes, Mutual Assured Destruction, and the Meltdown of a Nuclear Family, ISBN 978-0-19-955227-6

[23] "Whether you can observe a thing or not depends on the theory which you use. It is the theory which decides what can be observed." Albert Einstein to Werner Heisenberg, objecting to placing observables at the heart of the new quantum mechanics, during Heisenberg's 1926 lecture at Berlin; related by Heisenberg in 1968, quoted by Abdus Salam, *Unification of Fundamental Forces*, Cambridge University Press (1990) ISBN 0-521-37140-6, pp 98–101

3.13. NOTES

[24] N.P. Landsman, "The conclusion seems to be that no generally accepted derivation of the Born rule has been given to date, but this does not imply that such a derivation is impossible in principle.", in *Compendium of Quantum Physics* (eds.) F.Weinert, K. Hentschel, D.Greenberger and B. Falkenburg (Springer, 2008), ISBN 3-540-70622-4

[25] Adrian Kent (May 5, 2009), *One world versus many: the inadequacy of Everettian accounts of evolution, probability, and scientific confirmation*

[26] Kent, Adrian (1990). "Against Many-Worlds Interpretations". *Int.J.Mod.Phys* **A5**: 1745–1762. arXiv:gr-qc/9703089. Bibcode:1990IJMPA...5.1745K. doi:10.1142/S0217751X90000805.

[27] Gleason, A. M. (1957). "Measures on the closed subspaces of a Hilbert space". *Journal of Mathematics and Mechanics* **6**: 885–893. doi:10.1512/iumj.1957.6.56050. MR 0096113.

[28] James Hartle, Quantum Mechanics of Individual Systems, *American Journal of Physics*, 1968, vol 36 (#8), pp. 704–712

[29] E. Farhi, J. Goldstone & S. Gutmann. *How probability arises in quantum mechanics.*, Ann. Phys. (N.Y.) 192, 368–382 (1989).

[30] Pitowsky, I. (2005). "Quantum mechanics as a theory of probability". *Eprint arXiv:quant-ph/0510095*: 10095. arXiv:quant-ph/0510095. Bibcode:2005quant.ph.10095P.

[31] Deutsch, D. (1999). Quantum Theory of Probability and Decisions. *Proceedings of the Royal Society of London* A455, 3129–3137. .

[32] David Wallace: Quantum Probability and Decision Theory, Revisited

[33] David Wallace. Everettian Rationality: defending Deutsch's approach to probability in the Everett interpretation. Stud. Hist. Phil. Mod. Phys. 34 (2003), 415–438.

[34] David Wallace (2003), Quantum Probability from Subjective Likelihood: improving on Deutsch's proof of the probability rule

[35] David Wallace, 2009,A formal proof of the Born rule from decision-theoretic assumptions

[36] Simon Saunders: Derivation of the Born rule from operational assumptions. Proc. Roy. Soc. Lond. A460, 1771–1788 (2004).

[37] Simon Saunders, 2004: What is Probability?

[38] David J Baker, Measurement Outcomes and Probability in Everettian Quantum Mechanics, *Studies In History and Philosophy of Science Part B: Studies In History and Philosophy of Modern Physics*, Volume 38, Issue 1, March 2007, Pages 153–169

[39] H. Barnum, C. M. Caves, J. Finkelstein, C. A. Fuchs, R. Schack: Quantum Probability from Decision Theory? *Proc. Roy. Soc. Lond.* A456, 1175–1182 (2000).

[40] Merali, Zeeya (2007-09-21). "Parallel universes make quantum sense". *New Scientist* (2622). Retrieved 2013-11-22. (Summary only).

[41] Breitbart.com, Parallel universes exist – study, Sept 23 2007

[42] Perimeter Institute, Seminar overview, Probability in the Everett interpretation: state of play, David Wallace – Oxford University, 21 Sept 2007

[43] Perimeter Institute, Many worlds at 50 conference, September 21–24, 2007

[44] Wojciech H. Zurek: Probabilities from entanglement, Born's rule from envariance, *Phys. Rev.* A71, 052105 (2005).

[45] Schlosshauer, M.; Fine, A. (2005). "On Zurek's derivation of the Born rule". *Found. Phys.* **35**: 197–213. arXiv:quant-ph/0312058. Bibcode:2005FoPh...35..197S. doi:10.1007/s10701-004-1941-6.

[46] Lutz Polley, Position eigenstates and the statistical axiom of quantum mechanics, contribution to conference *Foundations of Probability and Physics*, Vaxjo, Nov 27 – Dec 1, 2000

[47] Lutz Polley, Quantum-mechanical probability from the symmetries of two-state systems

[48] Armando V.D.B. Assis (2011). "Assis, Armando V.D.B. On the nature of $a_k^* a_k$ and the emergence of the Born rule. Annalen der Physik, 2011.". *Annalen der Physik (Berlin)* **523**: 883–897. arXiv:1009.1532. Bibcode:2011AnP...523..883A. doi:10.1002/andp.201100062.

[49] Mark A. Rubin, Locality in the Everett Interpretation of Heisenberg-Picture Quantum Mechanics, *Foundations of Physics Letters*, 14, (2001), pp. 301–322, arXiv:quant-ph/0103079

[50] Paul C.W. Davies, *Other Worlds*, chapters 8 & 9 *The Anthropic Principle & Is the Universe an accident?*, (1980) ISBN 0-460-04400-1

[51] Paul C.W. Davies, *The Accidental Universe*, (1982) ISBN 0-521-28692-1

[52] Everett FAQ "Does many-worlds violate Ockham's Razor?"

[53] Vaidman, Lev. "Many-Worlds Interpretation of Quantum Mechanics". The Stanford Encyclopedia of Philosophy.

[54] Deutsch, D., (1986) 'Three experimental implications of the Everett interpretation', in R. Penrose and C.J. Isham (eds.), Quantum Concepts of Space and Time, Oxford: The Clarendon Press, pp. 204–214.

[55] Page, D., (2000) 'Can Quantum Cosmology Give Observational Consequences of Many-Worlds Quantum Theory?'

[56] Plaga, R. (1997). "On a possibility to find experimental evidence for the many-worlds interpretation of quantum mechanics". *Foundations of Physics* **27**: 559–577. arXiv:quant-ph/9510007. Bibcode:1997FoPh...27..559P. doi:10.1007/BF02550677.

[57] Page, Don N. (2000). "Can Quantum Cosmology Give Observational Consequences of Many-Worlds Quantum Theory?". arXiv:gr-qc/0001001. doi:10.1063/1.1301589.

[58] Bryce Seligman DeWitt, Quantum Mechanics and Reality: Could the solution to the dilemma of indeterminism be a universe in which all possible outcomes of an experiment actually occur?, *Physics Today*, 23(9) pp 30–40 (September 1970); see equation 10

[59] Penrose, R. *The Road to Reality*, §21.11

[60] Tegmark, Max The Interpretation of Quantum Mechanics: Many Worlds or Many Words?, 1998. To quote: "What Everett does NOT postulate: *"At certain magic instances, the world undergoes some sort of metaphysical 'split' into two branches that subsequently never interact."* This is not only a misrepresentation of the MWI, but also inconsistent with the Everett postulate, since the subsequent time evolution could in principle make the two terms...interfere. According to the MWI, there is, was and always will be only one wavefunction, and only decoherence calculations, not postulates, can tell us when it is a good approximation to treat two terms as non-interacting."

[61] Paul C.W. Davies, J.R. Brown, *The Ghost in the Atom* (1986) ISBN 0-521-31316-3, pp. 34–38: "The Many-Universes Interpretation", pp 83–105 for David Deutsch's test of MWI and reversible quantum memories

[62] Christoph Simon, 2009, *Conscious observers clarify many worlds*

[63] Joseph Gerver, The past as backward movies of the future, *Physics Today*, letters followup, 24(4), (April 1971), pp 46–7

[64] Bryce Seligman DeWitt, *Physics Today*,letters followup, 24(4), (April 1971), pp 43

[65] Arnold Neumaier's comments on the Everett FAQ, 1999 & 2003

[66] Everett [1956] 1973, "*Theory of the Universal Wavefunction*", chapter V, section 4 "Approximate Measurements", pp. 100–103 (e)

[67] Stapp, Henry (2002). "The basis problem in many-world theories" (PDF). *Canadian Journal of Physics* **80**: 1043–1052. arXiv:quant-ph/0110148. Bibcode:2002CaJPh..80.1043S. doi:10.1139/p02-068.

[68] Brown, Harvey R; Wallace, David (2005). "Solving the measurement problem: de Broglie–Bohm loses out to Everett" (PDF). *Foundations of Physics* **35**: 517–540. arXiv:quant-ph/0403094. Bibcode:2005FoPh...35..517B. doi:10.1007/s10701-004-2009-3.

[69] Mark A Rubin (2005), There Is No Basis Ambiguity in Everett Quantum Mechanics, *Foundations of Physics Letters*, Volume 17, Number 4 / August, 2004, pp 323–341

[70] Penrose, Roger (August 1991). "Roger Penrose Looks Beyond the Classic-Quantum Dichotomy". Sciencewatch. Retrieved 2007-10-21.

[71] Everett FAQ "Does many-worlds violate conservation of energy?"

[72] Everett FAQ "How do probabilities emerge within many-worlds?"

[73] Everett FAQ "When does Schrodinger's cat split?"

[74] Jeffrey A. Barrett, *The Quantum Mechanics of Minds and Worlds*, Oxford University Press, 1999. According to Barrett (loc. cit. Chapter 6) "There are many many-worlds interpretations."

[75] Barrett, Jeffrey A. (2010). Zalta, Edward N., ed. "Everett's Relative-State Formulation of Quantum Mechanics" (Fall 2010 ed.). The Stanford Encyclopedia of Philosophy. Again, according to Barrett "It is... unclear precisely how this was supposed to work."

[76] Aldhous, Peter (2007-11-24). "Parallel lives can never touch". *New Scientist* (2631). Retrieved 2007-11-21.

[77] Eugene Shikhovtsev's Biography of Everett, in particular see "*Keith Lynch remembers 1979–1980*"

[78] David Deutsch, *The Fabric of Reality: The Science of Parallel Universes And Its Implications*, Penguin Books (1998), ISBN 0-14-027541-X

[79] Deutsch, David (1985). "Quantum theory, the Church–Turing principle and the universal quantum computer". *Proceedings of the Royal Society of London A* **400**: 97–117. Bibcode:1985RSPSA.400...97D. doi:10.1098/rspa.1985.0070.

[80] A response to Bryce DeWitt, Martin Gardner, May 2002

[81] Award winning 1995 Channel 4 documentary "Reality on the rocks: Beyond our Ken" where, in response to Ken Campbell's question "all these trillions of Universes of the Multiverse, are they as real as this one seems to be to me?" Hawking states, "Yes.... According to Feynman's idea, every possible history (of Ken) is equally real."

[82] Gardner, Martin (2003). *Are universes thicker than blackberries?*. W.W. Norton. p. 10. ISBN 978-0-393-05742-3.

[83] Ferris, Timothy (1997). *The Whole Shebang*. Simon & Schuster. pp. 345. ISBN 978-0-684-81020-1.

[84] Hawking, Stephen; Roger Penrose (1996). *The Nature of Space and Time*. Princeton University Press. pp. 121. ISBN 978-0-691-03791-2.

[85] Elvridge., Jim (2008-01-02). *The Universe – Solved!*. pp. 35–36. ISBN 978-1-4243-3626-5. OCLC 247614399. 58% believed that the Many Worlds Interpretation (MWI) was true, including Stephen Hawking and Nobel Laureates Murray Gell-Mann and Richard Feynman

[86] Bruce., Alexandra. "How does reality work?". *Beyond the bleep : the definitive unauthorized guide to What the bleep do we know!?*. p. 33. ISBN 978-1-932857-22-1. [the poll was] published in the French periodical *Sciences et Avenir* in January 1998

[87] Stenger, V.J. (1995). *The Unconscious Quantum: Metaphysics in Modern Physics and Cosmology*. Prometheus Books. p. 176. ISBN 978-1-57392-022-3. LCCN lc95032599. Gell-Mann and collaborator James Hartle, along with a score of others, have been working to develop a more palatable interpretation of quantum mechanics that is free of the problems that plague all the interpretations we have considered so far. This new interpretation is called, in its various incarnations, **post-Everett quantum mechanics**, alternate histories, consistent histories, or decoherent histories. I will not be overly concerned with the detailed differences between these characterizations and will use the terms more or less interchangeably.

[88] Max Tegmark on many-worlds (contains MWI poll)

[89] Caroll, Sean (1 April 2004). "Preposterous Universe". Archived from the original on 8 September 2004.

[90] Nielsen, Michael (3 April 2004). "Michael Nielsen: The Interpretation of Quantum Mechanics". Archived from the original on 20 May 2004.

[91] Interpretation of Quantum Mechanics class survey

[92] "A Snapshot of Foundational Attitudes Toward Quantum Mechanics", Schlosshauer et al 2013

[93] "The Many Minds Approach". 25 October 2010. Retrieved 7 December 2010. This idea was first proposed by Austrian mathematician Hans Moravec in 1987...

[94] Moravec, Hans (1988). "The Doomsday Device". *Mind Children: The Future of Robot and Human Intelligence*. Harvard: Harvard University Press. p. 188. ISBN 978-0-674-57618-6. (If MWI is true, apocalyptic particle accelerators won't function as advertised).

[95] Marchal, Bruno (1988). "Informatique théorique et philosophie de l'esprit" [Theoretical Computer Science and Philosophy of Mind]. *Acte du 3ème colloque international Cognition et Connaissance [Proceedings of the 3rd International Conference Cognition and Knowledge]* (Toulouse): 193–227.

[96] Marchal, Bruno (1991). De Glas, M.; Gabbay, D., eds. "Mechanism and personal identity" (PDF). *Proceedings of WOCFAI 91* (Paris. Angkor.): 335–345.

[97] Tegmark, Max The Interpretation of Quantum Mechanics: Many Worlds or Many Words?, 1998

[98] Tegmark, Max (November 1998). "Quantum immortality". Retrieved 25 October 2010.

[99] W.M.Itano et al., Phys.Rev. A47,3354 (1993).

[100] M.SargentIII,M.O.Scully and W.E.Lamb, Laser physics (Addison-Wesley, Reading, 1974), p.27.

[101] M.O.Scully and H.Walther, Phys.Rev. A39,5229 (1989).

[102] J.Polchinski, Phys.Rev.Lett. 66,397 (1991).

[103] M.Gell-Mann and J.B.Hartle, Equivalent Sets of Histories and Multiple Quasiclassical Domains, preprint University of California at Santa Barbara UCSBTH-94-09 (1994).

[104] H.D.Zeh, Found.Phys. 3,109 (1973).

[105] H.D.Zeh, Phys.Lett.A 172,189 (1993).

[106] A.Albrecht, Phys.Rev. D48,3768 (1993).

[107] D.Deutsch, Int.J.theor.Phys. 24,1 (1985).

3.14 Further reading

- Jeffrey A. Barrett, *The Quantum Mechanics of Minds and Worlds*, Oxford University Press, Oxford, 1999.

- Peter Byrne, *The Many Worlds of Hugh Everett III: Multiple Universes, Mutual Assured Destruction, and the Meltdown of a Nuclear Family*, Oxford University Press, 2010.

- Jeffrey A. Barrett and Peter Byrne, eds., "The Everett Interpretation of Quantum Mechanics: Collected Works 1955–1980 with Commentary", Princeton University Press, 2012.

- Julian Brown, *Minds, Machines, and the Multiverse*, Simon & Schuster, 2000, ISBN 0-684-81481-1

- Paul C.W. Davies, *Other Worlds*, (1980) ISBN 0-460-04400-1

- James P. Hogan, *The Proteus Operation* (science fiction involving the many-worlds interpretation, time travel and World War 2 history), Baen, Reissue edition (August 1, 1996) ISBN 0-671-87757-7

- Adrian Kent, One world versus many: the inadequacy of Everettian accounts of evolution, probability, and scientific confirmation

- Andrei Linde and Vitaly Vanchurin, How Many Universes are in the Multiverse?

- Osnaghi, Stefano; Freitas, Fabio; Olival Freire, Jr (2009). "The Origin of the Everettian Heresy" (PDF). *Studies in History and Philosophy of Modern Physics* **40**: 97–123. doi:10.1016/j.shpsb.2008.10.002. A study of the painful three-way relationship between Hugh Everett, John A Wheeler and Niels Bohr and how this affected the early development of the many-worlds theory.

- Asher Peres, *Quantum Theory: Concepts and Methods*, Kluwer, Dordrecht, 1993.

- Mark A. Rubin, Locality in the Everett Interpretation of Heisenberg-Picture Quantum Mechanics, *Foundations of Physics Letters*, 14, (2001), pp. 301–322, arXiv:quant-ph/0103079

- David Wallace, Harvey R. Brown, Solving the measurement problem: de Broglie–Bohm loses out to Everett, *Foundations of Physics*, arXiv:quant-ph/0403094

- David Wallace, Worlds in the Everett Interpretation, *Studies in the History and Philosophy of Modern Physics*, 33, (2002), pp. 637–661, arXiv:quant-ph/0103092

- John A. Wheeler and Wojciech Hubert Zurek (eds), *Quantum Theory and Measurement*, Princeton University Press, (1983), ISBN 0-691-08316-9

- Sean M. Carroll, Charles T. Sebens, *Many Worlds, the Born Rule, and Self-Locating Uncertainty*, arXiv:1405.7907

3.15 External links

- Everett's Relative-State Formulation of Quantum Mechanics – Jeffrey A. Barrett's article on Everett's formulation of quantum mechanics in the Stanford Encyclopedia of Philosophy.

- Many-Worlds Interpretation of Quantum Mechanics – Lev Vaidman's article on the many-worlds interpretation of quantum mechanics in the Stanford Encyclopedia of Philosophy.

- Hugh Everett III Manuscript Archive (UC Irvine) – Jeffrey A. Barrett, Peter Byrne, and James O. Weatherall (eds.).

- Michael C Price's Everett FAQ – a clear FAQ-style presentation of the theory.

- The Many-Worlds Interpretation of Quantum Mechanics – a description for the lay reader with links.

- Against Many-Worlds Interpretations by Adrian Kent

- Many-Worlds is a "lost cause" according to R. F. Streater

- The many worlds of quantum mechanics John Sankey

- Max Tegmark's web page

- Henry Stapp's critique of MWI, focusing on the basis problem Canadian J. Phys. 80,1043–1052 (2002).

- Everett hit count on arxiv.org

- Many Worlds 50th anniversary conference at Oxford

- "Many Worlds at 50" conference at Perimeter Institute

- Scientific American report on the Many Worlds 50th anniversary conference at Oxford

- Highfield, Roger (September 21, 2007). "Parallel universe proof boosts time travel hopes". The Daily Telegraph. Archived from the original on 2007-10-20. Retrieved 2007-10-26..

- HowStuffWorks article

- Physicists Calculate Number of Parallel Universes Physorg.com October 16, 2009.

- TED-Education video – How many universes are there?.

Chapter 4

Eternal inflation

Eternal inflation is a hypothetical inflationary universe model, which is itself an outgrowth or extension of the Big Bang theory. In theories of eternal inflation, the inflationary phase of the universe's expansion lasts forever in at least some regions of the universe. Because these regions expand exponentially rapidly, most of the volume of the universe at any given time is inflating. All models of eternal inflation produce a hypothetically infinite multiverse, typically a fractal.

In 1983, Paul Steinhardt presented the first example of eternal inflation and Alexander Vilenkin showed that it is generic. [1] [2]

Eternal inflation was found to be predicted by many different models of cosmic inflation. MIT professor Alan Guth proposed an inflation model involving a "false vacuum" phase with positive vacuum energy. Parts of the universe in that phase inflate, and only occasionally decay to lower-energy, non-inflating phases or the ground state. In chaotic inflation, proposed by physicist Andrei Linde, the peaks in the evolution of a scalar field (determining the energy of the vacuum) correspond to regions of rapid inflation which dominate. Chaotic inflation usually eternally inflates,[3] since the expansions of the inflationary peaks exhibit positive feedback and come to dominate the large-scale dynamics of the universe.

Alan Guth's 2007 paper, "Eternal inflation and its implications",[3] details what is now known on the subject, and demonstrates that this particular flavor of inflationary universe theory is relatively current, or is still considered viable, more than 20 years after its inception.[4] [5][6]

4.1 Inflation and the multiverse

Both Linde and Guth believe that inflationary models of the early universe most likely lead to a multiverse but more proof is required.

> It's hard to build models of inflation that don't lead to a multiverse. It's not impossible, so I think there's still certainly research that needs to be done. But most models of inflation do lead to a multiverse, and evidence for inflation will be pushing us in the direction of taking [the idea of a] multiverse seriously. Alan Guth[7]

> It's possible to invent models of inflation that do not allow [a] multiverse, but it's difficult. Every experiment that brings better credence to inflationary theory brings us much closer to hints that the multiverse is real. Andrei Linde [7]

Polarization in the cosmic microwave background radiation suggests inflationary models for the early universe are more likely but confirmation is needed.[7]

4.2 History

Inflation, or the inflationary universe theory, was developed as a way to overcome the few remaining problems with what was otherwise considered a successful theory of cosmology, the Big Bang model. Although Alexei Starobinsky of the L.D. Landau Institute of Theoretical Physics in Moscow developed the first realistic inflation theory in 1979[8][9] he did not articulate its relevance to modern cosmological problems.

In 1979, Alan Guth developed an inflationary model independently, which offered a mechanism for inflation to begin: the decay of a so-called false vacuum into "bubbles" of "true vacuum" that expanded at the speed of light. Guth coined the term "inflation", and he was the first to discuss the theory with other scientists worldwide. But this formulation was problematic, as there was no consistent way to bring an end to the inflationary epoch and end up with the isotropic, homogeneous universe observed today. (See False vacuum: Development of theories). In 1982, this "graceful exit problem" was solved by Andrei Linde in the new inflationary scenario. A few months later, the same result was also obtained by Andreas Albrecht and Paul J. Steinhardt.

In 1986, Linde published an alternative model of inflation that also reproduced the same successes of new inflation entitled "Eternally Existing Self-Reproducing Chaotic Inflationary Universe",[10] which provides a detailed description of what has become known as the Chaotic Inflation theory or eternal inflation. The Chaotic Inflation theory is in some ways similar to Fred Hoyle's steady state theory, as it employs the concept of a universe that is eternally existing, and thus does not require a unique beginning or an ultimate end of the cosmos.

4.3 Quantum fluctuations of the inflation field

Chaotic Inflation theory models quantum fluctuations in the rate of inflation.[11] Those regions with a higher rate of inflation expand faster and dominate the universe, despite the natural tendency of inflation to end in other regions. This allows inflation to continue forever, to produce future-eternal inflation.

> Within the framework of established knowledge of physics and cosmology, our universe could be one of many in a super-universe or multiverse. Linde (1990, 1994) has proposed that a background space-time "foam" empty of matter and radiation will experience local quantum fluctuations in curvature, forming many bubbles of false vacuum that individually inflate into mini-universes with random characteristics. Each universe within the multiverse can have a different set of constants and physical laws. Some might have life of a form different from ours; others might have no life at all or something even more complex or so different that we cannot even imagine it. Obviously we are in one of those universes with life.[12]
>
> — Victor J. Stenger

Past-eternal models have been proposed which adhere to the perfect cosmological principle and have features of the steady state cosmos.[13][14][15]

A 2014 paper by Kohli and Haslam [16] analyzed Linde's chaotic inflation theory in which the quantum fluctuations are modeled as Gaussian white noise. They showed that in this popular scenario, eternal inflation in fact cannot be eternal, and the random noise leads to spacetime being filled with singularities. This was demonstrated by showing that solutions to the Einstein field equations diverge in a finite time. Their paper therefore concluded that the theory of eternal inflation based on random quantum fluctuations would not be a viable theory, and the resulting existence of a multiverse is "still very much an open question that will require much deeper investigation".

4.4 Differential decay

In **standard inflation**, inflationary expansion occurred while the universe was in a false vacuum state, halting when the universe decayed to a true vacuum state and became a general and inclusive phenomenon with homogeneity throughout, yielding a single expanding universe which is "our general reality" wherein the laws of physics are consistent throughout. In this case, the physical laws "just happen" to be compatible with the evolution of life.

The **bubble universe model** proposes that different regions of this inflationary universe (termed a multiverse) decayed to a true vacuum state at different times, with decaying regions corresponding to "sub"- universes not in causal contact with each other and existing in discrete regions that are subject to truly random "selection", determining each region's components based upon the persistence of the quantum components within that region. The end result will be a finite number of universes with physical laws consistent within each region of spacetime.

4.5 False vacuum and true vacuum

Variants of the bubble universe model postulate multiple false vacuum states, which result in lower-energy false-vacuum "progeny" universes spawned, which in turn produce true vacuum state progeny universes within themselves.

4.5.1 Evidence from the fluctuation level in our universe

New inflation does not produce a perfectly symmetric universe; tiny quantum fluctuations in the inflaton are created. These tiny fluctuations form the primordial seeds for all structure created in the later universe. These fluctuations were first calculated by Viatcheslav Mukhanov and G. V. Chibisov in the Soviet Union in analyzing Starobinsky's similar model.[17][18][19] In the context of inflation, they were worked out independently of the work of Mukhanov and Chibisov at the three-week 1982 Nuffield Workshop on the Very Early Universe at Cambridge University.[20] The fluctuations were calculated by four groups working separately over the course of the workshop: Stephen Hawking;[21] Starobinsky;[22] Guth and So-Young Pi;[23] and James M. Bardeen, Paul Steinhardt and Michael Turner.[24]

The fact that these models are consistent with WMAP data adds weight to the idea that the universe could be created in such a way. As a result, many physicists in the field agree it is possible, but needs further support to be accepted.[25]

4.6 See also

- Astrophysics
- Inflation
- Cosmology
- Fractal cosmology
- Physical cosmology
- Shape of the universe

4.7 References

[1] Guth, Alan H. (2000). "Inflation and Eternal Inflation". *Phys.Rept.* **333**: 11. arXiv:astro-ph/0002156. Bibcode:2000PhR...333..555G. doi:10.1016/S0370-1573(00)00037-5.

[2] Vilenkin, Alexander (1983). "Birth of Inflationary Universes". *Phys.Rev.D* **27** (12): 2848–2855. Bibcode:1983PhRvD..27.2848V. doi:10.1103/PhysRevD.27.2848.

[3] Guth, Alan; Eternal inflation and its implications arXiv: hep-th/0702178

[4] Holt, Jim. "The Big Lab Experiment. Was our universe created by design?". *Slate*.

[5] Jones, Douglas S. "Many worlds interpretation".

[6] Guth, Alan. "Eternal inflation: Successes and questions".

[7] Our Universe May Exist in a Multiverse, Cosmic Inflation Discovery Suggests

[8] Starobinsky, A. A. (1979). "Spectrum of Relict Gravitational Radiation and The Early State of the Universe" (PDF). *JETP Lett.* 30, 682 (Pisma Zh. Eksp. Teor. Fiz. 30, 719).

[9] Linde, Andrei (November 1994). "The Self-Reproducing Inflationary Universe" (PDF). *Scientific American*: page 51.

[10] Linde, A.D. (August 1986). "Eternally Existing Self-Reproducing Chaotic Inflationary Universe" (PDF). *Physics Letters B* **175** (4): 395–400. Bibcode:1986PhLB..175..395L. doi:10.1016/0370-2693(86)90611-8.

[11] Linde, A. (1986). "Eternal Chaotic Inflation". *Mod. Phys. Lett.* **A1** (2): 81. Bibcode:1986MPLA....1...81L. doi:10.1142/S0217732386000129.

[12] Stenger, Victor J. "Is the Universe fine-tuned for us?" (PDF).

[13] Aguirre, Anthony and Gratton, Steven n (2003). "Inflation without a beginning: A null boundary proposal". *Phys. Rev. D* **67** (8). arXiv:gr-qc/0301042. Bibcode:2003PhRvD..67h3515A. doi:10.1103/PhysRevD.67.083515.

[14] Aguirre, Anthony, and Gratton, Steven (2002). "Steady-State Eternal Inflation". *Phys. Rev. D* **65** (8). arXiv:astro-ph/0111191. Bibcode:2002PhRvD..65h3507A. doi:10.1103/PhysRevD.65.083507.

[15] Gribbin, John. "Inflation for Beginners".

[16] http://arxiv.org/pdf/1408.2249.pdf

[17] See Linde (1990) and Mukhanov (2005).

[18] Mukhanov, Viatcheslav F.; Chibisov, G. V. (1981). "Quantum fluctuation and "nonsingular" universe". *JETP Lett.* **33**: 532–5. Bibcode:1981JETPL..33..532M.

[19] Mukhanov, Viatcheslav F. (1982). "The vacuum energy and large scale structure of the universe". *Sov. Phys. JETP* **56**: 258–65.

[20] See Guth (1997) for a popular description of the workshop, or *The Very Early Universe*, ISBN 0521316774 eds Hawking, Gibbon & Siklos for a more detailed report

[21] Hawking, S.W. (1982). "The development of irregularities in a single bubble inflationary universe". *Phys. Lett.* **B115**: 295. Bibcode:1982PhLB..115..295H. doi:10.1016/0370-2693(82)90373-2.

[22] Starobinsky, Alexei A. (1982). "Dynamics of phase transition in the new inflationary universe scenario and generation of perturbations". *Phys. Lett.* **B117**: 175–8. Bibcode:1982PhLB..117..175S. doi:10.1016/0370-2693(82)90541-X.

[23] Guth, A.H. (1982). "Fluctuations in the new inflationary universe". *Phys. Rev. Lett.* **49** (15): 1110–3. Bibcode:1982PhRvL..49.1110G. doi:10.1103/PhysRevLett.49.1110.

[24] Bardeen, James M. (1983). "Spontaneous creation Of almost scale-free density perturbations in an inflationary universe". *Phys. Rev.* **D28**: 679. Bibcode:1983PhRvD..28..679B. doi:10.1103/PhysRevD.28.679.

[25] Weinberg, Steven (2006-11-05). "Beyond Belief: Science, Reason, Religion & Survival, Session 1". Salk Institute: The Science Network. 13:00–14:10. Retrieved 2012-08-27. Just in recent years, through developments in the theory of the very early universe — in particular, the theory of chaotic inflation due to Andrei Linde — we now have a picture which is, I would say, plausible but not yet well established, that our Big Bang... is just one episode in a much larger multiverse, in which Big Bangs — or maybe I should say, not-so-Big Bangs are popping off all the time.

4.8 External links

- 'Multiverse' theory suggested by microwave background BBC News, 3 August 2011 about testing eternal inflation.

Chapter 5

Cosmological principle

See also: Friedmann–Lemaître–Robertson–Walker metric and Observable universe § Large-scale structure

In modern physical cosmology, the **cosmological principle** is the notion that the distribution of matter in the universe is homogeneous and isotropic when viewed on a large enough scale, since the forces are expected to act uniformly throughout the universe, and should, therefore, produce no observable irregularities in the large scale structuring over the course of evolution of the matter field that was initially laid down by the Big Bang.

Astronomer William Keel explains:

> The cosmological principle is usually stated formally as 'Viewed on a sufficiently large scale, the properties of the universe are the same for all observers.' This amounts to the strongly philosophical statement that the part of the universe which we can see is a fair sample, and that the same physical laws apply throughout. In essence, this in a sense says that the universe is knowable and is playing fair with scientists.[1]

The cosmological principle depends on a definition of "observer," and contains an implicit qualification and two testable consequences.

"Observers" means any observer at any location in the universe, not simply any human observer at any location on Earth: as Andrew Liddle puts it, "the cosmological principle [means that] the universe looks the same whoever and wherever you are."[2]

The qualification is that variation in physical structures can be overlooked, provided this does not imperil the uniformity of conclusions drawn from observation: the Sun is different from the Earth, our galaxy is different from a black hole, some galaxies advance toward rather than recede from us, and the universe has a "foamy" texture of galaxy clusters and voids, but none of these different structures appears to violate the basic laws of physics.

The two testable structural consequences of the cosmological principle are homogeneity and isotropy. Homogeneity means that the same observational evidence is available to observers at different locations in the universe ("the part of the universe which we can see is a fair sample"). Isotropy means that the same observational evidence is available by looking in any direction in the universe ("the same physical laws apply throughout"). The principles are distinct but closely related, because a universe that appears isotropic from any two (for a spherical geometry, three) locations must also be homogeneous.

5.1 Origin

The cosmological principle is first clearly asserted in the *Philosophiæ Naturalis Principia Mathematica* (1687) of Isaac Newton. In contrast to earlier classical or medieval cosmologies, in which Earth rested at the center of universe, Newton conceptualized the Earth as a sphere in orbital motion around the Sun within an empty space that extended uniformly in all directions to immeasurably large distances. He then showed, through a series of mathematical proofs on detailed observational data of the motions of planets and comets, that their motions could be explained by a single principle of "universal gravitation" that applied as well to the orbits of the Galilean moons around Jupiter, the Moon around the Earth, the Earth around the Sun, and to falling bodies on Earth. That is, he asserted the equivalent material nature of all bodies within the Solar System, the identical nature of the Sun and distant stars ("the light of the fixed stars is of the same nature with the light of the Sun, ... and lest the systems of the fixed stars should, by their gravity, fall on each other, [God] hath placed those systems at immense distances from one another"), and thus the uniform extension of the physical laws of motion to a great distance beyond the observational location of Earth itself.

5.2 Implications

The cosmological principle represents both the principle on which cosmological theory and observation can proceed and a "null" hypothesis of uniformity that is an area of active research inquiry.[3] Many important advances in astronomy and cosmology, and the formulation of new cosmological theories, have occurred through the resolution of apparent violations of the cosmological principle. For example, the original discovery that far galaxies appeared to have higher spectral redshifts than near galaxies (an apparent violation of homogeneity) led to the discovery of Hubble flow, the metric expansion of space that occurs equally in all locations (restoring homogeneity).

The universe is now described as having a history, starting with the Big Bang and proceeding through distinct epochs of stellar and galaxy formation. Because this history is currently described (after the first fraction of a second after the origin) almost entirely in terms of known physical processes and particle physics, the cosmological principle is extended to assert the homogeneity of cosmological evolution across the anisotropy of time:

> ... all points in space ought to experience the same physical development, correlated in time in such a way that all points at a certain distance from an observer appear to be at the same stage of development. In that sense, all spatial conditions in the universe must appear to be homogeneous and isotropic to an observer at all times in the future and in the past.[4]

That is, earlier times are identical to the "distance from the observer" in spacetime, which is assessed as the redshift of the light arriving from the observed celestial object: the cosmological principle is preserved because the same sequence of evolution is observed in all directions from Earth, and is inferred to be identical to the sequence that would be observed from any other location in the universe.

Observations show that more distant galaxies are closer together and have lower content of chemical elements heavier than lithium.[5] Applying the cosmological principle, this suggests that heavier elements were not created in the Big Bang but were produced by nucleosynthesis in giant stars and expelled across a series of supernovae explosions and new star formation from the supernovae remnants, which means heavier elements would accumulate over time. Another observation is that the furthest galaxies (earlier time) are often more fragmentary, interacting and unusually shaped than local galaxies (recent time), suggesting evolution in galaxy structure as well.

A related implication of the cosmological principle is that the largest discrete structures in the universe are in mechanical equilibrium. Homogeneity and isotropy of matter at the largest scales would suggest that the largest discrete structures are parts of a single indiscrete form, like the crumbs which make up the interior of a cake. At extreme cosmological distances, the property of mechanical equilibrium in surfaces lateral to the line of sight can be empirically tested; however, under the assumption of the cosmological principle, it cannot be detected parallel to the line of sight (see timeline of the universe).

Cosmologists agree that in accordance with observations of distant galaxies, a universe must be non-static if it follows the cosmological principle. In 1923, Alexander Friedmann set out a variant of Einstein's equations of general relativity that describe the dynamics of a homogeneous isotropic universe.[6][7] Independently, Georges Lemaître derived in 1927 the equations of an expanding universe from the General Relativity equations.[8] Thus, a non-static universe is also implied, independent of observations of distant galaxies, as the result of applying the cosmological principle to general relativity.

5.3 Justification

Although the universe can seem inhomogeneous at smaller scales, it *is* statistically homogeneous on scales larger than 250 million light years. The cosmic microwave background is isotropic, that is to say that its intensity is about the same whichever direction we look at.[9] However, the European Space Agency has concluded, based on data from the Planck Mission showing hemispheric bias in 2 respects: one with respect to average temperature, the second with respect to larger variations in the degree of perturbations. i.e. temperature fluctuations, i.e. densities, that these anisotropies are, in fact, statistically significant and can no longer be ignored.[10]

5.4 Criticism

Karl Popper criticized the cosmological principle on the grounds that it makes "our *lack* of knowledge a principle of *knowing something*". He summarized his position as follows:[11]

> the "cosmological principles" were, I fear, dogmas that should not have been proposed.

The *cosmological principle* implies that at a sufficiently large scale, the universe is homogeneous; different places will appear similar to one another. Whilst Yadav *et al.* have suggested a maximum scale of 260/h Mpc for structures within the universe according to this heuristic, other authors have suggested values as low as 60/h Mpc.[12] Yadav's calculation suggests that the maximum size of a structure can be about 370 Mpc.[13]

The Clowes–Campusano LQG, discovered in 1991, has a length of 580 Mpc, and is marginally larger than the consistent scale.

The Sloan Great Wall, discovered in 2003, has a length of 423 Mpc,[14] which is only just consistent with the cosmological principle.

U1.11, a large quasar group discovered in 2011, has a length of 780 Mpc, and is two times larger than the upper limit of the homogeneity scale.

The Huge-LQG, discovered in 2012, is three times longer than, and twice as wide as is predicted possible according to these current models, and so challenges our understanding of the universe on large scales.

In November 2013, a new structure 10 billion light years away measuring 2000-3000 Mpc (more than six times that of the SGW) has been discovered, the Hercules–Corona Borealis Great Wall, putting further doubt on the validity of the cosmological principle.[15]

5.5 See also

- Background independence
- Copernican principle
- End of Greatness
- Friedmann–Lemaître–Robertson–Walker metric
- Large scale structure of the cosmos
- Metric expansion of space
- Perfect Cosmological Principle
- Redshift

5.6 References

[1] William C. Keel (2007). *The Road to Galaxy Formation (2nd ed.)*. Springer-Praxis. ISBN 978-3-540-72534-3.. p. 2.

[2] Andrew Liddle (2003). *An Introduction to Modern Cosmology (2nd ed.)*. John Wiley & Sons. ISBN 978-0-470-84835-7.. p. 2.

[3] GFR Ellis (1975). "Cosmology and verifiability". *Quarterly Journal of the Royal Astronomical Society* **16**: 245–264. Bibcode:1975QJRAS..16..245E.

[4] Klaus Mainzer and J Eisinger (2002). *The Little Book of Time*. Springer. ISBN 0-387-95288-8.. P. 55.

[5] Image:CMB Timeline75.jpg - NASA (public domain image)

[6] Alexander Friedmann (1923). *Die Welt als Raum und Zeit (The World as Space and Time)*. Ostwalds Klassiker der exakten Wissenschaften. ISBN 3-8171-3287-5..

[7] Éduard Abramovich Tropp, Viktor Ya. Frenkel, Artur Davidovich Chernin (1993). *Alexander A. Friedmann: The Man who Made the Universe Expand*. Cambridge University Press. p. 219. ISBN 0-521-38470-2.

[8] Lemaître, Georges (1927). "Un univers homogène de masse constante et de rayon croissant rendant compte de la vitesse radiale des nébuleuses extra-galactiques". *Annales de la Société Scientifique de Bruxelles* **A47**: 49–56. Bibcode:1927ASSB...47...49L. *translated by A. S. Eddington*: Lemaître, Georges (1931). "Expansion of the universe, A homogeneous universe of constant mass and increasing radius accounting for the radial velocity of extra-galactic nebulæ". *Monthly Notices of the Royal Astronomical Society* **91**: 483–490. Bibcode:1931MNRAS..91..483L. doi:10.1093/mnras/91.5.483.

[9]

[10]

[11] Helge Kragh: "The most philosophically of all the sciences": Karl Popper and physical cosmology (2012)

[12] Yadav, Jaswant; J. S. Bagla; Nishikanta Khandai (25 February 2010). "Fractal dimension as a measure of the scale of homogeneity". *Monthly Notices of the Royal Astronomical Society* **405** (3): 2009–2015. arXiv:1001.0617. Bibcode:2010MNRAS.405.2009Y. doi:10.1111/j.1365-2966.2010.16612.x. Retrieved 15 January 2013.

[13] "A structure in the early universe at z ~ 1.3 that exceeds the homogeneity scale of the R-W concordance cosmology". *Cosmological Principle*. Cornell university. Retrieved 5 February 2013. |first1= missing |last1= in Authors list (help)

[14] Gott, J. Richard, III; et al. (May 2005). "A Map of the Universe". *The Astrophysical Journal* **624** (2): 463–484. arXiv:astro-ph/0310571. Bibcode:2005ApJ...624..463G. doi:10.1086/428890

[15] http://arxiv.org/abs/1311.1104

Chapter 6

Inflation (cosmology)

"Inflation model" and "Inflation theory" redirect here. For a general rise in the price level, see Inflation. For other uses, see Inflation (disambiguation).

In physical cosmology, **cosmic inflation**, **cosmological inflation**, or just **inflation** is a theory of exponential expansion of space in the early universe. The inflationary epoch lasted from 10^{-36} seconds after the Big Bang to sometime between 10^{-33} and 10^{-32} seconds. Following the inflationary period, the Universe continues to expand, but at a less rapid rate.[1]

Inflation theory was developed in the early 1980s. It explains the origin of the large-scale structure of the cosmos. Quantum fluctuations in the microscopic inflationary region, magnified to cosmic size, become the seeds for the growth of structure in the Universe (see galaxy formation and evolution and structure formation).[2] Many physicists also believe that inflation explains why the Universe appears to be the same in all directions (isotropic), why the cosmic microwave background radiation is distributed evenly, why the Universe is flat, and why no magnetic monopoles have been observed.

While the detailed particle physics mechanism responsible for inflation is not known, the basic picture makes a number of predictions that have been confirmed by observation.[3] The hypothetical field thought to be responsible for inflation is called the inflaton.[4]

In 2002, three of the original architects of the theory were recognized for their major contributions; physicists Alan Guth of M.I.T., Andrei Linde of Stanford and Paul Steinhardt of Princeton shared the prestigious Dirac Prize "for development of the concept of inflation in cosmology".[5]

6.1 Overview

Main article: Metric expansion of space

An expanding universe generally has a cosmological horizon, which, by analogy with the more familiar horizon caused by the curvature of the Earth's surface, marks the boundary of the part of the Universe that an observer can see. Light (or other radiation) emitted by objects beyond the cosmological horizon never reaches the observer, because the space in between the observer and the object is expanding too rapidly.

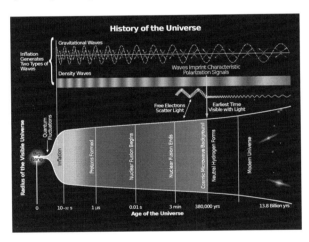

History of the Universe - gravitational waves are hypothesized to arise from cosmic inflation, a faster-than-light expansion just after the Big Bang (17 March 2014).[6][7][8]

The observable universe is one *causal patch* of a much larger unobservable universe; other parts of the Universe cannot communicate with Earth yet. These parts of the Universe are outside our current cosmological horizon. In the standard hot big bang model, without inflation, the cosmological horizon moves out, bringing new regions into view. Yet as a local observer sees such a region for the first time, it looks no different from any other region of space the local observer has already seen: its background radiation is at nearly the same temperature as the background radiation of other regions, and its space-time curvature is evolving lock-step with the others. This presents a mystery: how did these new regions know what temperature and curvature they were supposed to have? They couldn't have learned it by getting signals, because they were not previously in communication with our past light cone.[9][10]

Inflation answers this question by postulating that all the regions come from an earlier era with a big vacuum energy, or cosmological constant. A space with a cosmological constant is qualitatively different: instead of moving outward, the cosmological horizon stays put. For any one observer, the distance to the cosmological horizon is constant. With exponentially expanding space, two nearby observers are separated very quickly; so much so, that the distance between them quickly exceeds the limits of communications. The spatial slices are expanding very fast to cover huge volumes. Things are constantly moving beyond the cosmological horizon, which is a fixed distance away, and everything becomes homogeneous.

As the inflationary field slowly relaxes to the vacuum, the cosmological constant goes to zero and space begins to expand normally. The new regions that come into view during the normal expansion phase are exactly the same regions that were pushed out of the horizon during inflation, and so they are at nearly the same temperature and curvature, because they come from the same originally small patch of space.

The theory of inflation thus explains why the temperatures and curvatures of different regions are so nearly equal. It also predicts that the total curvature of a space-slice at constant global time is zero. This prediction implies that the total ordinary matter, dark matter and residual vacuum energy in the Universe have to add up to the critical density, and the evidence supports this. More strikingly, inflation allows physicists to calculate the minute differences in temperature of different regions from quantum fluctuations during the inflationary era, and many of these quantitative predictions have been confirmed.[11][12]

6.1.1 Space expands

To say that space expands exponentially means that two inertial observers are moving farther apart with accelerating velocity. In stationary coordinates for one observer, a patch of an inflating universe has the following polar metric:[13][14]

$$ds^2 = -(1 - \Lambda r^2)\,dt^2 + \frac{1}{1 - \Lambda r^2}\,dr^2 + r^2\,d\Omega^2.$$

This is just like an inside-out black hole metric—it has a zero in the dt component on a fixed radius sphere called the cosmological horizon. Objects are drawn away from the observer at $r = 0$ towards the cosmological horizon, which they cross in a finite proper time. This means that any inhomogeneities are smoothed out, just as any bumps or matter on the surface of a black hole horizon are swallowed and disappear.

Since the space–time metric has no explicit time dependence, once an observer has crossed the cosmological horizon, observers closer in take its place. This process of falling outward and replacement points closer in are always steadily replacing points further out—an exponential expansion of space–time.

This steady-state exponentially expanding spacetime is called a de Sitter space, and to sustain it there must be a cosmological constant, a vacuum energy proportional to Λ everywhere. In this case, the equation of state is $p = -\rho$. The physical conditions from one moment to the next are stable: the rate of expansion, called the Hubble parameter, is nearly constant, and the scale factor of the Universe is proportional to e^{Ht}. Inflation is often called a period of *accelerated expansion* because the distance between two fixed observers is increasing exponentially (i.e. at an accelerating rate as they move apart), while Λ can stay approximately constant (see deceleration parameter).

6.1.2 Few inhomogeneities remain

Cosmological inflation has the important effect of smoothing out inhomogeneities, anisotropies and the curvature of space. This pushes the Universe into a very simple state, in which it is completely dominated by the inflaton field, the source of the cosmological constant, and the only significant inhomogeneities are the tiny quantum fluctuations in the inflaton. Inflation also dilutes exotic heavy particles, such as the magnetic monopoles predicted by many extensions to the Standard Model of particle physics. If the Universe was only hot enough to form such particles *before* a period of inflation, they would not be observed in nature, as they would be so rare that it is quite likely that there are none in the observable universe. Together, these effects are called the inflationary "no-hair theorem"[15] by analogy with the no hair theorem for black holes.

The "no-hair" theorem works essentially because the cosmological horizon is no different from a black-hole horizon, except for philosophical disagreements about what is on the other side. The interpretation of the no-hair theorem is that the Universe (observable and unobservable) expands by an enormous factor during inflation. In an expanding universe, energy densities generally fall, or get diluted, as the volume of the Universe increases. For example, the density of ordinary "cold" matter (dust) goes down as the inverse of the volume: when linear dimensions double, the energy density goes down by a factor of eight; the radiation energy density goes down even more rapidly as the Universe expands since the wavelength of each photon is stretched (redshifted), in addition to the photons being dispersed by the expansion. When linear dimensions are doubled, the energy density in radiation falls by a factor of sixteen (see the solution of the energy density continuity equation for an ultra-relativistic fluid). During inflation, the energy density in the inflaton field is roughly constant. However, the energy density in everything else, including inhomogeneities, curvature, anisotropies, exotic particles, and standard-model particles is falling, and through sufficient inflation these all become negligible. This leaves the Universe flat and symmetric, and (apart from the homogeneous inflaton field) mostly empty, at the moment inflation ends and reheating begins.[16]

6.1.3 Duration

A key requirement is that inflation must continue long enough to produce the present observable universe from a single, small inflationary Hubble volume. This is necessary to ensure that the Universe appears flat, homogeneous and isotropic at the largest observable scales. This requirement is generally thought to be satisfied if the Universe expanded by a factor of at least 10^{26} during inflation.[17]

6.1.4 Reheating

Inflation is a period of supercooled expansion, when the temperature drops by a factor of 100,000 or so. (The exact drop is model dependent, but in the first models it was typically from 10^{27} K down to 10^{22} K.[18]) This relatively low temperature is maintained during the inflationary phase. When inflation ends the temperature returns to the pre-inflationary temperature; this is called *reheating* or thermalization because the large potential energy of the inflaton field decays into particles and fills the Universe with Standard Model particles, including electromagnetic radiation, starting the radiation dominated phase of the Universe. Because the nature of the inflation is not known, this process is still poorly understood, although it is believed to take place through a parametric resonance.[19][20]

6.2 Motivations

Inflation resolves several problems in Big Bang cosmology that were discovered in the 1970s.[21] Inflation was first proposed by Guth while investigating the problem of why no magnetic monopoles are seen today; he found that a positive-energy false vacuum would, according to general relativity, generate an exponential expansion of space. It was very quickly realised that such an expansion would resolve many other long-standing problems. These problems arise from the observation that to look like it does *today*, the Universe would have to have started from very finely tuned, or "special" initial conditions at the Big Bang. Inflation attempts to resolve these problems by providing a dynamical mechanism that drives the Universe to this special state, thus making a universe like ours much more likely in the context of the Big Bang theory.

6.2.1 Horizon problem

Main article: Horizon problem

The horizon problem is the problem of determining why the Universe appears statistically homogeneous and isotropic in accordance with the cosmological principle.[22][23][24] For example, molecules in a canister of gas are distributed homogeneously and isotropically because they are in thermal equilibrium: gas throughout the canister has had enough time to interact to dissipate inhomogeneities and anisotropies. The situation is quite different in the big bang model without inflation, because gravitational expansion does not give the early universe enough time to equilibrate. In a big bang with only the matter and radiation known in the Standard Model, two widely separated regions of the observable universe cannot have equilibrated because they move apart from each other faster than the speed of light and thus have never come into causal contact. In the early Universe, it was not possible to send a light signal between the two regions. Because they have had no interaction, it is difficult to explain why they have the same temperature (are thermally equilibrated). Historically, proposed solutions included the *Phoenix universe* of Georges Lemaître,[25] the related oscillatory universe of Richard Chase Tolman,[26] and the Mixmaster universe of Charles Misner. Lemaître and Tolman proposed that a universe undergoing a number of cycles of contraction and expansion could come into thermal equilibrium. Their models failed, however, because of the buildup of entropy over several cycles. Misner made the (ultimately incorrect) conjecture that the Mixmaster mechanism, which made the Universe *more* chaotic, could lead to statistical homogeneity and isotropy.[23][27]

6.2.2 Flatness problem

Main article: Flatness problem

The flatness problem is sometimes called one of the Dicke coincidences (along with the cosmological constant problem).[28][29] It became known in the 1960s that the density of matter in the Universe was comparable to the critical density necessary for a flat universe (that is, a universe whose large scale geometry is the usual Euclidean geometry, rather than a non-Euclidean hyperbolic or spherical geometry).[30]:61

Therefore, regardless of the shape of the universe the contribution of spatial curvature to the expansion of the Universe could not be much greater than the contribution of matter. But as the Universe expands, the curvature redshifts away more slowly than matter and radiation. Extrapolated into the past, this presents a fine-tuning problem because the contribution of curvature to the Universe must be exponentially small (sixteen orders of magnitude less than the density of radiation at big bang nucleosynthesis, for example). This problem is exacerbated by recent observations of the cosmic microwave background that have demonstrated that the Universe is flat to within a few percent.[31]

6.2.3 Magnetic-monopole problem

The magnetic monopole problem, sometimes called the exotic-relics problem, says that if the early universe were very hot, a large number of very heavy, stable magnetic monopoles would have been produced. This is a problem with Grand Unified Theories, which propose that at high temperatures (such as in the early universe) the electromagnetic force, strong, and weak nuclear forces are not actually fundamental forces but arise due to spontaneous symmetry breaking from a single gauge theory.[32] These theories predict a number of heavy, stable particles that have not been observed in nature. The most notorious is the magnetic monopole, a kind of stable, heavy "charge" of magnetic field.[33][34] Monopoles are predicted to be copiously produced following Grand Unified Theories at high temperature,[35][36] and they should have persisted to the present day, to such an extent that they would become the primary constituent of the Universe.[37][38] Not only is that not the case, but all searches for them have failed, placing stringent limits on the density of relic magnetic monopoles in the Universe.[39] A period of inflation that occurs below the temperature where magnetic monopoles can be produced would offer a possible resolution of this problem: monopoles would be separated from each other as the Universe around them expands, potentially lowering their observed density by many orders of magnitude. Though, as cosmologist Martin Rees has written, "Skeptics about exotic physics might not be hugely impressed by a theoretical argument to explain the absence of particles that are themselves only hypothetical. Preventive medicine can readily seem 100 percent effective against a disease that doesn't exist!"[40]

6.3 History

6.3.1 Precursors

In the early days of General Relativity, Albert Einstein introduced the cosmological constant to allow a static solution, which was a three-dimensional sphere with a uniform density of matter. Later, Willem de Sitter found a highly symmetric inflating universe, which described a universe with a cosmological constant that is otherwise empty.[41] It was discovered that Einstein's universe is unstable, and that small fluctuations cause it to collapse or turn into a de Sitter universe.

In the early 1970s Zeldovich noticed the flatness and horizon problems of Big Bang cosmology; before his work, cosmology was presumed to be symmetrical on purely philosophical grounds. In the Soviet Union, this and other considerations led Belinski and Khalatnikov to analyze the chaotic BKL singularity in General Relativity. Misner's Mixmaster universe attempted to use this chaotic behavior to solve the cosmological problems, with limited success.

In the late 1970s, Sidney Coleman applied the instanton techniques developed by Alexander Polyakov and collaborators to study the fate of the false vacuum in quantum field theory. Like a metastable phase in statistical mechanics—water below the freezing temperature or above the boiling point—a quantum field would need to nucleate a large enough bubble of the new vacuum, the new phase, in order to make a transition. Coleman found the most likely decay pathway for vacuum decay and calculated the inverse lifetime per unit volume. He eventually noted that gravitational effects would be significant, but he did not calculate these effects and did not apply the results to cosmology.

In the Soviet Union, Alexei Starobinsky noted that quantum corrections to general relativity should be important for the early universe. These generically lead to curvature-squared corrections to the Einstein–Hilbert action and a form of $f(R)$ modified gravity. The solution to Einstein's equations in the presence of curvature squared terms, when the curvatures are large, leads to an effective cosmological constant. Therefore, he proposed that the early universe went through an inflationary de Sitter era.[42] This resolved the cosmology problems and led to specific predictions for the corrections to the microwave background radiation, corrections that were then calculated in detail.

In 1978, Zeldovich noted the monopole problem, which was an unambiguous quantitative version of the horizon problem, this time in a subfield of particle physics, which led to several speculative attempts to resolve it. In 1980 Alan Guth realized that false vacuum decay in the early universe would solve the problem, leading him to propose a scalar-driven inflation. Starobinsky's and Guth's scenarios both predicted an initial deSitter phase, differing only in mechanistic details.

6.3.2 Early inflationary models

Guth proposed inflation in January 1980 to explain the nonexistence of magnetic monopoles;[43][44] it was Guth who coined the term "inflation".[45] At the same time, Starobinsky argued that quantum corrections to gravity would replace the initial singularity of the Universe with an exponentially expanding deSitter phase.[46] In October 1980, Demosthenes Kazanas suggested that exponential expansion could eliminate the particle horizon and perhaps solve the horizon problem,[47] while Sato suggested that an exponential expansion could eliminate domain walls (another kind of exotic relic).[48] In 1981 Einhorn and Sato[49] published a model similar to Guth's and showed that it would resolve the puzzle of the magnetic monopole abundance in Grand Unified Theories. Like Guth, they concluded that such a model not only required fine tuning of the cosmological constant, but also would likely lead to a much too granular universe, i.e., to large density variations resulting from bubble wall collisions.

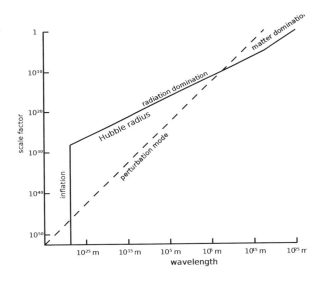

The physical size of the Hubble radius (solid line) as a function of the linear expansion (scale factor) of the universe. During cosmological inflation, the Hubble radius is constant. The physical wavelength of a perturbation mode (dashed line) is also shown. The plot illustrates how the perturbation mode grows larger than the horizon during cosmological inflation before coming back inside the horizon, which grows rapidly during radiation domination. If cosmological inflation had never happened, and radiation domination continued back until a gravitational singularity, then the mode would never have been outside the horizon in the very early universe, and no causal mechanism could have ensured that the universe was homogeneous on the scale of the perturbation mode.

Guth proposed that as the early universe cooled, it was trapped in a false vacuum with a high energy density, which is much like a cosmological constant. As the very early universe cooled it was trapped in a metastable state (it was supercooled), which it could only decay out of through the process of bubble nucleation via quantum tunneling. Bubbles of true vacuum spontaneously form in the sea of false vacuum and rapidly begin expanding at the speed of light. Guth recognized that this model was problematic because the model did not reheat properly: when the bubbles nucleated, they did not generate any radiation. Radiation could only be generated in collisions between bubble walls. But if inflation lasted long enough to solve the initial conditions problems, collisions between bubbles became exceedingly rare. In any one causal patch it is likely that only one bubble would nucleate.

6.3.3 Slow-roll inflation

The bubble collision problem was solved by Linde[50] and independently by Andreas Albrecht and Paul Steinhardt[51] in a model named *new inflation* or *slow-roll inflation* (Guth's model then became known as *old inflation*). In this model, instead of tunneling out of a false vacuum state, inflation occurred by a scalar field rolling down a potential energy hill. When the field rolls very slowly compared to the expansion of the Universe, inflation occurs. However, when the hill becomes steeper, inflation ends and reheating can occur.

6.3.4 Effects of asymmetries

Eventually, it was shown that new inflation does not produce a perfectly symmetric universe, but that quantum fluctuations in the inflaton are created. These fluctuations form the primordial seeds for all structure created in the later universe.[52] These fluctuations were first calculated by Viatcheslav Mukhanov and G. V. Chibisov in analyzing Starobinsky's similar model.[53][54][55] In the context of inflation, they were worked out independently of the work of Mukhanov and Chibisov at the three-week 1982 Nuffield Workshop on the Very Early Universe at Cambridge University.[56] The fluctuations were calculated by four groups working separately over the course of the workshop: Stephen Hawking;[57] Starobinsky;[58] Guth and So-Young Pi;[59] and Bardeen, Steinhardt and Turner.[60]

6.4 Observational status

Inflation is a mechanism for realizing the cosmological principle, which is the basis of the standard model of physical cosmology: it accounts for the homogeneity and isotropy of the observable universe. In addition, it accounts for the observed flatness and absence of magnetic monopoles. Since Guth's early work, each of these observations has received further confirmation, most impressively by the detailed observations of the cosmic microwave background made by the Wilkinson Microwave Anisotropy Probe (WMAP) spacecraft.[11] This analysis shows that the Universe is flat to within at least a few percent, and that it is homogeneous and isotropic to one part in 100,000.

In addition, inflation predicts that the structures visible in the Universe today formed through the gravitational collapse of perturbations that were formed as quantum mechanical fluctuations in the inflationary epoch. The detailed form of the spectrum of perturbations called a nearly-scale-invariant Gaussian random field (or Harrison–Zel'dovich spectrum) is very specific and has only two free parameters, the amplitude of the spectrum and the *spectral index*, which measures the slight deviation from scale invariance predicted by inflation (perfect scale invariance corresponds to the idealized de Sitter universe).[61] Inflation predicts that the observed perturbations should be in thermal equilibrium with each other (these are called *adiabatic* or *isentropic* perturbations). This structure for the perturbations has been confirmed by the WMAP spacecraft and other cosmic microwave background (CMB) experiments,[11] and galaxy surveys, especially the ongoing Sloan Digital Sky Survey.[62] These experiments have shown that the one part in 100,000 inhomogeneities observed have exactly the form predicted by theory. Moreover, there is evidence for a slight deviation from scale invariance. The *spectral index*, n_s is equal to one for a scale-invariant spectrum. The simplest inflation models predict that this quantity is between 0.92 and 0.98.[63][64][65][66] From WMAP data it can be inferred that $n_s = 0.963 \pm 0.012$,[67] implying that it differs from one at the level of two standard deviations (2σ). This is considered an important confirmation of the theory of inflation.[11]

Various inflation theories have been proposed that make radically different predictions, but they generally have much more fine tuning than should be necessary.[63][64] As a physical model, however, inflation is most valuable in that it robustly predicts the initial conditions of the Universe based on only two adjustable parameters: the spectral index (that can only change in a small range) and the amplitude of the perturbations. Except in contrived models, this is true regardless of how inflation is realized in particle physics.

Occasionally, effects are observed that appear to contradict the simplest models of inflation. The first-year WMAP data suggested that the spectrum might not be nearly scale-invariant, but might instead have a slight curvature.[68] However, the third-year data revealed that the effect was a statistical anomaly.[11] Another effect remarked upon since the first cosmic microwave background satellite, the Cosmic Background Explorer is that the amplitude of the quadrupole moment of the CMB is unexpectedly low and the other low multipoles appear to be preferentially aligned with the ecliptic plane. Some have claimed that this is a signature of non-Gaussianity and thus contradicts the simplest models of inflation. Others have suggested that the effect may be due to other new physics, foreground contamination, or even publication bias.[69]

An experimental program is underway to further test inflation with more precise CMB measurements. In particular, high precision measurements of the so-called "B-modes" of the polarization of the background radiation could provide evidence of the gravitational radiation produced by inflation, and could also show whether the energy scale of inflation predicted by the simplest models (10^{15}–10^{16} GeV) is correct.[64][65] In March 2014, it was announced that B-mode CMB polarization consistent with that predicted from inflation had been demonstrated by a South Pole experiment.[6][7][8][70][71][72] However, on 19 June 2014, lowered confidence in confirming the findings was reported;[71][73][74] on 19 September 2014, a further reduction in confidence was reported[75][76] and, on 30 January 2015, even less confidence yet was reported.[77][78]

Other potentially corroborating measurements are expected from the Planck spacecraft, although it is unclear if the signal will be visible, or if contamination from foreground sources will interfere.[79] Other forthcoming measurements, such as those of 21 centimeter radiation (radiation emitted and absorbed from neutral hydrogen before the first stars turned on), may measure the power spectrum with even greater resolution than the CMB and galaxy surveys, although it is not known if these measurements will be possible or if interference with radio sources on Earth and in the galaxy will be too great.[80]

Dark energy is broadly similar to inflation and is thought to be causing the expansion of the present-day universe to accelerate. However, the energy scale of dark energy is much lower, 10^{-12} GeV, roughly 27 orders of magnitude less than the scale of inflation.

6.5 Theoretical status

In Guth's early proposal, it was thought that the inflaton was the Higgs field, the field that explains the mass of the elementary particles.[44] It is now believed by some that the inflaton cannot be the Higgs field[81] although the recent discovery of the Higgs boson has increased the number of works considering the Higgs field as inflaton.[82] One problem of this identification is the current tension with experimental data at the electroweak scale,[83] which is currently under study at the Large Hadron Collider (LHC). Other models of inflation relied on the properties of Grand Unified Theories.[51] Since the simplest models of grand unification have failed, it is now thought by many physicists that inflation will be included in a supersymmetric theory such as string theory or a supersymmetric grand unified theory. At present, while inflation is understood principally by its detailed predictions of the initial conditions for the hot early universe, the particle physics is largely *ad hoc* modelling. As such, although predictions of inflation have been consistent with the results of observational tests, many open questions remain.

6.5.1 Fine-tuning problem

One of the most severe challenges for inflation arises from the need for fine tuning. In new inflation, the *slow-roll conditions* must be satisfied for inflation to occur. The slow-roll conditions say that the inflaton potential must be flat (compared to the large vacuum energy) and that the inflaton particles must have a small mass.[84] New inflation requires the Universe to have a scalar field with an especially flat potential and special initial conditions. However, explanations for these fine-tunings have been proposed. For example, classically scale invariant field theories, where scale invariance is broken by quantum effects, provide an explanation of the flatness of inflationary potentials, as long as the theory can be studied through perturbation theory.[85]

Andrei Linde

Linde proposed a theory known as *chaotic inflation* in which he suggested that the conditions for inflation were actually satisfied quite generically. Inflation will occur in virtually any universe that begins in a chaotic, high energy state that has a scalar field with unbounded potential energy.[86] However, in his model the inflaton field necessarily takes values larger than one Planck unit: for this reason, these are often called *large field* models and the competing new inflation models are called *small field* models. In this situation, the predictions of effective field theory are thought to be invalid, as renormalization should cause large corrections that could prevent inflation.[87] This problem has not yet been resolved and some cosmologists argue that the small field models, in which inflation can occur at a much lower energy scale, are better models.[88] While inflation depends on quantum field theory (and the semiclassical approximation to quantum gravity) in an important way, it has not been completely reconciled with these theories.

Brandenberger commented on fine-tuning in another situation.[89] The amplitude of the primordial inhomogeneities produced in inflation is directly tied to the energy scale of inflation. This scale is suggested to be around 10^{16} GeV or 10^{-3} times the Planck energy. The natural scale is naïvely the Planck scale so this small value could be seen as another form of fine-tuning (called a hierarchy problem): the energy density given by the scalar potential is down by 10^{-12} compared to the Planck density. This is not usually considered to be a critical problem, however, because the scale of inflation corresponds naturally to the scale of gauge unification.

6.5.2 Eternal inflation

Main article: Eternal inflation

In many models, the inflationary phase of the Universe's expansion lasts forever in at least some regions of the Universe. This occurs because inflating regions expand very rapidly, reproducing themselves. Unless the rate of decay to the non-inflating phase is sufficiently fast, new inflating regions are produced more rapidly than non-inflating regions. In such models most of the volume of the Universe at any given time is inflating. All models of eternal inflation produce an infinite multiverse, typically a fractal.

Although new inflation is classically rolling down the potential, quantum fluctuations can sometimes lift it to previous levels. These regions in which the inflaton fluctuates upwards expand much faster than regions in which the inflaton has a lower potential energy, and tend to dominate in terms of physical volume. This steady state, which first developed by Vilenkin,[90] is called "eternal inflation". It has been shown that any inflationary theory with an unbounded potential is eternal.[91] It is a popular conclusion among physicists that this steady state cannot continue forever into the past.[92][93][94] Inflationary spacetime, which is similar to de Sitter space, is incomplete without a contracting region. However, unlike de Sitter space, fluctuations in a contracting inflationary space collapse to form a gravitational singularity, a point where densities become infinite. Therefore, it is necessary to have a theory for the Universe's initial conditions. Linde, however, believes inflation may be past eternal.[95]

In eternal inflation, regions with inflation have an exponentially growing volume, while regions that are not inflating don't. This suggests that the volume of the inflating part of the Universe in the global picture is always unimaginably larger than the part that has stopped inflating, even though inflation eventually ends as seen by any single preinflationary observer. Scientists disagree about how to assign a probability distribution to this hypothetical anthropic landscape. If the probability of different regions is counted by volume, one should expect that inflation will never end or applying boundary conditions that a local observer exists to observe it, that inflation will end as late as possible. Some physicists believe this paradox can be resolved by weighting observers by their pre-inflationary volume.

6.5.3 Initial conditions

Some physicists have tried to avoid the initial conditions problem by proposing models for an eternally inflating universe with no origin.[96][97][98][99] These models propose that while the Universe, on the largest scales, expands exponentially it was, is and always will be, spatially infinite and has existed, and will exist, forever.

Other proposals attempt to describe the ex nihilo creation of the Universe based on quantum cosmology and the following inflation. Vilenkin put forth one such scenario.[90] Hartle and Hawking offered the no-boundary proposal for the initial creation of the Universe in which inflation comes about naturally.[100]

Guth described the inflationary universe as the "ultimate free lunch":[101][102] new universes, similar to our own, are continually produced in a vast inflating background. Gravitational interactions, in this case, circumvent (but do not violate) the first law of thermodynamics (energy conservation) and the second law of thermodynamics (entropy and the arrow of time problem). However, while there is consensus that this solves the initial conditions problem, some have disputed this, as it is much more likely that the Universe came about by a quantum fluctuation. Don Page was an outspoken critic of inflation because of this anomaly.[103] He stressed that the thermodynamic arrow of time necessitates low entropy initial conditions, which would be highly unlikely. According to them, rather than solving this problem, the inflation theory aggravates it – the reheating at the end of the inflation era increases entropy, making it necessary for the initial state of the Universe to be even more orderly than in other Big Bang theories with no inflation phase.

Hawking and Page later found ambiguous results when they attempted to compute the probability of inflation in the Hartle-Hawking initial state.[104] Other authors have argued that, since inflation is eternal, the probability doesn't matter as long as it is not precisely zero: once it starts, inflation perpetuates itself and quickly dominates the Universe.[105][106]:223–225 However, Albrecht and Lorenzo Sorbo argued that the probability of an inflationary cosmos, consistent with today's observations, emerging by a random fluctuation from some pre-existent state is much higher than that of a non-inflationary cosmos. This is because the "seed" amount of non-gravitational energy required for the inflationary cosmos is so much less than that for a non-inflationary alternative, which outweighs any entropic considerations.[107]

Another problem that has occasionally been mentioned is the trans-Planckian problem or trans-Planckian effects.[108] Since the energy scale of inflation and the Planck scale are relatively close, some of the quantum fluctuations that have made up the structure in our universe were smaller than the Planck length before inflation. Therefore, there ought to be corrections from Planck-scale physics, in particular the unknown quantum theory of gravity. Some disagreement remains about the magnitude of this effect: about whether it is just on the threshold of detectability or completely undetectable.[109]

6.5.4 Hybrid inflation

Another kind of inflation, called *hybrid inflation*, is an extension of new inflation. It introduces additional scalar fields, so that while one of the scalar fields is responsible for normal slow roll inflation, another triggers the end of inflation: when inflation has continued for sufficiently long, it becomes favorable to the second field to decay into a much lower energy state.[110]

In hybrid inflation, one scalar field is responsible for most of the energy density (thus determining the rate of expansion), while another is responsible for the slow roll (thus determining the period of inflation and its termination). Thus fluctuations in the former inflaton would not affect inflation termination, while fluctuations in the latter would not affect the rate of expansion. Therefore, hybrid inflation is not eternal.[111][112] When the second (slow-rolling) inflaton reaches the bottom of its potential, it changes the location of the minimum of the first inflaton's potential, which leads to a fast roll of the inflaton down its potential, leading to termination of inflation.

6.5.5 Inflation and string cosmology

The discovery of flux compactifications opened the way for reconciling inflation and string theory.[113] *Brane inflation* suggests that inflation arises from the motion of D-branes[114] in the compactified geometry, usually towards a stack of anti-D-branes. This theory, governed by the *Dirac-Born-Infeld action*, is different from ordinary inflation. The dynamics are not completely understood. It appears that special conditions are necessary since inflation occurs in tunneling between two vacua in the string landscape. The process of tunneling between two vacua is a form of old inflation, but new inflation must then occur by some other mechanism.

6.5.6 Inflation and loop quantum gravity

When investigating the effects the theory of loop quantum gravity would have on cosmology, a loop quantum cosmology model has evolved that provides a possible mechanism for cosmological inflation. Loop quantum gravity assumes a quantized spacetime. If the energy density is larger than can be held by the quantized spacetime, it is thought to bounce back.[115]

6.6 Alternatives

Other models explain some of the observations explained by inflation. However none of these "alternatives" has the same breadth of explanation and still require inflation for a more complete fit with observation. They should therefore be regarded as adjuncts to inflation, rather than as alternatives.

6.6.1 Big bounce

The flatness and horizon problems are naturally solved in the Einstein-Cartan-Sciama-Kibble theory of gravity, without needing an exotic form of matter or free parameters.[116][117] This theory extends general relativity by removing a constraint of the symmetry of the affine connection and regarding its antisymmetric part, the torsion tensor, as a dynamical variable. The minimal coupling between torsion and Dirac spinors generates a spin-spin interaction that is significant in fermionic matter at extremely high densities. Such an interaction averts the unphysical Big Bang singularity, replacing it with a cusp-like bounce at a finite minimum scale factor, before which the Universe was contracting. The rapid expansion immediately after the Big Bounce explains why the present Universe at largest scales appears spatially flat, homogeneous and isotropic. As the density of the Universe decreases, the effects of torsion weaken and the Universe smoothly enters the radiation-dominated era.

6.6.2 String theory

String theory requires that, in addition to the three observable spatial dimensions, additional dimensions exist that are curled up or compactified (see also Kaluza–Klein theory). Extra dimensions appear as a frequent component of supergravity models and other approaches to quantum gravity. This raised the contingent question of why four space-time dimensions became large and the rest became unobservably small. An attempt to address this question, called *string gas cosmology*, was proposed by Robert Brandenberger and Cumrun Vafa.[118] This model focuses on the dynamics of the early universe considered as a hot gas of strings. Brandenberger and Vafa show that a dimension of spacetime can only expand if the strings that wind around it can efficiently annihilate each other. Each string is a one-dimensional object, and the largest number of dimensions in which two strings will generically intersect (and, presumably, annihilate) is three. Therefore, the most likely number of non-compact (large) spatial dimensions is three. Current work on this model centers on whether it can succeed in stabilizing the size of the compactified dimensions and produce the correct spectrum of primordial density perturbations.[119] Supporters admit that their model "does not solve the entropy and flatness problems of standard cosmology ..... and we can provide no explanation for why the current universe is so close to being spatially flat".[120]

6.6.3 Ekpyrotic and cyclic models

The ekpyrotic and cyclic models are also considered adjuncts to inflation. These models solve the horizon problem through an expanding epoch well *before* the Big Bang, and then generate the required spectrum of primordial density perturbations during a contracting phase leading to a Big Crunch. The Universe passes through the Big Crunch and emerges in a hot Big Bang phase. In this sense they are reminiscent of Richard Chace Tolman's oscillatory universe; in Tolman's model, however, the total age of the Universe is necessarily finite, while in these models this is not necessarily so. Whether the correct spectrum of density fluctuations can be produced, and whether the Universe can successfully navigate the Big Bang/Big Crunch transition, remains a topic of controversy and current research. Ekpyrotic models avoid the magnetic monopole problem as long as the temperature at the Big Crunch/Big Bang transition remains below the Grand Unified Scale, as this is the temperature required to produce magnetic monopoles in the first place. As things stand, there is no evidence of any 'slowing down' of the expansion, but this is not surprising as each cycle is expected to last on the order of a trillion years.

6.6.4 Varying C

Another adjunct, the varying speed of light model was offered by Jean-Pierre Petit in 1988, John Moffat in 1992 as well Albrecht and João Magueijo in 1999, instead of superluminal expansion the speed of light was 60 orders of magnitude faster than its current value solving the horizon and homogeneity problems in the early universe.

6.7 Criticisms

Since its introduction by Alan Guth in 1980, the inflationary paradigm has become widely accepted. Nevertheless, many physicists, mathematicians, and philosophers of science have voiced criticisms, claiming untestable predictions and a lack of serious empirical support.[105] In 1999, John Earman and Jesús Mosterín published a thorough critical review of inflationary cosmology, concluding, "we do not think that there are, as yet, good grounds for admitting any of the models of inflation into the standard core of cosmology."[121]

In order to work, and as pointed out by Roger Penrose from 1986 on, inflation requires extremely specific initial conditions of its own, so that the problem (or pseudo-problem) of initial conditions is not solved: "There is something fundamentally misconceived about trying to explain the uniformity of the early universe as resulting from a thermalization process. [...] For, if the thermalization is actually doing anything [...] then it represents a definite increasing of the entropy. Thus, the universe would have been even more special before the thermalization than after."[122] The problem of specific or "fine-tuned" initial conditions would not have been solved; it would have gotten worse. At a conference in 2015, Penrose said that "inflation isn't falsifiable, it's falsified. [...] BICEP did a wonderful service by bringing all the Inflation-ists out of their shell, and giving them a black eye."[123]

A recurrent criticism of inflation is that the invoked inflation field does not correspond to any known physical field, and that its potential energy curve seems to be an ad hoc contrivance to accommodate almost any data obtainable. Paul Steinhardt, one of the founding fathers of inflationary cosmology, has recently become one of its sharpest critics. He calls 'bad inflation' a period of accelerated expansion whose outcome conflicts with observations, and 'good inflation' one compatible with them: "Not only is bad inflation more likely than good inflation, but no inflation is more likely than either.... Roger Penrose considered all the possible configurations of the inflaton and gravitational fields. Some of these configurations lead to inflation ... Other configurations lead to a uniform, flat universe directly – without inflation. Obtaining a flat universe is unlikely overall. Penrose's shocking conclusion, though, was that obtaining a flat universe without inflation is much more likely than with inflation – by a factor of 10 to the googol (10 to the 100) power!"[105][106] Together with Anna Ijjas and Abraham Loeb, he wrote articles claiming that the inflationary paradigm is in trouble in view of the data from the Planck satellite.[124][125] Counter-arguments were presented by Alan Guth, David Kaiser, and Yasunori Nomura[126] and by Andrei Linde,[127] saying that "cosmic inflation is on a stronger footing than ever before".[126]

6.8 See also

- Brane cosmology
- Conservation of angular momentum
- Cosmology
- Dark flow
- Doughnut theory of the universe
- Hubble's law
- Non-minimally coupled inflation
- Nonlinear optics
- Varying speed of light
- Warm inflation

6.9 Notes

[1] "First Second of the Big Bang". *How The Universe Works 3*. 2014. Discovery Science.

[2] Tyson, Neil deGrasse and Donald Goldsmith (2004), *Origins: Fourteen Billion Years of Cosmic Evolution*, W. W. Norton & Co., pp. 84–5.

[3] Tsujikawa, Shinji (28 Apr 2003). "Introductory review of cosmic inflation". p. 4257. arXiv:hep-ph/0304257. Bibcode:2003hep.ph....4257T. In fact temperature anisotropies observed by the COBE satellite in 1992 exhibit nearly scale-invariant spectra as predicted by the inflationary paradigm. Recent observations of WMAP also show strong evidence for inflation.

[4] Guth, Alan H. (1997). *The Inflationary Universe: The Quest for a New Theory of Cosmic Origins*. Basic Books. pp. 233–234. ISBN 0201328402.

[5] "The Medallists: A list of past Dirac Medallists". *ictp.it*.

[6] Staff (17 March 2014). "BICEP2 2014 Results Release". *National Science Foundation*. Retrieved 18 March 2014.

[7] Clavin, Whitney (17 March 2014). "NASA Technology Views Birth of the Universe". *NASA*. Retrieved 17 March 2014.

[8] Overbye, Dennis (17 March 2014). "Space Ripples Reveal Big Bang's Smoking Gun". *The New York Times*. Retrieved 17 March 2014.

[9] Using Tiny Particles To Answer Giant Questions. Science Friday, 3 April 2009.

[10] See also Faster than light#Universal expansion.

[11] Spergel, D.N. (2006). "Three-year Wilkinson Microwave Anisotropy Probe (WMAP) observations: Implications for cosmology". WMAP... confirms the basic tenets of the inflationary paradigm...

[12] "Our Baby Universe Likely Expanded Rapidly, Study Suggests". *Space.com*.

[13] Melia, Fulvio (2007). "The Cosmic Horizon". *Monthly Notices of the Royal Astronomical Society* **382** (4): 1917–1921. arXiv:0711.4181. Bibcode:2007MNRAS.382.1917M. doi:10.1111/j.1365-2966.2007.12499.x.

[14] Melia, Fulvio; et al. (2009). "The Cosmological Spacetime". *International Journal of Modern Physics D* **18** (12): 1889–1901. arXiv:0907.5394. Bibcode:2009IJMPD..18.1889M. doi:10.1142/s0218271809015746.

[15] Kolb and Turner (1988).

[16] Barbara Sue Ryden (2003). *Introduction to cosmology*. Addison-Wesley. ISBN 978-0-8053-8912-8. Not only is inflation very effective at driving down the number density of magnetic monopoles, it is also effective at driving down the number density of every other type of particle, including photons.:202–207

[17] This is usually quoted as 60 e-folds of expansion, where $e^{60} \approx 10^{26}$. It is equal to the amount of expansion since reheating, which is roughly $E_{\text{inflation}}/T_0$, where $T_0 = 2.7$ K is the temperature of the cosmic microwave background today. See, *e.g.* Kolb and Turner (1998) or Liddle and Lyth (2000).

[18] Guth, *Phase transitions in the very early universe*, in The Very Early Universe, ISBN 0-521-31677-4 eds Hawking, Gibbon & Siklos

[19] See Kolb and Turner (1988) or Mukhanov (2005).

[20] Kofman, Lev; Linde, Andrei; Starobinsky, Alexei (1994). "Reheating after inflation". *Physical Review Letters* **73** (5): 3195–3198. arXiv:hep-th/9405187. Bibcode:1986CQGra...3..811K. doi:10.1088/0264-9381/3/5/011.

[21] Much of the historical context is explained in chapters 15–17 of Peebles (1993).

[22] Misner, Charles W.; Coley, A A; Ellis, G F R; Hancock, M (1968). "The isotropy of the universe". *Astrophysical Journal* **151** (2): 431. Bibcode:1998CQGra..15..331W. doi:10.1088/0264-9381/15/2/008.

[23] Misner, Charles; Thorne, Kip S. and Wheeler, John Archibald (1973). *Gravitation*. San Francisco: W. H. Freeman. pp. 489–490, 525–526. ISBN 0-7167-0344-0.

[24] Weinberg, Steven (1971). *Gravitation and Cosmology*. John Wiley. pp. 740, 815. ISBN 0-471-92567-5.

[25] Lemaître, Georges (1933). "The expanding universe". *Annales de la Société Scientifique de Bruxelles* **47A**: 49., English in *Gen. Rel. Grav.* **29**:641–680, 1997.

[26] R. C. Tolman (1934). *Relativity, Thermodynamics, and Cosmology*. Oxford: Clarendon Press. ISBN 0-486-65383-8. LCCN 34032023. Reissued (1987) New York: Dover ISBN 0-486-65383-8.

[27] Misner, Charles W.; Leach, P G L (1969). "Mixmaster universe". *Physical Review Letters* **22** (15): 1071–74. Bibcode:2008JPhA...41o5201A. doi:10.1088/1751-8113/41/15/155201.

[28] Dicke, Robert H. (1970). *Gravitation and the Universe*. Philadelphia: American Philosophical Society.

[29] Dicke, Robert H.; P. J. E. Peebles (1979). "The big bang cosmology – enigmas and nostrums". In ed. S. W. Hawking and W. Israel. *General Relativity: an Einstein Centenary Survey*. Cambridge University Press.

[30] Alan P. Lightman (1 January 1993). *Ancient Light: Our Changing View of the Universe*. Harvard University Press. ISBN 978-0-674-03363-4.

[31] "WMAP- Content of the Universe". *nasa.gov*.

[32] Since supersymmetric Grand Unified Theory is built into string theory, it is still a triumph for inflation that it is able to deal with these magnetic relics. See, *e.g.* Kolb and Turner (1988) and Raby, Stuart (2006). ed. Bruce Hoeneisen, ed. "Grand Unified Theories". arXiv:hep-ph/0608183.

[33] 't Hooft, Gerard (1974). "Magnetic monopoles in Unified Gauge Theories". *Nuclear Physics B* **79** (2): 276–84. Bibcode:1974NuPhB..79..276T. doi:10.1016/0550-3213(74)90486-6.

[34] Polyakov, Alexander M. (1974). "Particle spectrum in quantum field theory". *JETP Letters* **20**: 194–5. Bibcode:1974JETPL..20..194P.

[35] Guth, Alan; Tye, S. (1980). "Phase Transitions and Magnetic Monopole Production in the Very Early Universe". *Physical Review Letters* **44** (10): 631–635; Erratum *ibid.*,**44**:963, 1980. Bibcode:1980PhRvL..44..631G. doi:10.1103/PhysRevLett.44.631.

[36] Einhorn, Martin B; Stein, D. L.; Toussaint, Doug (1980). "Are Grand Unified Theories Compatible with Standard Cosmology?". *Physical Review D* **21** (12): 3295–3298. Bibcode:1980PhRvD..21.3295E. doi:10.1103/PhysRevD.21.3295.

[37] Zel'dovich, Ya.; Khlopov, M. Yu. (1978). "On the concentration of relic monopoles in the universe". *Physics Letters B* **79** (3): 239–41. Bibcode:1978PhLB...79..239Z. doi:10.1016/0370-2693(78)90232-0.

[38] Preskill, John (1979). "Cosmological production of superheavy magnetic monopoles". *Physical Review Letters* **43** (19): 1365–1368. Bibcode:1979PhRvL..43.1365P. doi:10.1103/PhysRevLett.43.1365.

6.9. NOTES

[39] See, *e.g.* Yao, W.-M.; Amsler, C.; Asner, D.; Barnett, R. M.; Beringer, J.; Burchat, P. R.; Carone, C. D.; Caso, C.; Dahl, O.; d'Ambrosio, G.; De Gouvea, A.; Doser, M.; Eidelman, S.; Feng, J. L.; Gherghetta, T.; Goodman, M.; Grab, C.; Groom, D. E.; Gurtu, A.; Hagiwara, K.; Hayes, K. G.; Hernández-Rey, J. J.; Hikasa, K.; Jawahery, H.; Kolda, C.; Kwon, Y.; Mangano, M. L.; Manohar, A. V.; Masoni, A.; et al. (2006). "Review of Particle Physics". *J. Phys. G* **33** (1): 1–1232. arXiv:astro-ph/0601168. Bibcode:2006JPhG...33....1Y. doi:10.1088/0954-3899/33/1/001.

[40] Rees, Martin. (1998). *Before the Beginning* (New York: Basic Books) p. 185 ISBN 0-201-15142-1

[41] de Sitter, Willem (1917). "Einstein's theory of gravitation and its astronomical consequences. Third paper". *Monthly Notices of the Royal Astronomical Society* **78**: 3–28. Bibcode:1917MNRAS..78....3D. doi:10.1093/mnras/78.1.3.

[42] Starobinsky, A. A. (December 1979). "Spectrum Of Relict Gravitational Radiation And The Early State Of The Universe". *Journal of Experimental and Theoretical Physics Letters* **30**: 682. Bibcode:1979JETPL..30..682S.; Starobinskii, A. A. (December 1979). "Spectrum of relict gravitational radiation and the early state of the universe". *Pisma Zh. Eksp. Teor. Fiz. (Soviet Journal of Experimental and Theoretical Physics Letters)* **30**: 719. Bibcode:1979ZhPmR..30..719S.

[43] SLAC seminar, "10^{-35} seconds after the Big Bang", 23 January 1980. see Guth (1997), pg 186

[44] Guth, Alan H. (1981). "Inflationary universe: A possible solution to the horizon and flatness problems" (PDF). *Physical Review D* **23** (2): 347–356. Bibcode:1981PhRvD..23..347G. doi:10.1103/PhysRevD.23.347.

[45] Chapter 17 of Peebles (1993).

[46] Starobinsky, Alexei A. (1980). "A new type of isotropic cosmological models without singularity". *Physics Letters B* **91**: 99–102. Bibcode:1980PhLB...91...99S. doi:10.1016/0370-2693(80)90670-X.

[47] Kazanas, D. (1980). "Dynamics of the universe and spontaneous symmetry breaking". *Astrophysical Journal* **241**: L59–63. Bibcode:1980ApJ...241L..59K. doi:10.1086/183361.

[48] Sato, K. (1981). "Cosmological baryon number domain structure and the first order phase transition of a vacuum". *Physics Letters B* **33**: 66–70. Bibcode:1981PhLB...99...66S. doi:10.1016/0370-2693(81)90805-4.

[49] Einhorn, Martin B; Sato, Katsuhiko (1981). "Monopole Production In The Very Early Universe In A First Order Phase Transition". *Nuclear Physics B* **180** (3): 385–404. Bibcode:1981NuPhB.180..385E. doi:10.1016/0550-3213(81)90057-2.

[50] Linde, A (1982). "A new inflationary universe scenario: A possible solution of the horizon, flatness, homogeneity, isotropy and primordial monopole problems". *Physics Letters B* **108** (6): 389–393. Bibcode:1982PhLB..108..389L. doi:10.1016/0370-2693(82)91219-9.

[51] Albrecht, Andreas; Steinhardt, Paul (1982). "Cosmology for Grand Unified Theories with Radiatively Induced Symmetry Breaking" (PDF). *Physical Review Letters* **48** (17): 1220–1223. Bibcode:1982PhRvL..48.1220A. doi:10.1103/PhysRevLett.48.1220.

[52] J.B. Hartle (2003). *Gravity: An Introduction to Einstein's General Relativity* (1st ed.). Addison Wesley. p. 411. ISBN 0-8053-8662-9

[53] See Linde (1990) and Mukhanov (2005).

[54] Chibisov, Viatcheslav F.; Chibisov, G. V. (1981). "Quantum fluctuation and "nonsingular" universe". *JETP Letters* **33**: 532–5. Bibcode:1981JETPL..33..532M.

[55] Mukhanov, Viatcheslav F. (1982). "The vacuum energy and large scale structure of the universe". *Soviet Physics JETP* **56**: 258–65.

[56] See Guth (1997) for a popular description of the workshop, or *The Very Early Universe*, ISBN 0-521-31677-4 eds Hawking, Gibbon & Siklos for a more detailed report

[57] Hawking, S.W. (1982). "The development of irregularities in a single bubble inflationary universe". *Physics Letters B* **115** (4): 295–297. Bibcode:1982PhLB..115..295H. doi:10.1016/0370-2693(82)90373-2.

[58] Starobinsky, Alexei A. (1982). "Dynamics of phase transition in the new inflationary universe scenario and generation of perturbations". *Physics Letters B* **117** (3–4): 175–8. Bibcode:1982PhLB..117..175S. doi:10.1016/0370-2693(82)90541-X.

[59] Guth, A.H. (1982). "Fluctuations in the new inflationary universe". *Physical Review Letters* **49** (15): 1110–3. Bibcode:1982PhRvL..49.1110G. doi:10.1103/PhysRevLett.49.1110.

[60] Bardeen, James M.; Steinhardt, Paul J.; Turner, Michael S. (1983). "Spontaneous creation Of almost scale-free density perturbations in an inflationary universe". *Physical Review D* **28** (4): 679–693. Bibcode:1983PhRvD..28..679B. doi:10.1103/PhysRevD.28.679.

[61] Perturbations can be represented by Fourier modes of a wavelength. Each Fourier mode is normally distributed (usually called Gaussian) with mean zero. Different Fourier components are uncorrelated. The variance of a mode depends only on its wavelength in such a way that within any given volume each wavelength contributes an equal amount of power to the spectrum of perturbations. Since the Fourier transform is in three dimensions, this means that the variance of a mode goes as k^{-3} to compensate for the fact that within any volume, the number of modes with a given wavenumber k goes as k^3.

[62] Tegmark, M.; Eisenstein, Daniel J.; Strauss, Michael A.; Weinberg, David H.; Blanton, Michael R.; Frieman, Joshua A.; Fukugita, Masataka; Gunn, James E.; et al. (August 2006). "Cosmological constraints from the SDSS luminous red galaxies". *Physical Review D* **74** (12). arXiv:astro-ph/0608632. Bibcode:2006PhRvD..74l3507T. doi:10.1103/PhysRevD.74.123507.

[63] Steinhardt, Paul J. (2004). "Cosmological perturbations: Myths and facts". *Modern Physics Letters A* **19** (13 & 16): 967–82. Bibcode:2004MPLA...19..967S. doi:10.1142/S0217732304014252.

[64] Boyle, Latham A.; Steinhardt, PJ; Turok, N (2006). "Inflationary predictions for scalar and tensor fluctuations reconsidered". *Physical Review Letters* **96** (11): 111301. arXiv:astro-ph/0507455. Bibcode:2006PhRvL..96k1301B. doi:10.1103/PhysRevLett.96.111301. PMID 16605810.

[65] Tegmark, Max (2005). "What does inflation really predict?". *JCAP* **0504** (4): 001. arXiv:astro-ph/0410281. Bibcode:2005JCAP...04..001T. doi:10.1088/1475-7516/2005/04/001.

[66] This is known as a "red" spectrum, in analogy to redshift, because the spectrum has more power at longer wavelengths.

[67] Komatsu, E.; Smith, K. M.; Dunkley, J.; Bennett, C. L.; Gold, B.; Hinshaw, G.; Jarosik, N.; Larson, D.; et al. (January 2010). "Seven-Year Wilkinson Microwave Anisotropy Probe (WMAP) Observations: Cosmological Interpretation". *The Astrophysical Journal Supplement Series* **192** (2): 18. arXiv:1001.4538. Bibcode:2011ApJS..192...18K. doi:10.1088/0067-0049/192/2/18.

[68] Spergel, D. N.; Verde, L.; Peiris, H. V.; Komatsu, E.; Nolta, M. R.; Bennett, C. L.; Halpern, M.; Hinshaw, G.; et al. (2003). "First year Wilkinson Microwave Anisotropy Probe (WMAP) observations: determination of cosmological parameters". *Astrophysical Journal Supplement Series* **148** (1): 175–194. arXiv:astro-ph/0302209. Bibcode:2003ApJS..148..175S. doi:10.1086/377226.

[69] See cosmic microwave background#Low multipoles for details and references.

[70] Overbye, Dennis (24 March 2014). "Ripples From the Big Bang". *New York Times*. Retrieved 24 March 2014.

[71] Ade, P.A.R. (BICEP2 Collaboration); et al. (19 June 2014). "Detection of B-Mode Polarization at Degree Angular Scales by BICEP2". *Physical Review Letters* **112** (24): 241101. arXiv:1403.3985. Bibcode:2014PhRvL.112x1101A. doi:10.1103/PhysRevLett.112.241101. PMID 24996078.

[72] Woit, Peter (13 May 2014). "BICEP2 News". *Not Even Wrong*. Columbia University. Retrieved 19 January 2014.

[73] Overbye, Dennis (19 June 2014). "Astronomers Hedge on Big Bang Detection Claim". *New York Times*. Retrieved 20 June 2014.

[74] Amos, Jonathan (19 June 2014). "Cosmic inflation: Confidence lowered for Big Bang signal". *BBC News*. Retrieved 20 June 2014.

[75] Planck Collaboration Team (19 September 2014). "Planck intermediate results. XXX. The angular power spectrum of polarized dust emission at intermediate and high Galactic latitudes". *ArXiv*. arXiv:1409.5738. Bibcode:2014arXiv1409.5738P. Retrieved 22 September 2014.

[76] Overbye, Dennis (22 September 2014). "Study Confirms Criticism of Big Bang Finding". *New York Times*. Retrieved 22 September 2014.

[77] Clavin, Whitney (30 January 2015). "Gravitational Waves from Early Universe Remain Elusive". *NASA*. Retrieved 30 January 2015.

[78] Overbye, Dennis (30 January 2015). "Speck of Interstellar Dust Obscures Glimpse of Big Bang". *New York Times*. Retrieved 31 January 2015.

[79] Rosset, C.; PLANCK-HFI collaboration (2005). "Systematic effects in CMB polarization measurements". *Exploring the universe: Contents and structures of the universe (XXXIXth Rencontres de Moriond)*.

[80] Loeb, A.; Zaldarriaga, M (2004). "Measuring the small-scale power spectrum of cosmic density fluctuations through 21 cm tomography prior to the epoch of structure formation". *Physical Review Letters* **92** (21): 211301. arXiv:astro-ph/0312134. Bibcode:2004PhRvL..92u1301L. doi:10.1103/PhysRevLett.92.211301. PMID 15245272.

[81] Guth, Alan (1997). *The Inflationary Universe*. Addison–Wesley. ISBN 0-201-14942-7.

[82] Choi, Charles (Jun 29, 2012). "Could the Large Hadron Collider Discover the Particle Underlying Both Mass and Cosmic Inflation?". Scientific American. Retrieved Jun 25, 2014."The virtue of so-called Higgs inflation models is that they might explain inflation within the current Standard Model of particle physics, which successfully describes how most known particles and forces behave. Interest in the Higgs is running hot this summer because CERN, the lab in Geneva, Switzerland, that runs the LHC, has said it will announce highly anticipated findings regarding the particle in early July."

[83] Salvio, Alberto (2013-08-09). "Higgs Inflation at NNLO after the Boson Discovery". *Phys.Lett.* *B727 (2013) 234-239* **727**: 234–239. arXiv:1308.2244. Bibcode:2013PhLB..727..234S. doi:10.1016/j.physletb.2013.10.042.

[84] Technically, these conditions are that the logarithmic derivative of the potential, $\epsilon = (1/2)(V'/V)^2$ and second derivative $\eta = V''/V$ are small, where V is the potential and the equations are written in reduced Planck units. See, *e.g.* Liddle and Lyth (2000), pg 42-43.

[85] Salvio, Strumia (2014-03-17). "Agravity". *JHEP 1406 (2014) 080* **2014**. arXiv:1403.4226. Bibcode:2014JHEP...06..080S. doi:10.1007/JHEP06(2014)080.

[86] Linde, Andrei D. (1983). "Chaotic inflation". *Physics Letters B* **129** (3): 171–81. Bibcode:1983PhLB..129..177L. doi:10.1016/0370-2693(83)90837-7.

[87] Technically, this is because the inflaton potential is expressed as a Taylor series in φ/mP_l, where φ is the inflaton and mP_l is the Planck mass. While for a single term, such as the mass term $m_\varphi^4(\varphi/mP_l)^2$, the slow roll conditions can be satisfied for φ much greater than mP_l, this is precisely the situation in effective field theory in which higher order terms would be expected to contribute and destroy the conditions for inflation. The absence of these higher order corrections can be seen as another sort of fine tuning. See *e.g.* Alabidi, Laila; Lyth, David H (2006). "Inflation models and observation". *JCAP* **0605** (5): 016. arXiv:astro-ph/0510441. Bibcode:2006JCAP...05..016A. doi:10.1088/1475-7516/2006/05/016.

[88] See, *e.g.* Lyth, David H. (1997). "What would we learn by detecting a gravitational wave signal in the cosmic microwave background anisotropy?". *Physical Review Letters* **78** (10): 1861–3. arXiv:hep-ph/9606387. Bibcode:1997PhRvL..78.1861L. doi:10.1103/PhysRevLett.78.1861.

[89] Brandenberger, Robert H. (November 2004). "Challenges for inflationary cosmology". arXiv:astro-ph/0411671.

[90] Vilenkin, Alexander (1983). "The birth of inflationary universes". *Physical Review D* **27** (12): 2848–2855. Bibcode:1983PhRvD..27.2848V. doi:10.1103/PhysRevD.27.2848.

[91] A. Linde (1986). "Eternal chaotic inflation". *Modern Physics Letters A* **1** (2): 81–85. Bibcode:1986MPLA....1...81L. doi:10.1142/S0217732386000129. A. Linde (1986). "Eternally existing self-reproducing chaotic inflationary universe" (PDF). *Physics Letters B* **175** (4): 395–400. Bibcode:1986PhLB..175..395L. doi:10.1016/0370-2693(86)90611-8.

[92] A. Borde, A. Guth and A. Vilenkin (2003). "Inflationary space-times are incomplete in past directions". *Physical Review Letters* **90** (15): 151301. arXiv:gr-qc/0110012. Bibcode:2003PhRvL..90o1301B. doi:10.1103/PhysRevLett.90.151301. PMID 12732026.

[93] A. Borde (1994). "Open and closed universes, initial singularities and inflation". *Physical Review D* **50** (6): 3692–702. arXiv:gr-qc/9403049. Bibcode:1994PhRvD..50.3692B. doi:10.1103/PhysRevD.50.3692.

[94] A. Borde and A. Vilenkin (1994). "Eternal inflation and the initial singularity". *Physical Review Letters* **72** (21): 3305–9. arXiv:gr-qc/9312022. Bibcode:1994PhRvL..72.3305B. doi:10.1103/PhysRevLett.72.3305.

[95] Linde (2005, §V).

[96] Carroll, Sean M.; Chen, Jennifer (2005). "Does inflation provide natural initial conditions for the universe?". *Gen. Rel. Grav.* **37** (10): 1671–4. arXiv:gr-qc/0505037. Bibcode:2005GReGr..37.1671C. doi:10.1007/s10714-005-0148-2.

[97] Carroll, Sean M.; Jennifer Chen (2004). "Spontaneous inflation and the origin of the arrow of time". arXiv:hep-th/0410270.

[98] Aguirre, Anthony; Gratton, Steven (2003). "Inflation without a beginning: A null boundary proposal". *Physical Review D* **67** (8): 083515. arXiv:gr-qc/0301042. Bibcode:2003PhRvD..67h3515A. doi:10.1103/PhysRevD.67.083515.

[99] Aguirre, Anthony; Gratton, Steven (2002). "Steady-State Eternal Inflation". *Physical Review D* **65** (8): 083507. arXiv:astro-ph/0111191. Bibcode:2002PhRvD..65h3507A. doi:10.1103/PhysRevD.65.083507.

[100] Hartle, J.; Hawking, S. (1983). "Wave function of the universe". *Physical Review D* **28** (12): 2960–2975. Bibcode:1983PhRvD..28.2960H. doi:10.1103/PhysRevD.28.2960.; See also Hawking (1998).

[101] Hawking (1998), p. 129.

[102] Wikiquote

[103] Page, Don N. (1983). "Inflation does not explain time asymmetry". *Nature* **304** (5921): 39–41. Bibcode:1983Natur.304...39P. doi:10.1038/304039a0.; see also Roger Penrose's book The Road to Reality: A Complete Guide to the Laws of the Universe.

[104] Hawking, S. W.; Page, Don N. (1988). "How probable is inflation?". *Nuclear Physics B* **298** (4): 789–809. Bibcode:1988NuPhB.298..789H. doi:10.1016/0550-3213(88)90008-9.

[105] Steinhardt, Paul J. (2011). "The inflation debate: Is the theory at the heart of modern cosmology deeply flawed?" (*Scientific American*, April; pp. 18-25).

[106] Paul J. Steinhardt; Neil Turok (2007). *Endless Universe: Beyond the Big Bang*. Broadway Books. ISBN 978-0-7679-1501-4.

[107] Albrecht, Andreas; Sorbo, Lorenzo (2004). "Can the universe afford inflation?". *Physical Review D* **70** (6): 063528. arXiv:hep-th/0405270. Bibcode:2004PhRvD..70f3528A. doi:10.1103/PhysRevD.70.063528.

[108] Martin, Jerome; Brandenberger, Robert (2001). "The trans-Planckian problem of inflationary cosmology". *Physical Review D* **63** (12): 123501. arXiv:hep-th/0005209. Bibcode:2001PhRvD..63l3501M. doi:10.1103/PhysRevD.63.123501.

[109] Martin, Jerome; Ringeval, Christophe (2004). "Superimposed Oscillations in the WMAP Data?". *Physical Review D* **69** (8): 083515. arXiv:astro-ph/0310382. Bibcode:2004PhRvD..69h3515M. doi:10.1103/PhysRevD.69.083515.

[110] Robert H. Brandenberger, "A Status Review of Inflationary Cosmology", proceedings Journal-ref: BROWN-HET-1256 (2001), (available from arXiv:hep-ph/0101119v1 11 January 2001)

[111] Andrei Linde, "Prospects of Inflation", *Physica Scripta Online* (2004) (available from arXiv:hep-th/0402051)

[112] Blanco-Pillado et al., "Racetrack inflation", (2004) (available from arXiv:hep-th/0406230)

[113] Kachru, Shamit; Kallosh, Renata; Linde, Andrei; Maldacena, Juan; McAllister, Liam; Trivedi, Sandip P (2003). "Towards inflation in string theory". *JCAP* **0310** (10): 013. arXiv:hep-th/0308055. Bibcode:2003JCAP...10..013K. doi:10.1088/1475-7516/2003/10/013.

[114] G. R. Dvali, S. H. Henry Tye, *Brane inflation*, Phys.Lett. **B450**, 72-82 (1999), arXiv:hep-ph/9812483.

[115] Bojowald, Martin (October 2008). "Big Bang or Big Bounce?: New Theory on the Universe's Birth". Retrieved 2015-08-31.

[116] Poplawski, N. J. (2010). "Cosmology with torsion: An alternative to cosmic inflation". *Physics Letters B* **694** (3): 181–185. arXiv:1007.0587. Bibcode:2010PhLB..694..181P. doi:10.1016/j.physletb.2010.09.056.

[117] Poplawski, N. (2012). "Nonsingular, big-bounce cosmology from spinor-torsion coupling". *Physical Review D* **85** (10): 107502. arXiv:1111.4595. Bibcode:2012PhRvD..85j7502P. doi:10.1103/PhysRevD.85.107502.

[118] Brandenberger, R; Vafa, C. (1989). "Superstrings in the early universe". *Nuclear Physics B* **316** (2): 391–410. Bibcode:1989NuPhB.316..391B. doi:10.1016/0550-3213(89)90037-0.

[119] Battefeld, Thorsten; Watson, Scott (2006). "String Gas Cosmology". *Reviews Modern Physics* **78** (2): 435–454. arXiv:hep-th/0510022. Bibcode:2006RvMP...78..435B. doi:10.1103/RevModPhys.78.435.

[120] Brandenberger, Robert H.; Nayeri, ALI; Patil, Subodh P.; Vafa, Cumrun (2007). "String Gas Cosmology and Structure Formation". *International Journal of Modern Physics A* **22** (21): 3621–3642. arXiv:hep-th/0608121. Bibcode:2007IJMPA..22.3621B. doi:10.1142/S0217751X07037159.

[121] Earman, John; Mosterín, Jesús (March 1999). "A Critical Look at Inflationary Cosmology". *Philosophy of Science* **66**: 1–49. doi:10.2307/188736 (inactive 2015-01-14). JSTOR 188736.

[122] Penrose, Roger (2004). *The Road to Reality: A Complete Guide to the Laws of the Universe.* London: Vintage Books, p. 755. See also Penrose, Roger (1989). "Difficulties with Inflationary Cosmology". *Annals of the New York Academy of Sciences* **271**: 249–264. Bibcode:1989NYASA.571..249P. doi:10.1111/j.1749-6632.1989.tb50513.x.

[123] Hložek, Renée (12 June 2015). "CMB@50 day three". Retrieved 15 July 2015.
This is a collation of remarks from the third day of the "Cosmic Microwave Background @50" conference held at Princeton, 10–12 June 2015.

[124] Ijjas, Anna; Steinhardt, Paul J.; Loeb, Abraham. "Inflationary paradigm in trouble after Planck2013". *Physics Letters* **B723**: 261–266. arXiv:1304.2785. Bibcode:2013PhLB..723..261I. doi:10.1016/j.physletb.2013.05.023.

[125] Ijjas, Anna; Steinhardt, Paul J.; Loeb, Abraham. "Inflationary schism after Planck2013". *Physics Letters* **B736**: 142–146. Bibcode:2014PhLB..736..142I. doi:10.1016/j.physletb.2014.07.012.

[126] Guth, Alan H.; Kaiser, David I.; Nomura, Yasunori. "Inflationary paradigm after Planck 2013". *Physics Letters* **B733**: 112–119. arXiv:1312.7619. Bibcode:2014PhLB..733..112G. doi:10.1016/j.physletb.2014.03.020.

[127] Linde, Andrei. "Inflationary cosmology after Planck 2013". arXiv:1402.0526. Bibcode:2014arXiv1402.0526L.

6.10 References

- Guth, Alan (1997). *The Inflationary Universe: The Quest for a New Theory of Cosmic Origins.* Perseus. ISBN 0-201-32840-2.

- Hawking, Stephen (1998). *A Brief History of Time.* Bantam. ISBN 0-553-38016-8.

- Hawking, Stephen; Gary Gibbons (1983). *The Very Early Universe.* Cambridge University Press. ISBN 0-521-31677-4.

- Kolb, Edward; Michael Turner (1988). *The Early Universe.* Addison-Wesley. ISBN 0-201-11604-9.

- Linde, Andrei (1990). *Particle Physics and Inflationary Cosmology.* Chur, Switzerland: Harwood. arXiv:hep-th/0503203. ISBN 3-7186-0490-6.

- Linde, Andrei (2005) "Inflation and String Cosmology", eConf **C040802** (2004) L024; J. Phys. Conf. Ser. **24** (2005) 151–60; arXiv:hep-th/0503195 v1 2005-03-24.

- Liddle, Andrew; David Lyth (2000). *Cosmological Inflation and Large-Scale Structure*. Cambridge. ISBN 0-521-57598-2.

- Lyth, David H.; Riotto, Antonio (1999). "Particle physics models of inflation and the cosmological density perturbation". *Phys. Rept.* **314** (1–2): 1–146. arXiv:hep-ph/9807278. Bibcode:1999PhR...314....1L. doi:10.1016/S0370-1573(98)00128-8.

- Mukhanov, Viatcheslav (2005). *Physical Foundations of Cosmology*. Cambridge University Press. ISBN 0-521-56398-4.

- Vilenkin, Alex (2006). *Many Worlds in One: The Search for Other Universes*. Hill and Wang. ISBN 0-8090-9523-8.

- Peebles, P. J. E. (1993). *Principles of Physical Cosmology*. Princeton University Press. ISBN 0-691-01933-9.

6.11 External links

- Was Cosmic Inflation The 'Bang' Of The Big Bang?, by Alan Guth, 1997

- An Introduction to Cosmological Inflation by Andrew Liddle, 1999

- update 2004 by Andrew Liddle

- hep-ph/0309238 Laura Covi: Status of observational cosmology and inflation

- hep-th/0311040 David H. Lyth: Which is the best inflation model?

- The Growth of Inflation *Symmetry*, December 2004

- Guth's logbook showing the original idea

- WMAP Bolsters Case for Cosmic Inflation, March 2006

- NASA March 2006 WMAP press release

- Max Tegmark's *Our Mathematical Universe* (2014), "Chapter 5: Inflation"

Chapter 7

Spontaneous symmetry breaking

Spontaneous symmetry breaking[1][2][3] is a mode of realization of symmetry breaking in a physical system, where the underlying laws are invariant under a symmetry transformation, but the system as a whole changes under such transformations, in contrast to explicit symmetry breaking. It is a spontaneous process by which a system in a symmetrical state ends up in an asymmetrical state. It thus describes systems where the equations of motion or the Lagrangian obey certain symmetries, but the lowest-energy solutions do not exhibit that symmetry.

Consider a symmetrical upward dome with a trough circling the bottom. If a ball is put at the very peak of the dome, the system is symmetrical with respect to a rotation around the center axis. But the ball may *spontaneously break* this symmetry by rolling down the dome into the trough, a point of lowest energy. Afterward, the ball has come to a rest at some fixed point on the perimeter. The dome and the ball retain their individual symmetry, but the system does not.[4]

Most simple phases of matter and phase transitions, like crystals, magnets, and conventional superconductors can be simply understood from the viewpoint of spontaneous symmetry breaking. Notable exceptions include topological phases of matter like the fractional quantum Hall effect.

7.1 Spontaneous symmetry breaking in physics

7.1.1 Particle physics

In particle physics the force carrier particles are normally specified by field equations with gauge symmetry; their equations predict that certain measurements will be the same at any point in the field. For instance, field equations might predict that the mass of two quarks is constant. Solving the equations to find the mass of each quark might give two solutions. In one solution, quark A is heavier than quark B. In the second solution, quark B is heavier than quark A *by the same amount*. The symmetry of the equations is

Spontaneous symmetry breaking simplified: – *At high energy levels* (left) *the ball settles in the center, and the result is symmetrical. At lower energy levels* (right), *the overall "rules" remain symmetrical, but the "Mexican hat" potential comes into effect: "local" symmetry is inevitably broken since eventually the ball must roll one way (at random) and not another.*

not reflected by the individual solutions, but it is reflected by the range of solutions. An actual measurement reflects only one solution, representing a breakdown in the symmetry of the underlying theory. "Hidden" is perhaps a better term than "broken" because the symmetry is always there in these equations. This phenomenon is called *spontaneous symmetry breaking* because *nothing* (that we know) breaks the symmetry in the equations.[5]:194–195

Chiral symmetry

Main article: Chiral symmetry breaking

Chiral symmetry breaking is an example of spontaneous symmetry breaking affecting the chiral symmetry of the strong interactions in particle physics. It is a property of quantum chromodynamics, the quantum field theory describing these interactions, and is responsible for the bulk of the mass (over 99%) of the nucleons, and thus of all common matter, as it converts very light bound quarks into 100 times heavier constituents of baryons. The approximate Nambu–Goldstone bosons in this spontaneous symmetry breaking process are the pions, whose mass is an order of magnitude lighter than the mass of the nucleons. It served as the prototype and significant ingredient of the Higgs mechanism underlying the electroweak symmetry breaking.

Higgs mechanism

Main articles: Brout–Englert–Higgs mechanism and Yukawa interaction

The strong, weak, and electromagnetic forces can all be understood as arising from gauge symmetries. The Higgs mechanism, the spontaneous symmetry breaking of gauge symmetries, is an important component in understanding the superconductivity of metals and the origin of particle masses in the standard model of particle physics. One important consequence of the distinction between true symmetries and *gauge symmetries*, is that the spontaneous breaking of a gauge symmetry does not give rise to characteristic massless Nambu–Goldstone modes, but only massive modes, like the plasma mode in a superconductor, or the Higgs mode observed in particle physics.

In the standard model of particle physics, spontaneous symmetry breaking of the SU(2) × U(1) gauge symmetry associated with the electro-weak force generates masses for several particles, and separates the electromagnetic and weak forces. The W and Z bosons are the elementary particles that mediate the weak interaction, while the photon mediates the electromagnetic interaction. At energies much greater than 100 GeV all these particles behave in a similar manner. The Weinberg–Salam theory predicts that, at lower energies, this symmetry is broken so that the photon and the massive W and Z bosons emerge.[6] In addition, fermions develop mass consistently.

Without spontaneous symmetry breaking, the Standard Model of elementary particle interactions requires the existence of a number of particles. However, some particles (the W and Z bosons) would then be predicted to be massless, when, in reality, they are observed to have mass. To overcome this, spontaneous symmetry breaking is augmented by the Higgs mechanism to give these particles mass. It also suggests the presence of a new particle, the Higgs boson, reported as possibly identifiable with a boson detected in 2012. (If the Higgs boson were not confirmed to have been found, it would mean that the simplest implementation of the Higgs mechanism and spontaneous symmetry breaking *as they are currently formulated* require modification.)

Superconductivity of metals is a condensed-matter analog of the Higgs phenomena, in which a condensate of Cooper pairs of electrons spontaneously breaks the U(1) gauge "symmetry" associated with light and electromagnetism.

7.1.2 Condensed matter physics

Most phases of matter can be understood through the lens of spontaneous symmetry breaking. For example, crystals are periodic arrays of atoms that are not invariant under all translations (only under a small subset of translations by a lattice vector). Magnets have north and south poles that are oriented in a specific direction, breaking rotational symmetry. In addition to these examples, there are a whole host of other symmetry-breaking phases of matter including nematic phases of liquid crystals, charge- and spin-density waves, superfluids and many others.

There are several known examples of matter that cannot be described by spontaneous symmetry breaking, including: topologically ordered phases of matter like fractional quantum Hall liquids, and spin-liquids. These states do not break any symmetry, but are distinct phases of matter. Unlike the case of spontaneous symmetry breaking, there is not a general framework for describing such states.

Continuous symmetry

The ferromagnet is the canonical system which spontaneously breaks the continuous symmetry of the spins below the Curie temperature and at $h = 0$, where h is the external magnetic field. Below the Curie temperature the energy of the system is invariant under inversion of the magnetization $m(\mathbf{x})$ such that $m(\mathbf{x}) = -m(-\mathbf{x})$. The symmetry is spontaneously broken as $h \to 0$ when the Hamiltonian becomes invariant under the inversion transformation, but the expectation value is not invariant.

Spontaneously, symmetry broken phases of matter are characterized by an order parameter that describes the quantity which breaks the symmetry under consideration. For example, in a magnet, the order parameter is the local magnetization.

Spontaneously breaking of a continuous symmetry is inevitably accompanied by gapless (meaning that these modes do not cost any energy to excite) Nambu–Goldstone modes associated with slow long-wavelength fluctuations of the order parameter. For example, vibrational modes in a crystal, known as phonons, are associated with slow density fluctuations of the crystal's atoms. The associated Goldstone mode for magnets are oscillating waves of spin known as spin-waves. For symmetry-breaking states, whose order parameter is not a conserved quantity, Nambu–Goldstone modes are typically massless and propagate at a constant velocity.

An important theorem, due to Mermin and Wagner, states that, at finite temperature, thermally activated fluctuations of Nambu–Goldstone modes destroy the long-range order, and prevent spontaneous symmetry breaking in one- and two-dimensional systems. Similarly, quantum fluctuations of the order parameter prevent most types of continuous symmetry breaking in one-dimensional systems even at zero temperature (an important exception is ferromagnets, whose order parameter, magnetization, is an exactly conserved quantity and does not have any quantum fluctuations).

Other long-range interacting systems such as cylindrical curved surfaces interacting via the Coulomb potential or Yukawa potential has been shown to break translational and rotational symmetries.[7] It was shown, in the presence of a symmetric Hamiltonian, and in the limit of infinite volume, the system spontaneously adopts a chiral configuration, i.e. breaks mirror plane symmetry.

7.1.3 Dynamical symmetry breaking

Dynamical symmetry breaking (DSB) is a special form of spontaneous symmetry breaking where the ground state of the system has reduced symmetry properties compared to its theoretical description (Lagrangian).

Dynamical breaking of a global symmetry is a spontaneous symmetry breaking, that happens not at the (classical) tree level (i.e. at the level of the bare action), but due to quantum corrections (i.e. at the level of the effective action).

Dynamical breaking of a gauge symmetry is subtler. In the conventional spontaneous gauge symmetry breaking, there exists an unstable Higgs particle in the theory, which drives the vacuum to a symmetry-broken phase (see e.g. Electroweak interaction). In dynamical gauge symmetry breaking, however, no unstable Higgs particle operates in the theory, but the bound states of the system itself provide the unstable fields that render the phase transition. For example, Bardeen, Hill, and Lindner published a paper which attempts to replace the conventional Higgs mechanism in the standard model, by a DSB that is driven by a bound state of top-antitop quarks (such models, where a composite particle plays the role of the Higgs boson, are often referred to as "Composite Higgs models").[8] Dynamical breaking of gauge symmetries is often due to creation of a fermionic condensate; for example the quark condensate, which is connected to the dynamical breaking of chiral symmetry in quantum chromodynamics. Conventional superconductivity is the paradigmatic example from the condensed matter side, where phonon-mediated attractions lead electrons to become bound in pairs and then condense, thereby breaking the electromagnetic gauge symmetry.

7.2 Generalisation and technical usage

For spontaneous symmetry breaking to occur, there must be a system in which there are several equally likely outcomes. The system as a whole is therefore symmetric with respect to these outcomes. (If we consider any two outcomes, the probability is the same. This contrasts sharply to explicit symmetry breaking.) However, if the system is sampled (i.e. if the system is actually used or interacted with in any way), a specific outcome must occur. Though the system as a whole is symmetric, it is never encountered with this symmetry, but only in one specific asymmetric state. Hence, the symmetry is said to be spontaneously broken in that theory. Nevertheless, the fact that each outcome is equally likely is a reflection of the underlying symmetry, which is thus often dubbed "hidden symmetry", and has crucial formal consequences. (See the article on the Goldstone boson).

When a theory is symmetric with respect to a symmetry group, but requires that one element of the group be distinct, then spontaneous symmetry breaking has occurred. The theory must not dictate *which* member is distinct, only that *one is*. From this point on, the theory can be treated as if this element actually is distinct, with the proviso that any results found in this way must be resymmetrized, by taking the average of each of the elements of the group being the distinct one.

The crucial concept in physics theories is the order parameter. If there is a field (often a background field) which acquires an expectation value (not necessarily a *vacuum* expectation value) which is not invariant under the symmetry in question, we say that the system is in the ordered phase, and the symmetry is spontaneously broken. This is because other subsystems interact with the order parameter, which specifies a "frame of reference" to be measured against. In that case, the vacuum state does not obey the initial symmetry (which would keep it invariant, in the linearly realized **Wigner mode** in which it would be a singlet), and, instead changes under the (hidden) symmetry, now implemented in the (nonlinear) **Nambu–Goldstone mode**. Normally, in the absence of the Higgs mechanism, massless Goldstone bosons arise.

The symmetry group can be discrete, such as the space group of a crystal, or continuous (e.g., a Lie group), such as the rotational symmetry of space. However, if the system contains only a single spatial dimension, then only discrete symmetries may be broken in a vacuum state of the full quantum theory, although a classical solution may break a continuous symmetry.

7.3 A pedagogical example: the Mexican hat potential

In the simplest idealized relativistic model, the spontaneously broken symmetry is summarized through an illustrative scalar field theory. The relevant Lagrangian, which essentially dictates how a system behaves, can be split up into kinetic and potential terms,

7.4. OTHER EXAMPLES

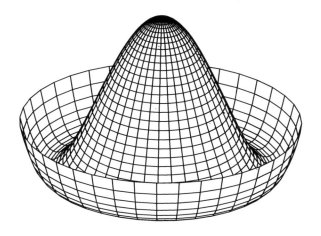

Graph of Goldstone's "Mexican hat" potential function V versus φ.

It is in this potential term $V(\Phi)$ that the symmetry breaking is triggered. An example of a potential, due to Jeffrey Goldstone[9] is illustrated in the graph at the right.

This potential has an infinite number of possible minima (vacuum states) given by

for any real θ between 0 and 2π. The system also has an unstable vacuum state corresponding to $\Phi = 0$. This state has a U(1) symmetry. However, once the system falls into a specific stable vacuum state (amounting to a choice of θ), this symmetry will appear to be lost, or "spontaneously broken".

In fact, any other choice of θ would have exactly the same energy, implying the existence of a massless Nambu–Goldstone boson, the mode running around the circle at the minimum of this potential, and indicating there is some memory of the original symmetry in the Lagrangian.

7.4 Other examples

- For ferromagnetic materials, the underlying laws are invariant under spatial rotations. Here, the order parameter is the magnetization, which measures the magnetic dipole density. Above the Curie temperature, the order parameter is zero, which is spatially invariant, and there is no symmetry breaking. Below the Curie temperature, however, the magnetization acquires a constant nonvanishing value, which points in a certain direction (in the idealized situation where we have full equilibrium; otherwise, translational symmetry gets broken as well). The residual rotational symmetries which leave the orientation of this vector invariant remain unbroken, unlike the other rotations which do not and are thus spontaneously broken.

- The laws describing a solid are invariant under the full Euclidean group, but the solid itself spontaneously breaks this group down to a space group. The displacement and the orientation are the order parameters.

- General relativity has a Lorentz symmetry, but in FRW cosmological models, the mean 4-velocity field defined by averaging over the velocities of the galaxies (the galaxies act like gas particles at cosmological scales) acts as an order parameter breaking this symmetry. Similar comments can be made about the cosmic microwave background.

- For the electroweak model, as explained earlier, a component of the Higgs field provides the order parameter breaking the electroweak gauge symmetry to the electromagnetic gauge symmetry. Like the ferromagnetic example, there is a phase transition at the electroweak temperature. The same comment about us not tending to notice broken symmetries suggests why it took so long for us to discover electroweak unification.

- In superconductors, there is a condensed-matter collective field ψ, which acts as the order parameter breaking the electromagnetic gauge symmetry.

- Take a thin cylindrical plastic rod and push both ends together. Before buckling, the system is symmetric under rotation, and so visibly cylindrically symmetric. But after buckling, it looks different, and asymmetric. Nevertheless, features of the cylindrical symmetry are still there: ignoring friction, it would take no force to freely spin the rod around, displacing the ground state in time, and amounting to an oscillation of vanishing frequency, unlike the radial oscillations in the direction of the buckle. This spinning mode is effectively the requisite Nambu–Goldstone boson.

- Consider a uniform layer of fluid over an infinite horizontal plane. This system has all the symmetries of the Euclidean plane. But now heat the bottom surface uniformly so that it becomes much hotter than the upper surface. When the temperature gradient becomes large enough, convection cells will form, breaking the Euclidean symmetry.

- Consider a bead on a circular hoop that is rotated about a vertical diameter. As the rotational velocity is increased gradually from rest, the bead will initially stay at its initial equilibrium point at the bottom of the hoop (intuitively stable, lowest gravitational potential). At a certain critical rotational velocity, this point will become unstable and the bead will jump to one of two other newly created equilibria, equidistant from the center. Initially, the system is symmetric with respect to the diameter, yet after passing the critical velocity, the bead ends up in one of the two new equilibrium points, thus breaking the symmetry.

7.5 Nobel Prize

On October 7, 2008, the Royal Swedish Academy of Sciences awarded the 2008 Nobel Prize in Physics to three scientists for their work in subatomic physics symmetry breaking. Yoichiro Nambu, of the University of Chicago, won half of the prize for the discovery of the mechanism of spontaneous broken symmetry in the context of the strong interactions, specifically chiral symmetry breaking. Physicists Makoto Kobayashi and Toshihide Maskawa shared the other half of the prize for discovering the origin of the explicit breaking of CP symmetry in the weak interactions.[10] This origin is ultimately reliant on the Higgs mechanism, but, so far understood as a "just so" feature of Higgs couplings, not a spontaneously broken symmetry phenomenon.

7.6 See also

- Autocatalytic reactions and order creation
- Catastrophe theory
- Chiral symmetry breaking
- CP-violation
- Explicit symmetry breaking
- Gauge gravitation theory
- Goldstone boson
- Grand unified theory
- Higgs mechanism
- Higgs boson
- Higgs field (classical)
- Irreversibility
- Magnetic catalysis of chiral symmetry breaking
- Mermin–Wagner theorem
- Quantum fluctuation
- Sakurai Prize for Theoretical Particle Physics
- Second-order phase transition
- Symmetry breaking
- Tachyon condensation
- Tachyonic field
- Wheeler–Feynman absorber theory
- 1964 PRL symmetry breaking papers

7.7 Notes

- ^ Note that (as in fundamental Higgs driven spontaneous gauge symmetry breaking) the term "symmetry breaking" is a misnomer when applied to gauge symmetries.

7.8 References

[1] *Dynamical Symmetry Breaking in Quantum Field Theories*. By Vladimir A. Miranskij. Pg 15.

[2] Patterns of Symmetry Breaking. Edited by Henryk Arodz, Jacek Dziarmaga, Wojciech Hubert Zurek. Pg 141.

[3] Bubbles, Voids and Bumps in Time: The New Cosmology. Edited by James Cornell. Pg 125.

[4] Gerald M. Edelman, Bright Air, Brilliant Fire: On the Matter of the Mind (New York: BasicBooks, 1992) 203.

[5] Steven Weinberg (20 April 2011). *Dreams of a Final Theory: The Scientist's Search for the Ultimate Laws of Nature*. Knopf Doubleday Publishing Group. ISBN 978-0-307-78786-6.

[6] A Brief History of Time, Stephen Hawking, Bantam; 10th anniversary edition (September 1, 1998). pp. 73–74.

[7] Kohlstedt, K.L.; Vernizzi, G.; Solis, F.J.; Olvera de la Cruz, M. (2007). "Spontaneous Chirality via Long-range Electrostatic Forces". *Physical Review Letters* **99**: 030602. arXiv:0704.3435. Bibcode:2007PhRvL..99c0602K. doi:10.1103/PhysRevLett.99.030602.

[8] William A. Bardeen; Christopher T. Hill; Manfred Lindner (1990). "Minimal dynamical symmetry breaking of the standard model". *Physical Review D* **41** (5): 1647–1660. Bibcode:1990PhRvD..41.1647B. doi:10.1103/PhysRevD.41.1647.

[9] Goldstone, J. (1961). "Field theories with " Superconductor " solutions". *Il Nuovo Cimento* **19**: 154–164. doi:10.1007/BF02812722.

[10] The Nobel Foundation. "The Nobel Prize in Physics 2008". *nobelprize.org*. Retrieved January 15, 2008.

7.9 External links

- Spontaneous symmetry breaking
- Physical Review Letters – 50th Anniversary Milestone Papers
- In CERN Courier, Steven Weinberg reflects on spontaneous symmetry breaking
- Englert–Brout–Higgs–Guralnik–Hagen–Kibble Mechanism on Scholarpedia
- History of Englert–Brout–Higgs–Guralnik–Hagen–Kibble Mechanism on Scholarpedia
- The History of the Guralnik, Hagen and Kibble development of the Theory of Spontaneous Symmetry Breaking and Gauge Particles
- International Journal of Modern Physics A: The History of the Guralnik, Hagen and Kibble development of the Theory of Spontaneous Symmetry Breaking and Gauge Particles
- Guralnik, G S; Hagen, C R and Kibble, T W B (1967). Broken Symmetries and the Goldstone Theorem. Advances in Physics, vol. 2 Interscience Publishers, New York. pp. 567–708 ISBN 0-470-17057-3
- Spontaneous Symmetry Breaking in Gauge Theories: a Historical Survey

Chapter 8

Cyclic model

A **cyclic model** (or **oscillating model**) is any of several cosmological models in which the universe follows infinite, or indefinite, self-sustaining cycles. For example, the oscillating universe theory briefly considered by Albert Einstein in 1930 theorized a universe following an eternal series of oscillations, each beginning with a big bang and ending with a big crunch; in the interim, the universe would expand for a period of time before the gravitational attraction of matter causes it to collapse back in and undergo a bounce.

8.1 Overview

In the 1920s, theoretical physicists, most notably Albert Einstein, considered the possibility of a cyclic model for the universe as an (everlasting) alternative to the model of an expanding universe. However, work by Richard C. Tolman in 1934 showed that these early attempts failed because of the cyclic problem: according to the Second Law of Thermodynamics, entropy can only increase.[1] This implies that successive cycles grow longer and larger. Extrapolating back in time, cycles before the present one become shorter and smaller culminating again in a Big Bang and thus not replacing it. This puzzling situation remained for many decades until the early 21st century when the recently discovered dark energy component provided new hope for a consistent cyclic cosmology.[2] In 2011, a five-year survey of 200,000 galaxies and spanning 7 billion years of cosmic time confirmed that "dark energy is driving our universe apart at accelerating speeds."[3][4]

One new cyclic model is a brane cosmology model of the creation of the universe, derived from the earlier ekpyrotic model. It was proposed in 2001 by Paul Steinhardt of Princeton University and Neil Turok of Cambridge University. The theory describes a universe exploding into existence not just once, but repeatedly over time.[5][6] The theory could potentially explain why a mysterious, repulsive form of energy known as the cosmological constant, which is accelerating the expansion of the universe, is several orders of magnitude smaller than predicted by the standard Big Bang model.

A different cyclic model relying on the notion of phantom energy was proposed in 2007 by Lauris Baum and Paul Frampton of the University of North Carolina at Chapel Hill.[7]

Other cyclic models include Conformal cyclic cosmology and Loop quantum cosmology.

8.2 The Steinhardt–Turok model

In this cyclic model, two parallel orbifold planes or M-branes collide periodically in a higher-dimensional space.[8] The visible four-dimensional universe lies on one of these branes. The collisions correspond to a reversal from contraction to expansion, or a big crunch followed immediately by a big bang. The matter and radiation we see today were generated during the most recent collision in a pattern dictated by quantum fluctuations created before the branes. After billions of years the universe reached the state we observe today; after additional billions of years it will ultimately begin to contract again. Dark energy corresponds to a force between the branes, and serves the crucial role of solving the monopole, horizon, and flatness problems. Moreover, the cycles can continue indefinitely into the past and the future, and the solution is an attractor, so it can provide a complete history of the universe.

As Richard C. Tolman showed, the earlier cyclic model failed because the universe would undergo inevitable thermodynamic heat death.[1] However, the newer cyclic model evades this by having a net expansion each cycle, preventing entropy from building up. However, there remain major open issues in the model. Foremost among them is that colliding branes are not understood by string theorists, and nobody knows if the scale invariant spectrum will be destroyed by the big crunch. Moreover, as with cosmic inflation, while the general character of the forces (in the ekpyrotic scenario, a force between branes) required to create the vacuum fluctuations is known, there is no candidate from particle physics. [9]

8.3 The Baum–Frampton model

This more recent cyclic model of 2007 makes a different technical assumption concerning the equation of state of the dark energy which relates pressure and density through a parameter w.[7][10] It assumes $w < -1$ (a condition called phantom energy) throughout a cycle, including at present. (By contrast, Steinhardt–Turok assume w is never less than -1.) In the Baum–Frampton model, a septillionth (or less) of a second (i.e. 10^{-24} seconds or less) before the would-be Big Rip, a turnaround occurs and only one causal patch is retained as our universe. The generic patch contains no quark, lepton or force carrier; only dark energy – and its entropy thereby vanishes. The adiabatic process of contraction of this much smaller universe takes place with constant vanishing entropy and with no matter including no black holes which disintegrated before turnaround.

The idea that the universe "comes back empty" is a central new idea of this cyclic model, and avoids many difficulties confronting matter in a contracting phase such as excessive structure formation, proliferation and expansion of black holes, as well as going through phase transitions such as those of QCD and electroweak symmetry restoration. Any of these would tend strongly to produce an unwanted premature bounce, simply to avoid violation of the second law of thermodynamics. The surprising $w < -1$ condition may be logically inevitable in a truly infinitely cyclic cosmology because of the entropy problem. Nevertheless, many technical back up calculations are necessary to confirm consistency of the approach. Although the model borrows ideas from string theory, it is not necessarily committed to strings, or to higher dimensions, yet such speculative devices may provide the most expeditious methods to investigate the internal consistency. The value of w in the Baum–Frampton model can be made arbitrarily close to, but must be less than, -1.

8.4 Other cyclic models

- Conformal cyclic cosmology - a general relativity based theory due to Roger Penrose in which the universe expands until all the matter decays and is turned to light - so there is nothing in the universe that has any time or distance scale associated with it. This permits it to become identical with the Big Bang, so starting the next cycle.

- Loop quantum cosmology which predicts a "quantum bridge" between contracting and expanding cosmological branches.

8.5 See also

Physical cosmologies:

- Big Bounce
- Conformal cyclic cosmology

Religion:

- Bhavacakra
- Cycles of time in Hinduism
- Eternal return
- Historic recurrence
- Kalachakra
- Wheel of time

8.6 Notes

[1] R.C. Tolman (1987) [1934]. *Relativity, Thermodynamics, and Cosmology*. New York: Dover. ISBN 0-486-65383-8. LCCN 34032023.

[2] P.H. Frampton (2006). "On Cyclic Universes". arXiv:astro-ph/0612243 [astro-ph].

[3] Dark Energy Is Driving Universe Apart: NASA's Galaxy Evolution Explorer Finds Dark Energy Repulsive

[4] The WiggleZ Dark Energy Survey: Direct constraints on blue galaxy intrinsic alignments at intermediate redshifts

[5] P.J. Steinhardt, N. Turok (2001). "Cosmic Evolution in a Cyclic Universe". *Phys. Rev. D* **65** (12). arXiv:hep-th/0111098. Bibcode:2002PhRvD..65l6003S. doi:10.1103/PhysRevD.65.126003.

[6] P.J. Steinhardt, N. Turok (2001). "A Cyclic Model of the Universe". *Science* **296** (5572): 1436–1439. arXiv:hep-th/0111030. Bibcode:2002Sci...296.1436S. doi:10.1126/science.1070462. PMID 11976408.

[7] L. Baum, P.H. Frampton (2007). "Entropy of Contracting Universe in Cyclic Cosmology". *Mod.Phys.Lett.A* **23**: 33. arXiv:hep-th/0703162. Bibcode:2008MPLA...23...33B. doi:10.1142/S0217732308026170.

[8] P.J. Steinhardt, N. Turok (2004). "The Cyclic Model Simplified". *New Astron.Rev.* **49** (2–6): 43–57. arXiv:astro-ph/0404480. Bibcode:2005NewAR..49...43S. doi:10.1016/j.newar.2005.01.003.

[9] P. Woit (2006). *Not Even Wrong*. London: Random House. ISBN 978-0-09-948864-4.

[10] L. Baum and P.H. Frampton (2007). "Turnaround in Cyclic Cosmology". *Physical Review Letters* **98** (7): 071301. arXiv:hep-th/0610213. Bibcode:2007PhRvL..98g1301B. doi:10.1103/PhysRevLett.98.071301. PMID 17359014.

8.7 Further reading

- P.J. Steinhardt, N. Turok (2007). *Endless Universe*. New York: Doubleday. ISBN 978-0-385-50964-0.

- R.C. Tolman (1987) [1934]. *Relativity, Thermodynamics, and Cosmology*. New York: Dover. ISBN 0-486-65383-8. LCCN 34032023.

- L. Baum and P.H. Frampton (2007). "Turnaround in Cyclic Cosmology". *Physical Review Letters* **98** (7): 071301. arXiv:hep-th/0610213. Bibcode:2007PhRvL..98g1301B. doi:10.1103/PhysRevLett.98.071301. PMID 17359014.

- R. H. Dicke, P. J. E. Peebles, P. G. Roll and D. T. Wilkinson, "Cosmic Black-Body Radiation," *Astrophysical Journal* **142** (1965), 414. This paper discussed the oscillatory universe as one of the main cosmological possibilities of the time.

- S. W. Hawking and G. F. R. Ellis, *The large-scale structure of space-time* (Cambridge, 1973).

- R. Penrose (2010). *Cycles of Time: an extraordinary new view of the universe*. London: The Bodley Head. ISBN 978-0-224-08036-1.

8.8 External links

- Paul J. Steinhardt, Department of Physics, Princeton University

- Paul H. Frampton, Department of Physics and Astronomy, The University of North Carolina at Chapel Hill

- "The Cyclic Universe": A Talk with Neil Turok

- Roger Penrose - Cyclical Universe Model

Chapter 9

Lee Smolin

Lee Smolin (/ˈsmoʊlɪn/; born 1955) is an American theoretical physicist, a faculty member at the Perimeter Institute for Theoretical Physics, an adjunct professor of physics at the University of Waterloo and a member of the graduate faculty of the philosophy department at the University of Toronto.

Smolin is best known for his contributions to quantum gravity theory, in particular the approach known as loop quantum gravity. He advocates that the two primary approaches to quantum gravity, loop quantum gravity and string theory, can be reconciled as different aspects of the same underlying theory. His research interests also include cosmology, elementary particle theory, the foundations of quantum mechanics, and theoretical biology.[2]

9.1 Early life

Smolin was born in New York City.[3] His brother, David M. Smolin, became a professor in the Cumberland School of Law in Birmingham, Alabama.[4]

9.2 Education and career

Smolin dropped out of Walnut Hills High School in Cincinnati, Ohio, and was educated at Hampshire College. He received his Ph.D in theoretical physics from Harvard University in 1979.[2] He held postdoctoral research positions at the Institute for Advanced Study in Princeton, New Jersey, the Kavli Institute for Theoretical Physics in Santa Barbara and the University of Chicago, before becoming a faculty member at Yale, Syracuse and Pennsylvania State Universities. He was a visiting scholar at the Institute for Advanced Study in 1995[5] and a visiting professor at Imperial College London (1999-2001) before becoming one of the founding faculty members at the Perimeter Institute in 2001.

9.3 Theories and work

9.3.1 Loop quantum gravity

Smolin contributed to the invention of loop quantum gravity (LQG) in collaborative work with Ted Jacobson, Carlo Rovelli, Louis Crane, Abhay Ashtekar and others. LQG is an approach to the unification of quantum mechanics with general relativity which utilizes a reformulation of general relativity in the language of gauge field theories, which allows the use of techniques from particle physics, particularly the expression of fields in terms of the dynamics of loops. (See main page loop quantum gravity.) With Rovelli he discovered the discreteness of areas and volumes and found their natural expression in terms of a discrete description of quantum geometry in terms of spin networks. In recent years he has focused on connecting LQG to phenomenology by developing implications for experimental tests of spacetime symmetries as well as investigating ways elementary particles and their interactions could emerge from spacetime geometry.

9.3.2 Background independent approaches to string theory

Between 1999 and 2002, Smolin made several proposals for the still open question of giving a fundamental formulation of string theory that does not depend on approximate descriptions involving classical background space-time models.

9.3.3 Experimental tests of quantum gravity

Smolin is among those theorists who have proposed that the effects of quantum gravity can be experimentally probed by searching for modifications in special relativity detected in observations of high energy astrophysical phenomena. These include very high energy cosmic rays and photons and neutrinos from gamma ray bursts. Among Smolin's contributions are the coinvention of doubly special relativity (with João Magueijo, independently of work by Giovanni Amelino-Camelia) and of relative locality (with Amelino-Camelia, Laurent Freidel and Jerzy Kowalski-Glikman).

9.3.4 Foundations of quantum mechanics

Smolin has worked since the early 1980s on a series of proposals for hidden variables theories, which would be non-local deterministic theories which would give a precise description of individual quantum phenomena. In recent years, he has pioneered two new approaches to the interpretation of quantum mechanics suggested by his work on the reality of time, called the real ensemble formulation and the principle of precedence.

9.3.5 Cosmological natural selection

Smolin's hypothesis of cosmological natural selection, also called the *fecund universes* theory, suggests that a process analogous to biological natural selection applies at the grandest of scales. Smolin published the idea in 1992 and summarized it in a book aimed at a lay audience called *The Life of the Cosmos*.

The theory surmises that a collapsing black hole causes the emergence of a new universe on the "other side", whose fundamental constant parameters (masses of elementary particles, Planck constant, elementary charge, and so forth) may differ slightly from those of the universe where the black hole collapsed. Each universe thus gives rise to as many new universes as it has black holes. The theory contains the evolutionary ideas of "reproduction" and "mutation" of universes, and so is formally analogous to models of population biology.

The resulting population of universes can be represented as a distribution of a landscape of parameters where the height of the landscape is proportional to the numbers of black holes that a universe with those parameters will have. Applying reasoning borrowed from the study of fitness landscapes in population biology, one can conclude that the population is dominated by universes whose parameters drive the production of black holes to a local peak in the landscape. This was the first use of the notion of a *landscape of parameters* in physics.

Leonard Susskind, who later promoted a similar string theory landscape, stated:

> "I'm not sure why Smolin's idea didn't attract much attention. I actually think it deserved far more than it got."[6]

However, Susskind also argued that, since Smolin's theory relies on information transfer from the parent universe to the baby universe through a black hole, it ultimately makes no sense as a theory of cosmological natural selection.[6] According to Susskind and many other physicists, the last decade of black hole physics has shown us that no information that goes into a black hole can be lost.[6] Indeed the debate over this issue has been resolved with Stephen Hawking, the largest proponent of the idea that information is lost in a black hole, reversing his position.[6] In this light, information transfer from the parent universe into the baby universe through a black hole is not conceivable.[6]

Smolin has noted that the string theory landscape is not Popper-falsifiable if other universes are not observable. This is the subject of the Smolin–Susskind debate concerning Smolin's argument: "[The] Anthropic Principle cannot yield any falsifiable predictions, and therefore cannot be a part of science."[6] There are then only two ways out: traversable wormholes connecting the different parallel universes, and "signal nonlocality", as described by Antony Valentini, a scientist at the Perimeter Institute.

In a critical review of *The Life of the Cosmos*, astrophysicist Joe Silk suggested that our universe falls short by about four orders of magnitude from being maximal for the production of black holes.[7] In his book *Questions of Truth*, particle physicist John Polkinghorne puts forward another difficulty with Smolin's thesis: one cannot impose the consistent multiversal time required to make the evolutionary dynamics work, since short-lived universes with few descendants would then dominate long-lived universes with many descendants.[8] Smolin responded to these criticisms in *Life of the Cosmos*, and later scientific papers.

When Smolin published the theory in 1992, he proposed as a prediction of his theory that no neutron star should exist with a mass of more than 1.6 times the mass of the sun. Later this figure was raised to two solar masses following more precise modeling of neutron star interiors by nuclear astrophysicists. If a more massive neutron star was ever observed, it would show that our universe's natural laws were not tuned for maximal black hole production, because the mass of the strange quark could be retuned to lower the mass threshold for production of a black hole. A 2-solar-mass pulsar was discovered in 2010.[9]

In 1992 Smolin also predicted that inflation, if true, must only be in its simplest form, governed by a single field and parameter. Both predictions have held up, and they demonstrate Smolin's main thesis: that the theory of cosmological natural selection is Popper falsifiable.

9.3.6 Contributions to philosophy of physics

Smolin has contributed to the philosophy of physics through a series of papers and books that advocate the relational, or Leibnizian, view of space and time. Since 2006, he has collaborated with the Brazilian philosopher and Harvard Law School professor Roberto Mangabeira Unger on the issues of the reality of time and the evolution of laws; in 2014 they published a book, its two parts being authored separately.[10]

A book length exposition of Smolin's philosophical views appeared in April 2013. *Time Reborn* argues that physical science has made time unreal while, as Smolin insists, it is the most fundamental feature of reality: "Space may be an illusion, but time must be real" (p. 179). An adequate description according to him would give a *Leibnizian universe*: indiscernibles would not be admitted and every difference should correspond to some other difference, as the principle of sufficient reason would have it. A few months later a more concise text has been made available in a paper with the title *Temporal Naturalism*.[11]

9.4 The Trouble with Physics

Smolin's 2006 book *The Trouble with Physics* explored the role of controversy and disagreement in the progress of science. It argued that science progresses fastest if the scientific community encourages the widest possible disagreement among trained and accredited professionals prior to the formation of consensus brought about by experimental confirmation of predictions of falsifiable theories. He proposed that this meant the fostering of diverse competing research programs, and that premature formation of paradigms not forced by experimental facts can slow the progress of science.

As a case study, *The Trouble with Physics* focused on the issue of the falsifiability of string theory due to the proposals that the anthropic principle be used to explain the properties of our universe in the context of the string landscape. The book was criticized by the physicists Joseph Polchinski[12] and other string theorists.

In his earlier book *Three Roads to Quantum Gravity* (2002), Smolin stated that loop quantum gravity and string theory were essentially the same concept seen from different perspectives. In that book, he also favored the holographic principle. *The Trouble with Physics*, on the other hand, was strongly critical of the prominence of string theory in contemporary theoretical physics, which he believes has suppressed research in other promising approaches. Smolin suggests that string theory suffers from serious deficiencies and has an unhealthy near-monopoly in the particle theory community. He called for a diversity of approaches to quantum gravity, and argued that more attention should be paid to loop quantum gravity, an approach Smolin has devised. Finally, *The Trouble with Physics* is also broadly concerned with the role of controversy and the value of diverse approaches in the ethics and process of science.

In the same year that *The Trouble with Physics* was published, Peter Woit published a book for nonspecialists whose conclusion was similar to Smolin's, namely that string theory was a fundamentally flawed research program.[13]

9.5 Views

Smolin's view on the nature of time:

> "More and more, I have the feeling that quantum theory and general relativity are both deeply wrong about the nature of time. It is not enough to combine them. There is a deeper problem, perhaps going back to the beginning of physics."[14]

Smolin does not believe that quantum mechanics is a "final theory":

> "I am convinced that quantum mechanics is not a final theory. I believe this because I have never encountered an interpretation of the present formulation of quantum mechanics that makes sense to me. I have studied most of them in depth and thought hard about them, and in the end I still can't make real sense of quantum theory as it stands."[15]

In a 2009 article, Smolin has articulated the following philosophical views (the sentences in italics are quotations):

- *There is only one universe. There are no others, nor is there anything isomorphic to it.* Smolin denies the existence of a "timeless" multiverse. Neither other universes nor copies of our universe — within or outside — exist. No copies can exist within the universe, because no subsystem can model precisely the larger system it is a part of. No copies can exist outside the universe, because the universe is by definition all there is. This principle also rules out the notion of a mathematical object isomorphic in every respect to the history of the entire universe, a notion more metaphysical than scientific.

- *All that is real is real in a moment, which is a succession of moments. Anything that is true is true of the present moment.* Not only is time real, but everything that is real is situated in time. Nothing exists timelessly.

- *Everything that is real in a moment is a process of change leading to the next or future moments. Anything that is true is then a feature of a process in this process causing or implying future moments.* This principle incorporates the notion that time is an aspect of causal relations. A reason for asserting it is that anything that existed for just one moment, without causing or implying some aspect of the world at a future moment, would be gone in the next moment. Things that persist must be thought of as processes leading to newly changed processes. An atom at one moment is a process leading to a different or a changed atom at the next moment.

- *Mathematics is derived from experience as a generalization of observed regularities, when time and particularity are removed.* Under this heading, Smolin distances himself from mathematical platonism, and gives his reaction to Eugene Wigner's "The Unreasonable Effectiveness of Mathematics in the Natural Sciences".

Smolin views rejecting the idea of a creator as essential to cosmology.[16] He also opposes the anthropic principle, which he claims "cannot help us to do science."[17] He also advocates "principles for an open future" which he claims underlie the work of both healthy scientific communities and democratic societies: "(1) When rational argument from public evidence suffices to decide a question, it must be considered to be so decided. (2) When rational argument from public evidence does not suffice to decide a question, the community must encourage a diverse range of viewpoints and hypotheses consistent with a good-faith attempt to develop convincing public evidence." (Time Reborn p 265.)

9.6 Publications

The following books are non-technical, and can be appreciated by those who are not physicists.

- 1999. *The Life of the Cosmos* ISBN 0195126645

- 2001. *Three Roads to Quantum Gravity* ISBN 0-465-07835-4

- 2006. *The Trouble With Physics: The Rise of String Theory, the Fall of a Science, and What Comes Next.* Houghton Mifflin. ISBN 978-0-618-55105-7

- 2013. *Time Reborn: From the Crisis in Physics to the Future of the Universe.* ISBN 978-0547511726

- 2014 *The Singular Universe and the Reality of Time: A Proposal in Natural Philosophy* by Lee Smolin and Roberto Mangabeira Unger, Cambridge University Press, ISBN 978-1107074064

9.7 Awards and honors

Smolin was named as #21 on Foreign Policy Magazine's list of Top 100 Public Intellectuals.[18] He is also one of many physicists dubbed the "New Einstein" by the media.[19] *The Trouble with Physics* was named by *Newsweek* magazine as number 17 on a list of 50 "Books for our Time", June 27, 2009. In 2007 he was awarded the Majorana Prize from the Electronic Journal of Theoretical Physics, and in 2009 the Klopsteg Memorial Award from the American Association of Physics Teachers (AAPT) for "extraordinary accomplishments in communicating the excitement of physics to the general public," He is a fellow of the Royal Society of Canada and the American Physical Society. In 2014 he was awarded the Buchalter Cosmology Prize for a work published in collaboration with Marina Cortês.[20]

9.8 Personal life

Smolin was born in New York City, USA. His father is Michael Smolin, an environmental and process engineer and his mother is the playwright Pauline Smolin. Lee Smolin has stayed involved with theatre becoming a scientific consultant for such plays as *A Walk in the Woods* by Lee Blessing, *Background Interference* by Drucilla Cornell and *Infinity* by Hannah Moscovitch.[21] He is married to Dina Graser, a lawyer and public servant in Toronto. His brother is law professor David M. Smolin.

9.9 References

[1] "The Institute for Advanced Study. Annual Report 1979/80", p. 58

[2] Smolin's faculty page, Perimeter Institute.

[3] "Perimeter Institute Profile". Perimeter Institute. Retrieved 2015-08-17.

[4] "David Smolin's Cumberland School of Law Faculty Page". Samford University. Retrieved 2015-08-17.

[5] "Institute for Advanced Study: Community of Scholars Profile". Institute for Advanced Study. Retrieved 2015-08-17.

[6] "Smolin vs. Susskind: The Anthropic Principle" *Edge* (August 18, 2004)

[7] Joe Silk (1997) "Holistic Cosmology," *Science* **277**: 644.

[8] John Polkinghorne and Nicholas Beale (2009) *Questions of Truth*. Westminster John Knox: 106-111.

[9] http://arxiv.org/abs/1010.5788

[10] Smolin L., and Roberto Mangabeira Unger R., (2014), *The Singular Universe and the Reality of Time: A Proposal in Natural Philosophy*, Cambridge University Press, ISBN 978-1107074064

[11] Smolin L., arXiv:1310.8539

[12] Joseph Polchinski (2007) "All Strung Out?" a review of *The Trouble with Physics* and *Not Even Wrong*, *American Scientist* 95(1):1.

[13] Woit, Peter (2006). *Not Even Wrong: The Failure of String Theory & the Continuing Challenge to Unify the Laws of Physics*. Jonathan Cape. ISBN 0-224-07605-1.

[14] Smolin, Lee. *The Trouble With Physics*, Houghton Mifflin Co, New York, 2006. (p. 256)

[15] Smolin's response to the question "What do you believe is true even though you cannot prove it?" "World Question Center 2005: Lee Smolin" *Edge*

[16] Huberman, Jack (2006). *The Quotable Atheist*. Nation Books. p. 282.

[17] *Scientific alternatives to the anthropic principle*, July 2004

[18] "Top 100 Public Intellectuals: The Final Rankings" *Foreign Policy Magazine* (June 27, 2008)

[19] Brockman, John. "Introduction" to Kauffman, Stuart and Smolin, Lee. "A Possible Solution For The Problem Of Time In Quantum Cosmology" on *Edge* (April 7, 1997)

[20] Announcement: First Prize for Dr. Marina Cortês, Dr. Lee Smolin The Universe as a Process of Unique Events, Phys. Rev. **D** 90, 084007 (2014)

[21] "Terragon Theatre Profile". Terragon Theatre. 2015. Retrieved 2015-08-17.

9.10 External links

- A partial list of Smolin's published work
- A debate of the merits of string theory between Smolin and Brian Greene, from National Public Radio (2006)
- "The Unique Universe": Smolin explains his skepticism re the multiverse (2009)
- Closer to the Truth: Series of interviews by Smolin on fundamental issues in physics
- "Time Reborn: a new theory of time - a new view of the world" on YouTube: Smolin's presentation at the Royal Society of Arts (2013)
- Lee Smolin at the Mathematics Genealogy Project

Chapter 10

Multiple histories

The concept of **multiple histories** is closely related to the many-worlds interpretation of quantum mechanics. In the same way that the many-worlds interpretation regards possible futures as having a real existence of their own, the theory of multiple histories reverses this in time to regard the many possible past histories of a given event as having real existence.

This concept was introduced by Richard Feynman, whose Feynman path integral is integrated over the set of all possible histories.

The idea of multiple histories has also been applied to cosmology, in a theoretical interpretation in which the universe has multiple possible cosmologies, and in which reasoning *backwards* from the current state of the universe to a quantum superposition of possible cosmic histories makes sense.

10.1 See also

- Consistent histories
- *Fabric of Reality*
- Interpretation of quantum mechanics
- Many-worlds interpretation
- Multiverse
- Quantum cosmology
- Hartle–Hawking state

Chapter 11

Many-minds interpretation

The **many-minds interpretation** of quantum mechanics extends the many-worlds interpretation by proposing that the distinction between worlds should be made at the level of the mind of an individual observer. The concept was first introduced in 1970 by H. Dieter Zeh as a variant of the Hugh Everett interpretation in connection with quantum decoherence, and later (in 1981) explicitly called a many or multi-consciousness interpretation. The name *many-minds interpretation* was first used by David Albert and Barry Loewer in their 1988 work *Interpreting the Many Worlds Interpretation*.

11.1 The central problems

One of the central problems in interpretation of quantum theory is the duality of time evolution of physical systems:

1. Unitary evolution by the Schrödinger equation

2. Nondeterministic, nonunitary change during measurement of physical observables, at which time the system "selects" a single value in the range of possible values for the observable. This process is known as wavefunction collapse. Moreover, the process of observation occurs outside the system, which presents a problem on its own if one considers the universe itself to be a quantum system. This is known as the measurement problem.

In the introduction to his paper, *The Problem Of Conscious Observation In Quantum Mechanical Description* (June 2000), H.D. Zeh offered an empirical basis for connecting the processes involved in (2) with conscious observation:

> John von Neumann seems to have first clearly pointed out the conceptual difficulties that arise when one attempts to formulate the physical process underlying subjective observation within quantum theory. He emphasized the latter's incompatibility with a psycho-physical parallelism, the traditional way of reducing the act of observation to a physical process. Based on the assumption of a physical reality in space and time, one either assumes a coupling (causal relationship — one-way or bidirectional) of matter and mind, or disregards the whole problem by retreating to pure behaviorism. However, even this may remain problematic when one attempts to describe classical behavior in quantum mechanical terms. Neither position can be upheld without fundamental modifications in a consistent quantum mechanical description of the physical world.

11.1.1 The many-worlds interpretation

Main article: Many-worlds interpretation

Hugh Everett described a way out of this problem by suggesting that the universe is in fact indeterminate as a whole. That is, if you were to measure the spin of a particle and find it to be "up", in fact there are two "yous" after the measurement, one who measured the spin up, the other spin down. Effectively by looking at the system in question, you take on its indeterminacy.

This relative state formulation, where all states (sets of measures) can only be measured relative to other such states, avoids a number of problems in quantum theory, including the original duality – no collapse takes place, the indeterminacy simply grows (or moves) to a larger system.

Everett claims that the universe has a single quantum state, which he called the universal wavefunction, that always evolves according to the Schrödinger equation or some relativistic equivalent; now the measurement problem suggests the universal wavefunction will be in a superposition corresponding to many different definite macroscopic realms ("macrorealms"); that one can recover the subjective appearance of a definite macrorealm by postulating that all the various definite macrorealms are actual – it seems to each observer that "we just happen to be in one rather than the others" because "we" are in all of them, but each are mutually unobservable.

11.1.2 Continuous infinity of minds

In Everett's conception the mind of an observer is split by the measuring process as a consequence of the decoherence induced by measurement. In many-minds each physical observer has a postulated associated continuous infinity of minds. The decoherence of the measuring event (observation) causes the infinity of minds associated with each observer to become categorized into distinct yet infinite subsets, each subset associated with each distinct outcome of the observation. No minds are split, in the many-minds view, because it is assumed that they are all already always distinct.

The idea of many-minds was suggested early on by Zeh in 1995. He argues that in a decohering no-collapse universe one can avoid the necessity of distinct macrorealms ("parallel worlds" in MWI terminology) by introducing a new psycho-physical parallelism, in which individual minds supervene on each non-interfering component in the physical state. Zeh indeed suggests that, given decoherence, this is the most natural interpretation of quantum mechanics.

The main difference between the many-minds and many-worlds interpretations then lies in the definition of the preferred quantity. The many-minds interpretation suggests that to solve the measurement problem, there is no need to secure a definite macrorealm: the only thing that's required is appearance of such. A bit more precisely: the idea is that the preferred quantity is whatever physical quantity, defined on brains (or brains and parts of their environments), has definite-valued states (eigenstates) that underpin such appearances, i.e. underpin the states of belief in, or sensory experience of, the familiar macroscopic realm.

In its original version (related to decoherence), there is no process of selection. The process of quantum decoherence explains in terms of the Schrödinger equation how certain components of the universal wave function become irreversibly dynamically independent of one another (*separate worlds* – even though there is but one quantum world that does not split). These components may (each) contain definite quantum states of observers, while the total quantum state may not. These observer states may then be assumed to correspond to definite states of awareness (*minds*), just as in a classical description of observation. States of different observers are consistently entangled with one another, thus warranting *objective* results of measurements.

However Albert and Loewer suggest that the mental does not supervene on the physical, because individual minds have trans-temporal identity of their own. The mind selects one of these identities to be its non-random reality, while the universe itself is unaffected. The process for selection of a single state remains unexplained. This is particularly problematic because it is not clear how different observers would thus end up agreeing on measurements, which happens all the time here in the real world. There is assumed to be a sort of feedback between the mental process that leads to selection and the universal wavefunction, thereby affecting other mental states as a matter of course. In order to make the system work, the "mind" must be separate from the body, an old duality of philosophy to replace the new one of quantum mechanics.

In general this interpretation has received little attention, largely for this last reason.

11.2 Objections

Objections that apply to the many-worlds interpretation also apply to the many-minds interpretation. On the surface both of these theories arguably violate Occam's Razor; proponents counter that in fact these solutions minimize entities by simplifying the rules that would be required to describe the universe.

Another serious objection is that workers in no collapse interpretations have produced no more than elementary models based on the definite existence of specific measuring devices. They have assumed, for example, that the Hilbert space of the universe splits naturally into a tensor product structure compatible with the measurement under consideration. They have also assumed, even when describing the behavior of macroscopic objects, that it is appropriate to employ models in which only a few dimensions of Hilbert space are used to describe all the relevant behavior.

In his *What is it like to be Schrödinger's cat?* (2000), Peter J. Lewis argues that the many-minds interpretation of quantum mechanics has absurd implications for agents facing life-or-death decisions.

In general, the many-minds theory holds that a conscious being who observes the outcome of a random zero-sum experiment will evolve into two successors in different observer states, each of whom observes one of the possible outcomes. Moreover, the theory advises you to favor choices in such situations in proportion to the probability that they will bring good results to your various successors. But in a life-or-death case like getting into the box with Schrödinger's cat, you will only have one successor, since one of the outcomes will ensure your death. So it seems that the many-minds interpretation advises you to get in the box with the cat, since it is certain that your only successor will emerge unharmed. See also quantum suicide and immortality.

Finally, it supposes that there is some physical distinction between a conscious observer and a non-conscious measuring device, so it seems to require eliminating the strong Church–Turing hypothesis or postulating a physical model for consciousness.

11.3 See also

- Consciousness
- Quantum immortality
- Quantum mind

11.4 External links

- Wikibook on consciousness
- On Many-Minds Interpretations of Quantum Theory
- Bibliography on the Many-minds interpretation

Chapter 12

Mathematical universe hypothesis

In physics and cosmology, the **mathematical universe hypothesis** (**MUH**), also known as the **Ultimate Ensemble**, is a speculative "theory of everything" (TOE) proposed by the cosmologist Max Tegmark.[1][2]

12.1 Description

Tegmark's mathematical universe hypothesis (**MUH**) is: *Our external physical reality is a mathematical structure.* That is, the physical universe *is* mathematics in a well-defined sense, and "in those [worlds] complex enough to contain self-aware substructures [they] will subjectively perceive themselves as existing in a physically 'real' world".[3][4] The hypothesis suggests that worlds corresponding to different sets of initial conditions, physical constants, or altogether different equations may be considered equally real. Tegmark elaborates the MUH into the **Computable Universe Hypothesis** (**CUH**), which posits that all computable mathematical structures (in Gödel's sense) exist.[5]

The theory can be considered a form of Pythagoreanism or Platonism in that it posits the existence of mathematical entities; a form of mathematical monism in that it denies that anything exists except mathematical objects; and a formal expression of ontic structural realism.

Tegmark claims that the hypothesis has no free parameters and is not observationally ruled out. Thus, he reasons, it is preferred over other theories-of-everything by Occam's Razor. He suggests conscious experience would take the form of mathematical "self-aware substructures" that exist in a physically "real" world.

The hypothesis is related to the anthropic principle and to Tegmark's categorization of four levels of the multiverse.[6]

Andreas Albrecht of Imperial College in London called it a "provocative" solution to one of the central problems facing physics. Although he "wouldn't dare" go so far as to say he believes it, he noted that "it's actually quite difficult to construct a theory where everything we see is all there is".[7]

12.2 Criticisms and responses

12.2.1 Definition of the Ensemble

Jürgen Schmidhuber[8] argues that "Although Tegmark suggests that '... all mathematical structures are a priori given equal statistical weight,' there is no way of assigning equal nonvanishing probability to all (infinitely many) mathematical structures." Schmidhuber puts forward a more restricted ensemble which admits only universe representations describable by constructive mathematics, that is, computer programs. He explicitly includes universe representations describable by non-halting programs whose output bits converge after finite time, although the convergence time itself may not be predictable by a halting program, due to Kurt Gödel's limitations.[9]

In response, Tegmark notes[3] (sec. V.E) that the measure over all universes has not yet been constructed for the String theory landscape either, so this should not be regarded as a "show-stopper".

12.2.2 Consistency with Gödel's theorem

It has also been suggested that the MUH is inconsistent with Gödel's incompleteness theorem. In a three-way debate between Tegmark and fellow physicists Piet Hut and Mark Alford,[10] the "secularist" (Alford) states that "the methods allowed by formalists cannot prove all the theorems in a sufficiently powerful system... The idea that math is 'out there' is incompatible with the idea that it consists of formal systems."

Tegmark's response in [10] (sec VI.A.1) is to offer a new hypothesis "that only Godel-complete (fully decidable) mathematical structures have physical existence. This drastically shrinks the Level IV multiverse, essentially placing an upper limit on complexity, and may have the attractive side effect of explaining the relative simplicity of our universe." Tegmark goes on to note that although conventional theories in physics are Godel-undecidable, the actual mathematical structure describing our world could still be Godel-complete, and "could in principle contain observers capable of thinking about Godel-incomplete mathematics, just as finite-state digital computers can prove certain theorems about Godel-incomplete formal systems like Peano arithmetic." In [3] (sec. VII) he gives a more detailed response, proposing as an alternative to MUH the more restricted "Computable Universe Hypothesis" (CUH) which only includes mathematical structures that are simple enough that Gödel's theorem does not require them to contain any undecidable or uncomputable theorems. Tegmark admits that this approach faces "serious challenges", including (a) it excludes much of the mathematical landscape; (b) the measure on the space of allowed theories may itself be uncomputable; and (c) "virtually all historically successful theories of physics violate the CUH".

12.2.3 Observability

Stoeger, Ellis, and Kircher[11] (sec. 7) note that in a true multiverse theory, "the universes are then completely disjoint and nothing that happens in any one of them is causally linked to what happens in any other one. This lack of any causal connection in such multiverses really places them beyond any scientific support". Ellis[12] (p29) specifically criticizes the MUH, stating that an infinite ensemble of completely disconnected universes is "completely untestable, despite hopeful remarks sometimes made, see, e.g., Tegmark (1998)." Tegmark maintains that MUH is testable, stating that it predicts (a) that "physics research will uncover mathematical regularities in nature", and (b) by assuming that we occupy a typical member of the multiverse of mathematical structures, one could "start testing multiverse predictions by assessing how typical our universe is" ([3] sec. VIII.C).

12.2.4 Plausibility of Radical Platonism

The MUH is based on the Radical Platonist view that math is an external reality ([3] sec V.C). However, Jannes[13] argues that "mathematics is at least in part a human construction", on the basis that if it is an external reality, then it should be found in some other animals as well: "Tegmark argues that, if we want to give a complete description of reality, then we will need a language independent of us humans, understandable for non-human sentient entities, such as aliens and future supercomputers. Brian Greene ([14] p. 299) argues similarly: "The deepest description of the universe should not require concepts whose meaning relies on human experience or interpretation. Reality transcends our existence and so shouldn't, in any fundamental way, depend on ideas of our making."

However, it is not clear why we should recur to aliens or supercomputers. We know many non-human entities, plenty of which are quite intelligent, and many of which can apprehend, memorise, compare and even approximately add numerical quantities. Several animals have also passed the mirror test of self-consciousness. But a few surprising examples of mathematical abstraction notwithstanding (for example, chimpanzees can be trained to carry out symbolic addition with digits, or the report of a parrot understanding a "zero-like concept"), all examples of animal intelligence with respect to mathematics are limited to basic counting abilities." He adds, "non-human intelligent beings should exist that understand the language of advanced mathematics. However, none of the non-human intelligent beings that we know of confirm the status of (advanced) mathematics as an objective language." In the paper "On Math, Matter and Mind"[10] the secularist viewpoint examined argues (sec. VI.A) that math is evolving over time, there is "no reason to think it is converging to a definite structure, with fixed questions and established ways to address them", and also that "The Radical Platonist position is just another metaphysical theory like solipsism... In the end the metaphysics just demands that we use a different language for saying what we already knew." Tegmark responds (sec VI.A.1) that "The notion of a mathematical structure is rigorously defined in any book on Model Theory", and that non-human mathematics would only differ from our own "because we are uncovering a different part of what is in fact a consistent and unified picture, so math is converging in this sense." In his 2014 book on the MUH,[15] Tegmark argues that the resolution is that we invent the language of mathematics but discover the structure of mathematics.

12.2.5 Coexistence of all mathematical structures

Don Page has argued[16] (sec 4) that "At the ultimate level, there can be only one world and, if mathematical structures are broad enough to include all possible worlds or at least our own, there must be one unique mathematical structure that describes ultimate reality. So I think it is logical nonsense to talk of Level 4 in the sense of the co-existence of all mathematical structures." Tegmark responds ([3] sec. V.E) that "this is less inconsistent with Level IV than it may sound, since many mathematical structures decompose into unrelated substructures, and separate ones can be unified."

12.2.6 Consistency with our "simple universe"

Alexander Vilenkin comments[17] (Ch. 19, p. 203) that "the number of mathematical structures increases with increasing complexity, suggesting that 'typical' structures should be horrendously large and cumbersome. This seems to be in conflict with the beauty and simplicity of the theories describing our world". He goes on to note (footnote 8, p. 222) that Tegmark's solution to this problem, the assigning of lower "weights" to the more complex structures ([6] sec. V.B) seems arbitrary ("Who determines the weights?") and may not be logically consistent ("It seems to introduce an additional mathematical structure, but all of them are supposed to be already included in the set").

12.2.7 Occam's razor

Tegmark has been criticized as misunderstanding the nature and application of Occam's razor; Massimo Pigliucci reminds us that "Occam's razor is just a useful heuristic, it should never be used as the final arbiter to decide which theory is to be favored".[18]

12.3 Major books

12.3.1 *Our Mathematical Universe*

Written by Max Tegmark and published on January 7, 2014, this book describes Tegmark's theory.

12.4 See also

- Church–Turing thesis
- Cosmology
- Digital physics
- Impossible world
- Modal realism
- Multiverse
- Ontology
- String theory
- Theory of everything
- The Unreasonable Effectiveness of Mathematics in the Natural Sciences

12.5 References

[1] Tegmark, Max (November 1998). "Is "the Theory of Everything" Merely the Ultimate Ensemble Theory?". *Annals of Physics* **270** (1): 1–51. arXiv:gr-qc/9704009. Bibcode:1998AnPhy.270....1T. doi:10.1006/aphy.1998.5855.

[2] M. Tegmark 2014, "Our Mathematical Universe", Knopf

[3] Tegmark, Max (February 2008). "The Mathematical Universe". *Foundations of Physics* **38** (2): 101–150. arXiv:0704.0646. Bibcode:2008FoPh...38..101T. doi:10.1007/s10701-007-9186-9.

[4] Tegmark (1998), p. 1.

[5] http://arxiv.org/abs/0704.0646

[6] Tegmark, Max (2003). "Parallel Universes". In Barrow, J.D.; Davies, P.C.W.' & Harper, C.L. *"Science and Ultimate Reality: From Quantum to Cosmos" honoring John Wheeler's 90th birthday*. Cambridge University Press. arXiv:astro-ph/0302131.

[7] Chown, Markus (June 1998). "Anything goes". *New Scientist* **158** (2157).

[8] J. Schmidhuber (2000) "Algorithmic Theories of Everything."

[9] Schmidhuber, J. (2002). "Hierarchies of generalized Kolmogorov complexities and nonenumerable universal measures computable in the limit". *International Journal of Foundations of Computer Science* **13** (4): 587–612. doi:10.1142/S0129054102001291.

[10] Hut, P.; Alford, M.; Tegmark, M. (2006). "On Math, Matter and Mind". *Foundations of Physics* **36**: 765–94. arXiv:physics/0510188. Bibcode:2006FoPh...36..765H. doi:10.1007/s10701-006-9048-x.

[11] W. R. Stoeger, G. F. R. Ellis, U. Kirchner (2006) "Multiverses and Cosmology: Philosophical Issues."

[12] G.F.R. Ellis, "83 years of general relativity and cosmology: Progress and problems", Class. Quant. Grav. 16, A37-A75, 1999

[13] Gil Jannes, "Some comments on 'The Mathematical Universe'", Found. Phys. 39, 397-406, 2009 arXiv:0904.0867

[14] B. Greene 2011, "*The Hidden Reality*"

[15] M. Tegmark 2014, "Our Mathematical Universe"

[16] D. Page, "Predictions and Tests of Multiverse Theories."

[17] A. Vilenkin (2006) *Many Worlds in One: The Search for Other Universes*. Hill and Wang, New York.

[18] Mathematical Universe? I Ain't Convinced

12.6 Further reading

- Schmidhuber, J. (1997) "A Computer Scientist's View of Life, the Universe, and Everything" in C. Freksa, ed., *Foundations of Computer Science: Potential - Theory - Cognition*. Lecture Notes in Computer Science, Springer: p.201-08.

- Tegmark, Max (1998). "Is the 'theory of everything' merely the ultimate ensemble theory?". *Annals of Physics* **270**: 1–51. arXiv:gr-qc/9704009. Bibcode:1998AnPhy.270....1T. doi:10.1006/aphy.1998.5855.

- Tegmark, Max (2008), "The Mathematical Universe", *Foundations of Physics* **38**:101–50.

- Tegmark, Max (2014), *Our Mathematical Universe: My Quest for the Ultimate Nature of Reality*, ISBN 978-0-307-59980-3

- Woit, P. (17 January 2014), "Book Review: 'Our Mathematical Universe' by Max Tegmark", *The Wall Street Journal*.

12.7 External links

- Jürgen Schmidhuber "The ensemble of universes describable by constructive mathematics."

- Page maintained by Max Tegmark with links to his technical and popular writings.

- "The 'Everything' mailing list" (and archives). Discusses the idea that all possible universes exist.

- "Is the universe actually made of math?" Interview with Max Tegmark in *Discover Magazine*.

Chapter 13

Brane cosmology

Brane cosmology refers to several theories in particle physics and cosmology related to string theory, superstring theory and M-theory.

13.1 Brane and bulk

Main article: Brane

The central idea is that the visible, four-dimensional universe is restricted to a brane inside a higher-dimensional space, called the "bulk" (also known as "hyperspace"). If the additional dimensions are compact, then the observed universe contains the extra dimensions, and then no reference to the bulk is appropriate. In the bulk model, at least some of the extra dimensions are extensive (possibly infinite), and other branes may be moving through this bulk. Interactions with the bulk, and possibly with other branes, can influence our brane and thus introduce effects not seen in more standard cosmological models.

13.2 Why gravity is weak and the cosmological constant is small

Some versions of brane cosmology, based on the large extra dimension idea, can explain the weakness of gravity relative to the other fundamental forces of nature, thus solving the so-called hierarchy problem. In the brane picture, the other three forces (electromagnetism and the weak and strong nuclear forces) are localized on the brane, but gravity has no such constraint and propagates on the full spacetime, called bulk. Much of the gravitational attractive power "leaks" into the bulk. As a consequence, the force of gravity should appear significantly stronger on small (subatomic or at least sub-millimetre) scales, where less gravitational force has "leaked". Various experiments are currently under way to test this.[1] Extensions of the large extra dimension idea with supersymmetry in the bulk appears to be promising in addressing the so-called cosmological constant problem.[2][3][4]

13.3 Models of brane cosmology

One of the earliest documented attempts to apply brane cosmology as part of a conceptual theory is dated to 1983.[5]

The authors discussed the possibility that the Universe has $(3+N)+1$ dimensions, but ordinary particles are confined in a potential well which is narrow along N spatial directions and flat along three others, and proposed a particular five-dimensional model.

In 1998/99 Merab Gogberashvili published on arXiv a number of articles where he showed that if the Universe is considered as a thin shell (a mathematical synonym for "brane") expanding in 5-dimensional space then there is a possibility to obtain one scale for particle theory corresponding to the 5-dimensional cosmological constant and Universe thickness, and thus to solve the hierarchy problem.[6][7][8] It was also shown that the four-dimensionality of the Universe is the result of the stability requirement found in mathematics since the extra component of the Einstein field equations giving the confined solution for matter fields coincides with one of the conditions of stability.

In 1999 there were proposed the closely related Randall–Sundrum (RS1 and RS2; see *5 dimensional warped geometry theory* for a nontechnical explanation of RS1) scenarios. These particular models of brane cosmology have attracted a considerable amount of attention.

Later, the pre-big bang, ekpyrotic and cyclic proposals appeared. The ekpyrotic theory hypothesizes that the origin of the observable universe occurred when two parallel branes collided.[9]

13.4 Empirical tests

See also: Large extra dimension, Empirical tests

As of now, no experimental or observational evidence of large extra dimensions, as required by the Randall–Sundrum models, has been reported. An analysis of results from the Large Hadron Collider in December 2010 severely constrains theories with large extra dimensions.[10]

13.5 See also

- Kaluza–Klein theory
- Loop quantum cosmology
- M-theory
- String theory

13.6 References

[1] Session D9 - Experimental Tests of Short Range Gravitation.

[2] Aghababaie, Burgess, Parameswaran, Quevedo (2003-04-29). "Towards a naturally small cosmological constant from branes in 6-D supergravity". *Nucl.Phys. B680 (2004) 389-414.* arXiv:hep-th/0304256. Bibcode:2004NuPhB.680..389A. doi:10.1016/j.nuclphysb.2003.12.015.

[3] Burgess, van Nierop (2011-08-01). "Technically Natural Cosmological Constant From Supersymmetric 6D Brane Backreaction". *Phys.Dark Univ. 2 (2013) 1-16.* arXiv:1108.0345. Bibcode:2013PDU.....2....1B. doi:10.1016/j.dark.2012.10.001.

[4] Burgess, van Nierop, Parameswaran, Salvio, Williams (2012-10-19). "Accidental SUSY: Enhanced Bulk Supersymmetry from Brane Back-reaction". *JHEP 1302 (2013) 120.* arXiv:1210.5405. Bibcode:2013JHEP...02..120B. doi:10.1007/JHEP02(2013)120.

[5] V. A. Rubakov and M. E. Shaposhnikov,*Do we live inside a domain wall?*, Physics Letters B 125 (1983) 136–138.

[6] M. Gogberashvili, *Hierarchy problem in the shell universe model*, Arxiv:hep-ph/9812296.

[7] M. Gogberashvili, *Our world as an expanding shell*, Arxiv:hep-ph/9812365.

[8] M. Gogberashvili, *Four dimensionality in noncompact Kaluza–Klein model*, Arxiv:hep-ph/9904383.

[9] Musser, George; Minkel, JR (2002-02-11). "A Recycled Universe: Crashing branes and cosmic acceleration may power an infinite cycle in which our universe is but a phase". Scientific American Inc. Retrieved 2008-05-03.

[10] CMS Collaboration, "Search for Microscopic Black Hole Signatures at the Large Hadron Collider", http://arxiv.org/abs/1012.3375

13.7 External links

- Brax, Philippe; van de Bruck, Carsten (2003). "Cosmology and Brane Worlds: A Review". arXiv:hep-th/0303095. – Cosmological consequences of the brane world scenario are reviewed in a pedagogical manner.

- Langlois, David (2002). "Brane cosmology: an introduction". arXiv:hep-th/0209261. – These notes (32 pages) give an introductory review on brane cosmology.

- Papantonopoulos, Eleftherios (2002). "Brane Cosmology". arXiv:hep-th/0202044. – Lectures (24 pages) presented at the First Aegean Summer School on Cosmology, Samos, September 2001.

- Brane cosmology on arxiv.org

- Dimensional Shortcuts - evidence for sterile neutrino; (August 2007; Scientific American)

Chapter 14

Brane

For other uses, see Brane (disambiguation).

In string theory and related theories such as supergravity theories, a **brane** is a physical object that generalizes the notion of a point particle to higher dimensions. For example, a point particle can be viewed as a brane of dimension zero, while a string can be viewed as a brane of dimension one. It is also possible to consider higher-dimensional branes. In dimension p, these are called p-branes. The word brane comes from the word "membrane" which refers to a two-dimensional brane.[1]

Branes are dynamical objects which can propagate through spacetime according to the rules of quantum mechanics. They have mass and can have other attributes such as charge. A p-brane sweeps out a $(p+1)$-dimensional volume in spacetime called its *worldvolume*. Physicists often study fields analogous to the electromagnetic field which live on the worldvolume of a brane.[2]

In string theory, D-branes are an important class of branes that arise when one considers open strings. As an open string propagates through spacetime, its endpoints are required to lie on a D-brane. The letter "D" in D-brane refers to a certain mathematical condition on the system known as the Dirichlet boundary condition. The study of D-branes in string theory has led to important results such as the AdS/CFT correspondence, which has shed light on many problems in quantum field theory.

Branes are also frequently studied from a purely mathematical point of view since they are related to subjects such as homological mirror symmetry and noncommutative geometry. Mathematically, branes may be represented as objects of certain categories, such as the derived category of coherent sheaves on a Calabi–Yau manifold, or the Fukaya category.

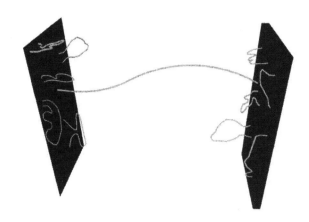

Open strings attached to a pair of D-branes

14.1 D-branes

Main article: D-brane

In string theory, a string may be open (forming a segment with two endpoints) or closed (forming a closed loop). D-branes are an important class of branes that arise when one considers open strings. As an open string propagates through spacetime, its endpoints are required to lie on a D-brane. The letter "D" in D-brane refers to a condition that it satisfies, the Dirichlet boundary condition.[3]

One crucial point about D-branes is that the dynamics on the D-brane worldvolume is described by a gauge theory, a kind of highly symmetric physical theory which is also used to describe the behavior of elementary particles in the standard model of particle physics. This connection has led to many important insights into gauge theory. For example, it led to the discovery of the AdS/CFT correspondence, a theoretical tool that physicists use to translate difficult problems in gauge theory into more mathematically tractable problems in string theory.[4]

14.2 Mathematical viewpoint

Mathematically, branes can be described using the notion of a category.[5] This is a mathematical structure consisting of *objects*, and for any pair of objects, a set of *morphisms* between them. In most examples, the objects are mathematical structures (such as sets, vector spaces, or topological spaces) and the morphisms are functions between these structures.[6] One can also consider categories where the objects are D-branes and the morphisms between two branes α and β are states of open strings stretched between α and β.[7]

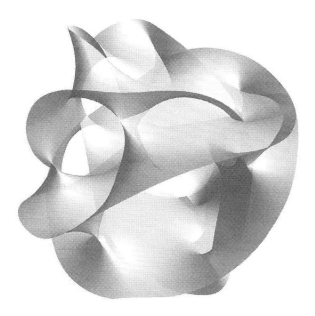

A cross section of a Calabi–Yau manifold

In one version of string theory known as the topological B-model, the D-branes are complex submanifolds of certain six-dimensional shapes called Calabi–Yau manifolds, together with additional data that arise physically from having charges at the endpoints of strings.[8] Intuitively, one can think of a submanifold as a surface embedded inside of a Calabi–Yau manifold, although submanifolds can also exist in dimensions different from two.[9] In mathematical language, the category having these branes as its objects is known as the derived category of coherent sheaves on the Calabi–Yau.[10] In another version of string theory called the topological A-model, the D-branes can again be viewed as submanifolds of a Calabi–Yau manifold. Roughly speaking, they are what mathematicians call special Lagrangian submanifolds.[11] This means among other things that they have half the dimension of the space in which they sit, and they are length-, area-, or volume-minimizing.[12] The category having these branes as its objects is called the Fukaya category.[13]

The derived category of coherent sheaves is constructed using tools from complex geometry, a branch of mathematics that describes geometric curves in algebraic terms and solves geometric problems using algebraic equations.[14] On the other hand, the Fukaya category is constructed using symplectic geometry, a branch of mathematics that arose from studies of classical physics. Symplectic geometry studies spaces equipped with a symplectic form, a mathematical tool that can be used to compute area in two-dimensional examples.[15]

The homological mirror symmetry conjecture of Maxim Kontsevich states that the derived category of coherent sheaves on one Calabi–Yau manifold is equivalent in a certain sense to the Fukaya category of a completely different Calabi–Yau manifold.[16] This equivalence provides an unexpected bridge between two branches of geometry, namely complex and symplectic geometry.[17]

14.3 See also

- Black brane
- Brane cosmology
- M2-brane
- M5-brane
- NS5-brane

14.4 Notes

[1] Moore 2005, p. 214

[2] Moore 2005, p. 214

[3] Moore 2005, p. 215

[4] Moore 2005, p. 215

[5] Aspinwall et al. 2009

[6] A basic reference on category theory is Mac Lane 1998.

[7] Zaslow 2008, p. 536

[8] Zaslow 2008, p. 536

[9] Yau and Nadis 2010, p. 165

[10] Aspinwal et al. 2009, p. 575

[11] Aspinwal et al. 2009, p. 575

[12] Yau and Nadis 2010, p. 175

[13] Aspinwal et al. 2009, p. 575

[14] Yau and Nadis 2010, pp. 180–1

[15] Zaslow 2008, p. 531

[16] Aspinwall et al. 2009, p. 616

[17] Yau and Nadis 2010, p. 181

14.5 References

- Aspinwall, Paul; Bridgeland, Tom; Craw, Alastair; Douglas, Michael; Gross, Mark; Kapustin, Anton; Moore, Gregory; Segal, Graeme; Szendrői, Balázs; Wilson, P.M.H., eds. (2009). *Dirichlet Branes and Mirror Symmetry*. American Mathematical Society. ISBN 978-0-8218-3848-8.

- Mac Lane, Saunders (1998). *Categories for the Working Mathematician*. ISBN 978-0-387-98403-2.

- Moore, Gregory (2005). "What is ... a Brane?" (PDF). *Notices of the AMS* **52**: 214. Retrieved June 2013.

- Yau, Shing-Tung; Nadis, Steve (2010). *The Shape of Inner Space: String Theory and the Geometry of the Universe's Hidden Dimensions*. Basic Books. ISBN 978-0-465-02023-2.

- Zaslow, Eric (2008). "Mirror Symmetry". In Gowers, Timothy. *The Princeton Companion to Mathematics*. ISBN 978-0-691-11880-2.

Chapter 15

D-brane

In string theory, **D-branes** are a class of extended objects upon which open strings can end with Dirichlet boundary conditions, after which they are named. D-branes were discovered by Dai, Leigh and Polchinski, and independently by Hořava in 1989. In 1995, Polchinski identified D-branes with black p-brane solutions of supergravity, a discovery that triggered the Second Superstring Revolution and led to both holographic and M-theory dualities.

D-branes are typically classified by their spatial dimension, which is indicated by a number written after the D. A D0-brane is a single point, a D1-brane is a line (sometimes called a "D-string"), a D2-brane is a plane, and a D25-brane fills the highest-dimensional space considered in bosonic string theory. There are also instantonic D(−1)-branes, which are localized in both space and time.

15.1 Theoretical background

The equations of motion of string theory require that the endpoints of an open string (a string with endpoints) satisfy one of two types of boundary conditions: The Neumann boundary condition, corresponding to free endpoints moving through spacetime at the speed of light, or the Dirichlet boundary conditions, which pin the string endpoint. Each coordinate of the string must satisfy one or the other of these conditions. There can also exist strings with mixed boundary conditions, where the two endpoints satisfy NN, DD, ND and DN boundary conditions. If p spatial dimensions satisfy the Neumann boundary condition, then the string endpoint is confined to move within a p-dimensional hyperplane. This hyperplane provides one description of a Dp-brane.

Although rigid in the limit of zero coupling, the spectrum of open strings ending on a D-brane contains modes associated with its fluctuations, implying that D-branes are dynamical objects. When N D-branes are nearly coincident, the spectrum of strings stretching between them becomes very rich. One set of modes produce a non-abelian gauge theory on the world-volume. Another set of modes is an $N \times N$ dimensional matrix for each transverse dimension of the brane. If these matrices commute, they may be diagonalized, and the eigenvalues define the position of the N D-branes in space. More generally, the branes are described by non-commutative geometry, which allows exotic behavior such as the Myers effect, in which a collection of Dp-branes expand into a D(p+2)-brane.

Tachyon condensation is a central concept in this field. Ashoke Sen has argued that in Type IIB string theory, tachyon condensation allows (in the absence of Neveu-Schwarz 3-form flux) an arbitrary D-brane configuration to be obtained from a stack of D9 and anti D9-branes. Edward Witten has shown that such configurations will be classified by the K-theory of the spacetime. Tachyon condensation is still very poorly understood. This is due to the lack of an exact string field theory that would describe the off-shell evolution of the tachyon.

15.2 Braneworld cosmology

This has implications for physical cosmology. Because string theory implies that the Universe has more dimensions than we expect—26 for bosonic string theories and 10 for superstring theories—we have to find a reason why the extra dimensions are not apparent. One possibility would be that the visible Universe is in fact a very large D-brane extending over three spatial dimensions. Material objects, made of open strings, are bound to the D-brane, and cannot move "at right angles to reality" to explore the Universe outside the brane. This scenario is called a brane cosmology. The force of gravity is *not* due to open strings; the gravitons which carry gravitational forces are vibrational states of *closed* strings. Because closed strings do not have to be attached to D-branes, gravitational effects could depend upon the extra dimensions orthogonal to the brane.

15.3 D-brane scattering

When two D-branes approach each other the interaction is captured by the one loop annulus amplitude of strings between the two branes. The scenario of two parallel branes approaching each other at a constant velocity can be mapped to the problem of two stationary branes that are rotated relative to each other by some angle. The annulus amplitude yields singularities that correspond to the on-shell production of open strings stretched between the two branes. This is true irrespective of the charge of the D-branes. At non-relativistic scattering velocities the open strings may be described by a low-energy effective action that contains two complex scalar fields that are coupled via a term $\phi^2\chi^2$. Thus, as the field ϕ (separation of the branes) changes, the mass of the field χ changes. This induces open string production and as a result the two scattering branes will be trapped.

15.4 Gauge theories

The arrangement of D-branes constricts the types of string states which can exist in a system. For example, if we have two parallel D2-branes, we can easily imagine strings stretching from brane 1 to brane 2 or vice versa. (In most theories, strings are *oriented* objects: each one carries an "arrow" defining a direction along its length.) The open strings permissible in this situation then fall into two categories, or "sectors": those originating on brane 1 and terminating on brane 2, and those originating on brane 2 and terminating on brane 1. Symbolically, we say we have the [1 2] and the [2 1] sectors. In addition, a string may begin and end on the same brane, giving [1 1] and [2 2] sectors. (The numbers inside the brackets are called *Chan-Paton indices*, but they are really just labels identifying the branes.) A string in either the [1 2] or the [2 1] sector has a minimum length: it cannot be shorter than the separation between the branes. All strings have some tension, against which one must pull to lengthen the object; this pull does work on the string, adding to its energy. Because string theories are by nature relativistic, adding energy to a string is equivalent to adding mass, by Einstein's relation $E = mc^2$. Therefore, the separation between D-branes controls the minimum mass open strings may have.

Furthermore, affixing a string's endpoint to a brane influences the way the string can move and vibrate. Because particle states "emerge" from the string theory as the different vibrational states the string can experience, the arrangement of D-branes controls the types of particles present in the theory. The simplest case is the [1 1] sector for a Dp-brane, that is to say the strings which begin and end on any particular D-brane of p dimensions. Examining the consequences of the Nambu-Goto action (and applying the rules of quantum mechanics to quantize the string), one finds that among the spectrum of particles is one resembling the photon, the fundamental quantum of the electromagnetic field. The resemblance is precise: a p-dimensional version of the electromagnetic field, obeying a p-dimensional analogue of Maxwell's equations, exists on every Dp-brane.

In this sense, then, one can say that string theory "predicts" electromagnetism: D-branes are a necessary part of the theory if we permit open strings to exist, and all D-branes carry an electromagnetic field on their volume.

Other particle states originate from strings beginning and ending on the same D-brane. Some correspond to massless particles like the photon; also in this group are a set of massless scalar particles. If a Dp-brane is embedded in a spacetime of d spatial dimensions, the brane carries (in addition to its Maxwell field) a set of $d - p$ massless scalars (particles which do not have polarizations like the photons making up light). Intriguingly, there are just as many massless scalars as there are directions perpendicular to the brane; the *geometry* of the brane arrangement is closely related to the *quantum field theory* of the particles existing on it. In fact, these massless scalars are Goldstone excitations of the brane, corresponding to the different ways the symmetry of empty space can be broken. Placing a D-brane in a universe breaks the symmetry among locations, because it defines a particular place, assigning a special meaning to a particular location along each of the $d - p$ directions perpendicular to the brane.

The quantum version of Maxwell's electromagnetism is only one kind of gauge theory, a **U**(1) gauge theory where the gauge group is made of unitary matrices of order 1. D-branes can be used to generate gauge theories of higher order, in the following way:

Consider a group of N separate Dp-branes, arranged in parallel for simplicity. The branes are labeled 1,2,...,N for convenience. Open strings in this system exist in one of many sectors: the strings beginning and ending on some brane i give that brane a Maxwell field and some massless scalar fields on its volume. The strings stretching from brane i to another brane j have more intriguing properties. For starters, it is worthwhile to ask which sectors of strings can interact with one another. One straightforward mechanism for a string interaction is for two strings to join endpoints (or, conversely, for one string to "split down the middle" and make two "daughter" strings). Since endpoints are restricted to lie on D-branes, it is evident that a [1 2] string may interact with a [2 3] string, but not with a [3 4] or a [4 17] one. The masses of these strings will be influenced by the separation between the branes, as discussed above, so for simplicity's sake we can imagine the branes squeezed closer and closer together, until they lie atop one another. If we regard two overlapping branes as distinct objects, then we still have all the sectors we had before, but without the effects due to the brane separations.

The zero-mass states in the open-string particle spectrum for a system of N coincident D-branes yields a set of interacting quantum fields which is exactly a $\mathbf{U}(N)$ gauge theory. (The string theory does contain other interactions, but they are only detectable at very high energies.) Gauge theories were not invented starting with bosonic or fermionic strings; they originated from a different area of physics, and have become quite useful in their own right. If nothing else, the relation between D-brane geometry and gauge theory offers a useful pedagogical tool for explaining gauge interactions, even if string theory fails to be the "theory of everything".

15.5 Black holes

Another important use of D-branes has been in the study of black holes. Since the 1970s, scientists have debated the problem of black holes having entropy. Consider, as a thought experiment, dropping an amount of hot gas into a black hole. Since the gas cannot escape from the hole's gravitational pull, its entropy would seem to have vanished from the universe. In order to maintain the second law of thermodynamics, one must postulate that the black hole gained whatever entropy the infalling gas originally had. Attempting to apply quantum mechanics to the study of black holes, Stephen Hawking discovered that a hole should emit energy with the characteristic spectrum of thermal radiation. The characteristic temperature of this Hawking radiation is given by

$$T_\text{H} = \frac{\hbar c^3}{8\pi G M k_B} \quad (\approx \frac{1.227 \times 10^{23} \; kg}{M} K)$$

where G is Newton's gravitational constant, M is the black hole's mass and k_B is Boltzmann's constant.

Using this expression for the Hawking temperature, and assuming that a zero-mass black hole has zero entropy, one can use thermodynamic arguments to derive the "Bekenstein entropy":

$$S_\text{B} = \frac{k_B 4\pi G}{\hbar c} M^2.$$

The Bekenstein entropy is proportional to the black hole mass squared; because the Schwarzschild radius is proportional to the mass, the Bekenstein entropy is proportional to the black hole's *surface area*. In fact,

$$S_\text{B} = \frac{A k_B}{4 l_\text{P}^2},$$

where l_P is the Planck length.

The concept of black hole entropy poses some interesting conundra. In an ordinary situation, a system has entropy when a large number of different "microstates" can satisfy the same macroscopic condition. For example, given a box full of gas, many different arrangements of the gas atoms can have the same total energy. However, a black hole was believed to be a featureless object (in John Wheeler's catchphrase, "Black holes have no hair"). What, then, are the "degrees of freedom" which can give rise to black hole entropy?

String theorists have constructed models in which a black hole is a very long (and hence very massive) string. This model gives rough agreement with the expected entropy of a Schwarzschild black hole, but an exact proof has yet to be found one way or the other. The chief difficulty is that it is relatively easy to count the degrees of freedom quantum strings possess *if they do not interact with one another*. This is analogous to the ideal gas studied in introductory thermodynamics: the easiest situation to model is when the gas atoms do not have interactions among themselves. Developing the kinetic theory of gases in the case where the gas atoms or molecules experience inter-particle forces (like the van der Waals force) is more difficult. However, a world without interactions is an uninteresting place: most significantly for the black hole problem, gravity is an interaction, and so if the "string coupling" is turned off, no black hole could ever arise. Therefore, calculating black hole entropy requires working in a regime where string interactions exist.

Extending the simpler case of non-interacting strings to the regime where a black hole could exist requires supersymmetry. In certain cases, the entropy calculation done for zero string coupling remains valid when the strings interact. The challenge for a string theorist is to devise a situation in which a black hole can exist which does not "break" supersymmetry. In recent years, this has been done by building black holes out of D-branes. Calculating the entropies of these hypothetical holes gives results which agree with the expected Bekenstein entropy. Unfortunately, the cases studied so far all involve higher-dimensional spaces — D5-branes in nine-dimensional space, for example. They do not directly apply to the familiar case, the Schwarzschild black holes observed in our own universe.

15.6 History

Dirichlet boundary conditions and D-branes had a long "pre-history" before their full significance was recognized. Mixed Dirichlet/Neumann boundary conditions were first considered by Warren Siegel in 1976 as a means of lowering the critical dimension of open string theory from 26 or

10 to 4 (Siegel also cites unpublished work by Halpern, and a 1974 paper by Chodos and Thorn, but a reading of the latter paper shows that it is actually concerned with linear dilation backgrounds, not Dirichlet boundary conditions). This paper, though prescient, was little-noted in its time (a 1985 parody by Siegel, "The Super-g String," contains an almost dead-on description of braneworlds). Dirichlet conditions for all coordinates including Euclidean time (defining what are now known as D-instantons) were introduced by Michael Green in 1977 as a means of introducing point-like structure into string theory, in an attempt to construct a string theory of the strong interaction. String compactifications studied by Harvey and Minahan, Ishibashi and Onogi, and Pradisi and Sagnotti in 1987-89 also employed Dirichlet boundary conditions.

The fact that T-duality interchanges the usual Neumann boundary conditions with Dirichlet boundary conditions was discovered independently by Horava and by Dai, Leigh, and Polchinski in 1989; this result implies that such boundary conditions must necessarily appear in regions of the moduli space of any open string theory. The Dai et al. paper also notes that the locus of the Dirichlet boundary conditions is dynamical, and coins the term Dirichlet-brane (D-brane) for the resulting object (this paper also coins orientifold for another object that arises under string T-duality). A 1989 paper by Leigh showed that D-brane dynamics are governed by the Dirac-Born-Infeld action. D-instantons were extensively studied by Green in the early 1990s, and were shown by Polchinski in 1994 to produce the $e^{-1/g}$ nonperturbative string effects anticipated by Shenker. In 1995 Polchinski showed that D-branes are the sources of electric and magnetic Ramond–Ramond fields that are required by string duality, leading to rapid progress in the nonperturbative understanding of string theory.

15.7 See also

- Bogomol'nyi–Prasad–Sommerfield bound

- M-theory

15.8 References

- Bachas, C. P. "Lectures on D-branes" (1998). arXiv:hep-th/9806199.

- Giveon, A. and Kutasov, D. "Brane dynamics and gauge theory", *Rev. Mod. Phys.* **71**, 983 (1999). arXiv:hep-th/9802067.

- Hashimoto, Koji, *D-Brane: Superstrings and New Perspective of Our World.* Springer (2012). ISBN 978-3-642-23573-3

- Johnson, Clifford (2003). *D-branes.* Cambridge: Cambridge University Press. ISBN 0-521-80912-6.

- Polchinski, Joseph, *TASI Lectures on D-branes*, arXiv:hep-th/9611050. Lectures given at TASI '96.

- Polchinski, Joseph, *Phys. Rev. Lett.* **75**, 4724 (1995). An article which established D-branes' significance in string theory.

- Zwiebach, Barton. *A First Course in String Theory.* Cambridge University Press (2004). ISBN 0-521-83143-1.

Chapter 16

M-theory

For a more accessible and less technical introduction to this topic, see Introduction to M-theory.

M-theory is a theory in physics that unifies all consistent versions of superstring theory. The existence of such a theory was first conjectured by Edward Witten at a string theory conference at the University of Southern California in the spring of 1995. Witten's announcement initiated a flurry of research activity known as the second superstring revolution.

Prior to Witten's announcement, string theorists had identified five versions of superstring theory. Although these theories appeared at first to be very different, work by several physicists showed that the theories were related in intricate and nontrivial ways. In particular, physicists found that apparently distinct theories could be unified by mathematical transformations called S-duality and T-duality. Witten's conjecture was based in part on the existence of these dualities and in part on the relationship of the string theories to a field theory called eleven-dimensional supergravity.

Although a complete formulation of M-theory is not known, the theory should describe two- and five-dimensional objects called branes and should be approximated by eleven-dimensional supergravity at low energies. Modern attempts to formulate M-theory are typically based on matrix theory or the AdS/CFT correspondence. According to Witten, M should stand for "magic", "mystery", or "membrane" according to taste, and the true meaning of the title should be decided when a more fundamental formulation of the theory is known.[1]

Investigations of the mathematical structure of M-theory have spawned important theoretical results in physics and mathematics. More speculatively, M-theory may provide a framework for developing a unified theory of all of the fundamental forces of nature. Attempts to connect M-theory to experiment typically focus on compactifying its extra dimensions to construct candidate models of our four-dimensional world, although so far none have been verified to give rise to physics as observed at, for instance, the Large Hadron Collider.

16.1 Background

16.1.1 Quantum gravity and strings

Main articles: Quantum gravity and String theory

One of the deepest problems in modern physics is the

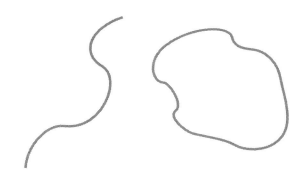

The fundamental objects of string theory are open and closed strings.

problem of quantum gravity. The current understanding of gravity is based on Albert Einstein's general theory of relativity, which is formulated within the framework of classical physics. However, nongravitational forces are described within the framework of quantum mechanics, a radically different formalism for describing physical phenomena based on probability.[lower-alpha 1] A quantum theory of gravity is needed in order to reconcile general relativity with the principles of quantum mechanics,[lower-alpha 2] but difficulties arise when one attempts to apply the usual prescriptions of quantum theory to the force of gravity.[lower-alpha 3]

String theory is a theoretical framework that attempts to reconcile gravity and quantum mechanics. In string theory, the point-like particles of particle physics are replaced by one-dimensional objects called strings. String theory describes how strings propagate through space and interact with each other. In a given version of string theory, there is only one kind of string, which may look like a small loop or segment of ordinary string, and it can vibrate in different ways. On distance scales larger than the string scale, a string will look just like an ordinary particle, with its mass, charge, and other properties determined by the vibrational state of the string. In this way, all of the different elementary particles may be viewed as vibrating strings. One of the vibrational states of a string gives rise to the graviton, a quantum mechanical particle that carries gravitational force.[lower-alpha 4]

There are several versions of string theory: type I, type IIA, type IIB, and two flavors of heterotic string theory ($SO(32)$ and $E_8 \times E_8$). The different theories allow different types of strings, and the particles that arise at low energies exhibit different symmetries. For example, the type I theory includes both open strings (which are segments with endpoints) and closed strings (which form closed loops), while types IIA and IIB include only closed strings.[2] Each of these five string theories arises as a special limiting case of M-theory. This theory, like its string theory predecessors, is an example of a quantum theory of gravity. It describes a force just like the familiar gravitational force subject to the rules of quantum mechanics.[3]

16.1.2 Number of dimensions

Main article: Compactification (physics)

In everyday life, there are three familiar dimensions of

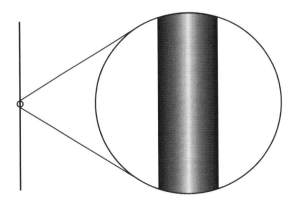

An example of compactification: At large distances, a two dimensional surface with one circular dimension looks one-dimensional.

space: height, width and depth. Einstein's general theory of relativity treats time as a dimension on par with the three spatial dimensions; in general relativity, space and time are not modeled as separate entities but are instead unified to a four-dimensional spacetime. In this framework, the phenomenon of gravity is viewed as a consequence of the geometry of spacetime.[4]

In spite of the fact that the universe is well described by four-dimensional spacetime, there are several reasons why physicists consider theories in other dimensions. In some cases, by modeling spacetime in a different number of dimensions, a theory becomes more mathematically tractable, and one can perform calculations and gain general insights more easily.[lower-alpha 5] There are also situations where theories in two or three spacetime dimensions are useful for describing phenomena in condensed matter physics.[5] Finally, there exist scenarios in which there could actually be more than four dimensions of spacetime which have nonetheless managed to escape detection.[6]

One notable feature of string theory and M-theory is that these theories require extra dimensions of spacetime for their mathematical consistency. In string theory, spacetime is ten-dimensional, while in M-theory it is eleven-dimensional. In order to describe real physical phenomena using these theories, one must therefore imagine scenarios in which these extra dimensions would not be observed in experiments.[7]

Compactification is one way of modifying the number of dimensions in a physical theory.[lower-alpha 6] In compactification, some of the extra dimensions are assumed to "close up" on themselves to form circles.[8] In the limit where these curled up dimensions become very small, one obtains a theory in which spacetime has effectively a lower number of dimensions. A standard analogy for this is to consider a multidimensional object such as a garden hose. If the hose is viewed from a sufficient distance, it appears to have only one dimension, its length. However, as one approaches the hose, one discovers that it contains a second dimension, its circumference. Thus, an ant crawling on the surface of the hose would move in two dimensions.[lower-alpha 7]

16.1.3 Dualities

Main articles: S-duality and T-duality

Theories that arise as different limits of M-theory turn out to be related in highly nontrivial ways. One of the relationships that can exist between these different physical theories is called S-duality. This is a relationship which says that a collection of strongly interacting particles in one theory can, in some cases, be viewed as a collection of weakly interacting particles in a completely different theory. Roughly speaking, a collection of particles is said to be strongly interacting if they combine and decay often and weakly interacting if they do so infrequently. Type I string theory turns out to be equivalent by S-duality to the $SO(32)$ heterotic string theory. Similarly, type IIB string theory is related to itself in a nontrivial way by S-duality.[10]

16.1. BACKGROUND

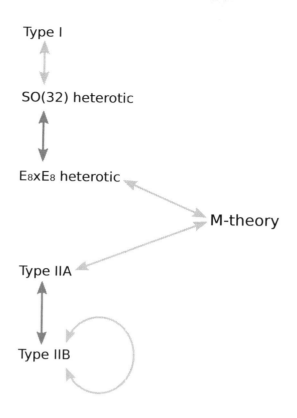

A diagram of string theory dualities. Yellow arrows indicate S-duality. Blue arrows indicate T-duality. These dualities may be combined to obtain equivalences of any of the five theories with M-theory.[9]

Another relationship between different string theories is T-duality. Here one considers strings propagating around a circular extra dimension. T-duality states that a string propagating around a circle of radius R is equivalent to a string propagating around a circle of radius $1/R$ in the sense that all observable quantities in one description are identified with quantities in the dual description. For example, a string has momentum as it propagates around a circle, and it can also wind around the circle one or more times. The number of times the string winds around a circle is called the winding number. If a string has momentum p and winding number n in one description, it will have momentum n and winding number p in the dual description. For example, type IIA string theory is equivalent to type IIB string theory via T-duality, and the two versions of heterotic string theory are also related by T-duality.[10]

In general, the term *duality* refers to a situation where two seemingly different physical systems turn out to be equivalent in a nontrivial way. If two theories are related by a duality, it means that one theory can be transformed in some way so that it ends up looking just like the other. The two theories are then said to be *dual* to one another under the transformation. Put differently, the two theories are mathematically different descriptions of the same phenomena.[11]

16.1.4 Supersymmetry

Main article: Supersymmetry

Another important theoretical idea that plays a role in M-theory is supersymmetry. This is a mathematical relation that exists in certain physical theories between a class of particles called bosons and a class of particles called fermions. Roughly speaking, fermions are the constituents of matter, while bosons mediate interactions between particles. In theories with supersymmetry, each boson has a counterpart which is a fermion, and vice versa. When supersymmetry is imposed as a local symmetry, one automatically obtains a quantum mechanical theory that includes gravity. Such a theory is called a supergravity theory.[12]

A theory of strings that incorporates the idea of supersymmetry is called a superstring theory. There are several different versions of superstring theory which are all subsumed within the M-theory framework. At low energies, the superstring theories are approximated by supergravity in ten spacetime dimensions. Similarly, M-theory is approximated at low energies by supergravity in eleven dimensions.[3]

16.1.5 Branes

Main article: Brane

In string theory and related theories such as supergravity theories, a brane is a physical object that generalizes the notion of a point particle to higher dimensions. For example, a point particle can be viewed as a brane of dimension zero, while a string can be viewed as a brane of dimension one. It is also possible to consider higher-dimensional branes. In dimension p, these are called p-branes. Branes are dynamical objects which can propagate through spacetime according to the rules of quantum mechanics. They can have mass and other attributes such as charge. A p-brane sweeps out a (p+1)-dimensional volume in spacetime called its *worldvolume*. Physicists often study fields analogous to the electromagnetic field which live on the worldvolume of a brane. The word brane comes from the word "membrane" which refers to a two-dimensional brane.[13]

In string theory, the fundamental objects that give rise to elementary particles are the one-dimensional strings. Although the physical phenomena described by M-theory are still poorly understood, physicists know that the theory describes two- and five-dimensional branes. Much of the current research in M-theory attempts to better understand the properties of these branes.[lower-alpha 8]

16.2 History and development

16.2.1 Kaluza–Klein theory

Main article: Kaluza–Klein theory

In the early 20th century, physicists and mathematicians including Albert Einstein and Hermann Minkowski pioneered the use of four-dimensional geometry for describing the physical world.[14] These efforts culminated in the formulation of Einstein's general theory of relativity, which relates gravity to the geometry of four-dimensional spacetime.[15]

The success of general relativity led to efforts to apply higher dimensional geometry to explain other forces. In 1919, work by Theodor Kaluza showed that by passing to five-dimensional spacetime, one can unify gravity and electromagnetism into a single force.[15] This idea was improved by physicist Oskar Klein, who suggested that the additional dimension proposed by Kaluza could take the form of a circle with radius around 10^{-30} cm.[16]

The Kaluza–Klein theory and subsequent attempts by Einstein to develop unified field theory were never completely successful. In part this was because Kaluza–Klein theory predicted a particle that has never been shown to exist, and in part because it was unable to correctly predict the ratio of an electron's mass to its charge. In addition, these theories were being developed just as other physicists were beginning to discover quantum mechanics, which would ultimately prove successful in describing known forces such as electromagnetism, as well as new nuclear forces that were being discovered throughout the middle part of the century. Thus it would take almost fifty years for the idea of new dimensions to be taken seriously again.[17]

16.2.2 Early work on supergravity

Main article: Supergravity
New concepts and mathematical tools provided fresh insights into general relativity, giving rise to a period in the 1960s and 70s now known as the golden age of general relativity.[18] In the mid-1970s, physicists began studying higher-dimensional theories combining general relativity with supersymmetry, the so-called supergravity theories.[19]

General relativity does not place any limits on the possible dimensions of spacetime. Although the theory is typ-

In the 1980s, Edward Witten contributed to the understanding of supergravity theories. In 1995, he introduced M-theory, sparking the second superstring revolution.

ically formulated in four dimensions, one can write down the same equations for the gravitational field in any number of dimensions. Supergravity is more restrictive because it places an upper limit on the number of dimensions.[12] In 1978, work by Werner Nahm showed that the maximum spacetime dimension in which one can formulate a consistent supersymmetric theory is eleven.[20] In the same year, Eugene Cremmer, Bernard Julia, and Joel Scherk of the École Normale Supérieure showed that supergravity not only permits up to eleven dimensions but is in fact most elegant in this maximal number of dimensions.[21][22]

Initially, many physicists hoped that by compactifying eleven-dimensional supergravity, it might be possible to construct realistic models of our four-dimensional world. The hope was that such models would provide a unified description of the four fundamental forces of nature: electromagnetism, the strong and weak nuclear forces, and gravity. Interest in eleven-dimensional supergravity soon waned as various flaws in this scheme were discovered. One of the problems was that the laws of physics appear to distinguish between clockwise and counterclockwise, a phenomenon known as chirality. Edward Witten and others observed this chirality property cannot be readily derived by compactifying from eleven dimensions.[22]

In the first superstring revolution in 1984, many physicists turned to string theory as a unified theory of particle physics and quantum gravity. Unlike supergravity theory, string theory was able to accommodate the chirality of the standard model, and it provided a theory of gravity consistent with quantum effects.[22] Another feature of string theory that many physicists were drawn to in the 1980s and 1990s was its high degree of uniqueness. In ordinary particle theories, one can consider any collection of elementary particles whose classical behavior is described by an arbitrary Lagrangian. In string theory, the possibilities are much more constrained: by the 1990s, physicists had argued that there were only five consistent supersymmetric versions of the theory.[22]

16.2.3 Relationships between string theories

Although there were only a handful of consistent superstring theories, it remained a mystery why there was not just one consistent formulation.[22] However, as physicists began to examine string theory more closely, they realized that these theories are related in intricate and nontrivial ways.[23]

In the late 1970s, Claus Montonen and David Olive had conjectured a special property of certain physical theories.[24] A sharpened version of their conjecture concerns a theory called $N=4$ supersymmetric Yang–Mills theory, which describes particles similar to the quarks and gluons that make up atomic nuclei. The strength with which the particles of this theory interact is measured by a number called the coupling constant. The result of Montonen and Olive, now known as Montonen–Olive duality, states that $N=4$ supersymmetric Yang–Mills theory with coupling constant g is equivalent to the same theory with coupling constant $1/g$. In other words, a system of strongly interacting particles (large coupling constant) has an equivalent description as a system of weakly interacting particles (small coupling constant) and vice versa.[25]

In the 1990s, several theorists generalized Montonen–Olive duality to the S-duality relationship, which connects different string theories. Ashoke Sen studied S-duality in the context of heterotic strings in four dimensions.[26][27] Chris Hull and Paul Townsend showed that type IIB string theory with a large coupling constant is equivalent via S-duality to the same theory with small coupling constant.[28] Theorists also found that different string theories may be related by T-duality. This duality implies that strings propagating on completely different spacetime geometries may be physically equivalent.[29]

16.2.4 Membranes and fivebranes

String theory extends ordinary particle physics by promoting zero-dimensional point particles to one-dimensional objects called strings. In the late 1980s, it was natural for theorists to attempt to formulate other extensions in which particles are replaced by two-dimensional supermembranes or by higher-dimensional objects called branes. Such objects had been considered as early as 1962 by Paul Dirac,[30] and they were reconsidered by a small but enthusiastic group of physicists in the 1980s.[22]

Supersymmetry severely restricts the possible number of dimensions of a brane. In 1987, Eric Bergshoeff, Ergin Sezgin, and Paul Townsend showed that eleven-dimensional supergravity includes two-dimensional branes.[31] Intuitively, these objects look like sheets or membranes propagating through the eleven-dimensional spacetime. Shortly after this discovery, Michael Duff, Paul Howe, Takeo Inami, and Kellogg Stelle considered a particular compactification of eleven-dimensional supergravity with one of the dimensions curled up into a circle.[32] In this setting, one can imagine the membrane wrapping around the circular dimension. If the radius of the circle is sufficiently small, then this membrane looks just like a string in ten-dimensional spacetime. In fact, Duff and his collaborators showed that this construction reproduces exactly the strings appearing in type IIA superstring theory.[25]

In 1990, Andrew Strominger published a similar result which suggested that strongly interacting strings in ten dimensions might have an equivalent description in terms of weakly interacting five-dimensional branes.[33] Initially, physicists were unable to prove this relationship for two important reasons. On the one hand, the Montonen–Olive duality was still unproven, and so Strominger's conjecture was even more tenuous. On the other hand, there were many technical issues related to the quantum properties of five-dimensional branes.[34] The first of these problems was solved in 1993 when Ashoke Sen established that certain physical theories require the existence of objects with both electric and magnetic charge which were predicted by the work of Montonen and Olive.[35]

In spite of this progress, the relationship between strings and five-dimensional branes remained conjectural because theorists were unable to quantize the branes. Starting in 1991, a team of researchers including Michael Duff, Ramzi Khuri, Jianxin Lu, and Ruben Minasian considered a special compactification of string theory in which four of the ten dimensions curl up. If one considers a five-dimensional brane wrapped around these extra dimensions, then the brane looks just like a one-dimensional string. In this way, the conjectured relationship between strings and branes was reduced to a relationship between strings and strings, and the latter could be tested using already established theoretical techniques.[29]

16.2.5 Second superstring revolution

Main article: Second superstring revolution

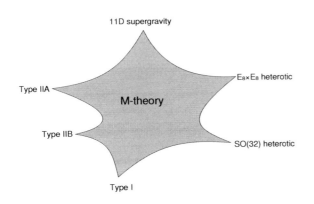

A schematic illustration of the relationship between M-theory, the five superstring theories, and eleven-dimensional supergravity. The shaded region represents a family of different physical scenarios that are possible in M-theory. In certain limiting cases corresponding to the cusps, it is natural to describe the physics using one of the six theories labeled there.

Speaking at the string theory conference at the University of Southern California in 1995, Edward Witten of the Institute for Advanced Study made the surprising suggestion that all five superstring theories were in fact just different limiting cases of a single theory in eleven spacetime dimensions. Witten's announcement drew together all of the previous results on S- and T-duality and the appearance of two- and five-dimensional branes in string theory.[36] In the months following Witten's announcement, hundreds of new papers appeared on the Internet confirming that the new theory involved membranes in an important way.[37] Today this flurry of work is known as the second superstring revolution.[38]

One of the important developments following Witten's announcement was Witten's work in 1996 with string theorist Petr Hořava.[39][40] Witten and Hořava studied M-theory on a special spacetime geometry with two ten-dimensional boundary components. Their work shed light on the mathematical structure of M-theory and suggested possible ways of connecting M-theory to real world physics.[41]

16.2.6 Origin of the term

Initially, some physicists suggested that the new theory was a fundamental theory of membranes, but Witten was skeptical of the role of membranes in the theory. In a paper from 1996, Hořava and Witten wrote

> As it has been proposed that the eleven-dimensional theory is a supermembrane theory but there are some reasons to doubt that interpretation, we will non-committally call it the M-theory, leaving to the future the relation of M to membranes.[39]

In the absence of an understanding of the true meaning and structure of M-theory, Witten has suggested that the *M* should stand for "magic", "mystery", or "membrane" according to taste, and the true meaning of the title should be decided when a more fundamental formulation of the theory is known.[1]

16.3 Matrix theory

16.3.1 BFSS matrix model

Main article: Matrix theory (physics)

In mathematics, a matrix is a rectangular array of numbers or other data. In physics, a matrix model is a particular kind of physical theory whose mathematical formulation involves the notion of a matrix in an important way. A matrix model describes the behavior of a set of matrices within the framework of quantum mechanics.[42][43]

One important example of a matrix model is the BFSS matrix model proposed by Tom Banks, Willy Fischler, Stephen Shenker, and Leonard Susskind in 1997. This theory describes the behavior of a set of nine large matrices. In their original paper, these authors showed, among other things, that the low energy limit of this matrix model is described by eleven-dimensional supergravity. These calculations led them to propose that the BFSS matrix model is exactly equivalent to M-theory. The BFSS matrix model can therefore be used as a prototype for a correct formulation of M-theory and a tool for investigating the properties of M-theory in a relatively simple setting.[42]

16.3.2 Noncommutative geometry

Main articles: Noncommutative geometry and Noncommutative quantum field theory

In geometry, it is often useful to introduce coordinates. For example, in order to study the geometry of the Euclidean plane, one defines the coordinates x and y as the distances between any point in the plane and a pair of axes. In ordinary geometry, the coordinates of a point are numbers, so they can be multiplied, and the product of two coordinates does not depend on the order of multiplication. That is, $xy = yx$. This property of multiplication is known as the commutative law, and this relationship between geometry and the commutative algebra of coordinates is the starting point for much of modern geometry.[44]

Noncommutative geometry is a branch of mathematics that attempts to generalize this situation. Rather than working with ordinary numbers, one considers some similar objects, such as matrices, whose multiplication does not satisfy the commutative law (that is, objects for which xy is not necessarily equal to yx). One imagines that these noncommuting objects are coordinates on some more general notion of "space" and proves theorems about these generalized spaces by exploiting the analogy with ordinary geometry.[45]

In a paper from 1998, Alain Connes, Michael R. Douglas, and Albert Schwarz showed that some aspects of matrix models and M-theory are described by a noncommutative quantum field theory, a special kind of physical theory in which the coordinates on spacetime do not satisfy the commutativity property.[43] This established a link between matrix models and M-theory on the one hand, and noncommutative geometry on the other hand. It quickly led to the discovery of other important links between noncommutative geometry and various physical theories.[46][47]

16.4 AdS/CFT correspondence

16.4.1 Overview

Main article: AdS/CFT correspondence

The application of quantum mechanics to physical objects such as the electromagnetic field, which are extended in space and time, is known as quantum field theory.[lower-alpha 9] In particle physics, quantum field theories form the basis for our understanding of elementary particles, which are modeled as excitations in the fundamental fields. Quantum field theories are also used throughout condensed matter physics to model particle-like objects called quasiparticles.[lower-alpha 10]

One approach to formulating M-theory and studying its properties is provided by the anti-de Sitter/conformal field theory (AdS/CFT) correspondence. Proposed by Juan Maldacena in late 1997, the AdS/CFT correspondence is a theoretical result which implies that M-theory is in some cases equivalent to a quantum field theory.[48] In addition to providing insights into the mathematical structure of string and M-theory, the AdS/CFT correspondence has shed light on many aspects of quantum field theory in regimes where traditional calculational techniques are ineffective.[49]

In the AdS/CFT correspondence, the geometry of spacetime is described in terms of a certain vacuum solution of Einstein's equation called anti-de Sitter space.[50] In very elementary terms, anti-de Sitter space is a mathematical model of spacetime in which the notion of distance between points (the metric) is different from the notion of distance in ordinary Euclidean geometry. It is closely related to hyperbolic space, which can be viewed as a disk as illustrated on the left.[51] This image shows a tessellation of a disk by triangles and squares. One can define the distance between points of this disk in such a way that all the triangles and squares are the same size and the circular outer boundary is infinitely far from any point in the interior.[52]

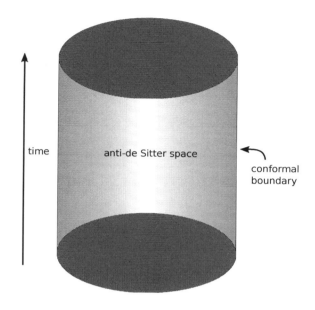

Three-dimensional anti-de Sitter space is like a stack of hyperbolic disks, each one representing the state of the universe at a given time. One can study theories of quantum gravity such as M-theory in the resulting spacetime.

Now imagine a stack of hyperbolic disks where each disk represents the state of the universe at a given time. The resulting geometric object is three-dimensional anti-de Sitter space.[51] It looks like a solid cylinder in which any cross section is a copy of the hyperbolic disk. Time runs along the vertical direction in this picture. The surface of this cylinder plays an important role in the AdS/CFT correspondence. As with the hyperbolic plane, anti-de Sitter space is curved in such a way that any point in the interior is actually infinitely far from this boundary surface.[52]

This construction describes a hypothetical universe with only two space dimensions and one time dimension, but it can be generalized to any number of dimensions. Indeed, hyperbolic space can have more than two dimensions and one can "stack up" copies of hyperbolic space to get higher-dimensional models of anti-de Sitter space.[51]

An important feature of anti-de Sitter space is its boundary (which looks like a cylinder in the case of three-dimensional anti-de Sitter space). One property of this boundary is that, within a small region on the surface around any given point, it looks just like Minkowski space, the model of spacetime used in nongravitational physics.[53] One can therefore consider an auxiliary theory in which "spacetime" is given by the boundary of anti-de Sitter space. This observation is the starting point for AdS/CFT correspondence, which states that the boundary of anti-de Sitter space can be regarded as the "spacetime" for a quantum field theory. The claim is that this quantum field theory is equivalent to the gravitational theory on the bulk anti-de Sitter space in the sense that there is a "dictionary" for translating entities and calculations in one theory into their counterparts in the other theory. For example, a single particle in the gravitational theory might correspond to some collection of particles in the boundary theory. In addition, the predictions in the two theories are quantitatively identical so that if two particles have a 40 percent chance of colliding in the gravitational theory, then the corresponding collections in the boundary theory would also have a 40 percent chance of colliding.[54]

16.4.2 6D (2,0) superconformal field theory

Main article: 6D (2,0) superconformal field theory
One particular realization of the AdS/CFT correspondence

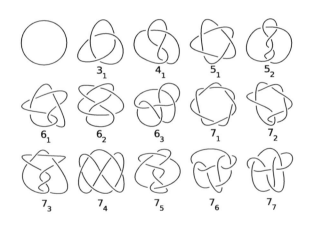

The six-dimensional (2,0)-theory has been used to understand results from the mathematical theory of knots.

states that M-theory on the product space $AdS_7 \times S^4$ is equivalent to the so-called (2,0)-theory on the six-dimensional boundary.[48] Here "(2,0)" refers to the particular type of supersymmetry that appears in the theory. In this example, the spacetime of the gravitational theory is effectively seven-dimensional (hence the notation AdS_7), and there are four additional "compact" dimensions (encoded by the S^4 factor). In the real world, spacetime is four-dimensional, at least macroscopically, so this version of the correspondence does not provide a realistic model of gravity. Likewise, the dual theory is not a viable model of any real-world system since it describes a world with six spacetime dimensions.[lower-alpha 11]

Nevertheless, the (2,0)-theory has proven to be important for studying the general properties of quantum field theories. Indeed, this theory subsumes many mathematically interesting effective quantum field theories and points to new dualities relating these theories. For example, Luis Alday, Davide Gaiotto, and Yuji Tachikawa showed that by compactifying this theory on a surface, one obtains a four-dimensional quantum field theory, and there is a duality known as the AGT correspondence which relates the physics of this theory to certain physical concepts associated with the surface itself.[55] More recently, theorists have extended these ideas to study the theories obtained by compactifying down to three dimensions.[56]

In addition to its applications in quantum field theory, the (2,0)-theory has spawned important results in pure mathematics. For example, the existence of the (2,0)-theory was used by Witten to give a "physical" explanation for a conjectural relationship in mathematics called the geometric Langlands correspondence.[57] In subsequent work, Witten showed that the (2,0)-theory could be used to understand a concept in mathematics called Khovanov homology.[58] Developed by Mikhail Khovanov around 2000, Khovanov homology provides a tool in knot theory, the branch of mathematics that studies and classifies the different shapes of knots.[59] Another application of the (2,0)-theory in mathematics is the work of Davide Gaiotto, Greg Moore, and Andrew Neitzke, which used physical ideas to derive new results in hyperkähler geometry.[60]

16.4.3 ABJM superconformal field theory

Main article: ABJM superconformal field theory

Another realization of the AdS/CFT correspondence states that M-theory on $AdS_4 \times S^7$ is equivalent to a quantum field theory called the ABJM theory in three dimensions. In this version of the correspondence, seven of the dimensions of M-theory are curled up, leaving four non-compact dimensions. Since the spacetime of our universe is four-dimensional, this version of the correspondence provides a somewhat more realistic description of gravity.[61]

The ABJM theory appearing in this version of the correspondence is also interesting for a variety of reasons. Introduced by Aharony, Bergman, Jafferis, and Maldacena, it is closely related to another quantum field theory called Chern–Simons theory. The latter theory was popularized by Witten in the late 1980s because of its applications to knot theory.[62] In addition, the ABJM theory serves as a semi-realistic simplified model for solving problems that arise in condensed matter physics.[61]

16.5 Phenomenology

16.5.1 Overview

Main article: String phenomenology

In addition to being an idea of considerable theoret-

A cross section of a Calabi–Yau manifold

ical interest, M-theory provides a framework for constructing models of real world physics that combine general relativity with the standard model of particle physics. Phenomenology is the branch of theoretical physics in which physicists construct realistic models of nature from more abstract theoretical ideas. String phenomenology is the part of string theory that attempts to construct realistic models of particle physics based on string and M-theory.[63]

Typically, such models are based on the idea of compactification.[lower-alpha 12] Starting with the ten- or eleven-dimensional spacetime of string or M-theory, physicists postulate a shape for the extra dimensions. By choosing this shape appropriately, they can construct models roughly similar to the standard model of particle physics, together with additional undiscovered particles,[64] usually supersymmetric partners to analogues of known particles. One popular way of deriving realistic physics from string theory is to start with the heterotic theory in ten dimensions and assume that the six extra dimensions of spacetime are shaped like a six-dimensional Calabi–Yau manifold. This is a special kind of geometric object named after mathematicians Eugenio Calabi and Shing-Tung Yau.[65] Calabi–Yau manifolds offer many ways of extracting realistic physics from string theory. Other similar methods can be used to construct models with physics resembling to some extent that of our four-dimensional world based on M-theory.[66]

Partly because of theoretical and mathematical difficulties and partly because of the extremely high energies (beyond what is technologically possible for the foreseeable future) needed to test these theories experimentally, there is so far no experimental evidence that would unambiguously point to any of these models being a correct fundamental description of nature. This has led some in the community to criticize these approaches to unification and question the value of continued research on these problems.[67]

16.5.2 Compactification on G_2 manifolds

In one approach to M-theory phenomenology, theorists assume that the seven extra dimensions of M-theory are shaped like a G_2 manifold. This is a special kind of seven-dimensional shape constructed by mathematician Dominic Joyce of the University of Oxford.[68] These G_2 manifolds are still poorly understood mathematically, and this fact has made it difficult for physicists to fully develop this approach to phenomenology.[69]

For example, physicists and mathematicians often assume that space has a mathematical property called smoothness, but this property cannot be assumed in the case of a G_2 manifold if one wishes to recover the physics of our four-dimensional world. Another problem is that G_2 manifolds are not complex manifolds, so theorists are unable to use tools from the branch of mathematics known as complex analysis. Finally, there are many open questions about the existence, uniqueness, and other mathematical properties of G_2 manifolds, and mathematicians lack a systematic way of searching for these manifolds.[69]

16.5.3 Heterotic M-theory

Because of the difficulties with G_2 manifolds, most attempts to construct realistic theories of physics based on M-theory have taken a more indirect approach to compactifying eleven-dimensional spacetime. One approach, pioneered by Witten, Hořava, Burt Ovrut, and others, is known as heterotic M-theory. In this approach, one imagines that one of the eleven dimensions of M-theory is shaped like a circle. If this circle is very small, then the spacetime becomes effectively ten-dimensional. One then assumes that six of the ten dimensions form a Calabi–Yau manifold. If this Calabi–Yau manifold is also taken to be small, one is left with a theory in four-dimensions.[69]

Heterotic M-theory has been used to construct models of brane cosmology in which the observable universe is thought to exist on a brane in a higher dimensional ambient space. It has also spawned alternative theories of the early universe that do not rely on the theory of cosmic inflation.[69]

16.6 References

16.6.1 Notes

[1] For a standard introduction to quantum mechanics, see Griffiths 2004.

[2] The necessity of a quantum mechanical description of gravity follows from the fact that one cannot consistently couple a classical system to a quantum one. See Wald 1984, p. 382.

[3] From a technical point of view, the problem is that the theory one gets in this way is not renormalizable and therefore cannot be used to make meaningful physical predictions. See Zee 2010, p. 72 for a discussion of this issue.

[4] For an accessible introduction to string theory, see Greene 2000.

[5] For example, in the context of the AdS/CFT correspondence, theorists often formulate and study theories of gravity in unphysical numbers of spacetime dimensions.

[6] Dimensional reduction is another way of modifying the number of dimensions.

[7] This analogy is used for example in Greene 2000, p. 186.

[8] For example, see the subsections on the 6D (2,0) superconformal field theory and ABJM superconformal field theory.

[9] A standard text is Peskin and Schroeder 1995.

[10] For an introduction to the applications of quantum field theory to condensed matter physics, see Zee 2010.

[11] For a review of the (2,0)-theory, see Moore 2012.

[12] Brane world scenarios provide an alternative way of recovering real world physics from string theory. See Randall and Sundrum 1999.

16.6.2 Citations

[1] Duff 1996, sec. 1

[2] Zwiebach 2009, p. 324

[3] Becker, Becker, and Schwarz 2007, p. 12

[4] Wald 1984, p. 4

[5] Zee 2010, Parts V and VI

[6] Zwiebach 2009, p. 9

[7] Zwiebach 2009, p. 8

[8] Yau and Nadis 2010, Ch. 6

[9] Becker, Becker, and Schwarz 2007, pp. 339–347

[10] Becker, Becker, and Schwarz 2007

[11] Zwiebach 2009, p. 376

[12] Duff 1998, p. 64

[13] Moore 2005

[14] Yau and Nadis 2010, p. 9

[15] Yau and Nadis 2010, p. 10

[16] Yau and Nadis 2010, p. 12

[17] Yau and Nadis 2010, p. 13

[18] Wald 1984, p. 3

[19] van Nieuwenhuizen 1981

[20] Nahm 1978

[21] Cremmer, Julia, and Scherk 1978

[22] Duff 1998, p. 65

[23] Duff 1998

[24] Montonen and Olive 1977

[25] Duff 1998, p. 66

[26] Sen 1994a

[27] Sen 1994b

[28] Hull and Townsend 1995

[29] Duff 1998, p. 67

[30] Dirac 1962

[31] Bergshoeff, Sezgin, and Townsend 1987

[32] Duff et al. 1987

[33] Strominger 1990

[34] Duff 1998, pp 66–67

[35] Sen 1993

[36] Witten 1995

[37] Duff 1998, pp. 67–68

[38] Becker, Becker, and Schwarz 2007, p. 296

[39] Hořava and Witten 1996a

[40] Hořava and Witten 1996b

[41] Duff 1998, p. 68

[42] Banks et al. 1997

[43] Connes, Douglas, and Schwarz 1998

[44] Connes 1994, p. 1

[45] Connes 1994

[46] Nekrasov and Schwarz 1998

[47] Seiberg and Witten 1999

[48] Maldacena 1998

[49] Klebanov and Maldacena 2009

[50] Klebanov and Maldacena 2009, p. 28

[51] Maldacena 2005, p. 60

[52] Maldacena 2005, p. 61

[53] Zwiebach 2009, p. 552

[54] Maldacena 2005, pp. 61–62

[55] Alday, Gaiotto, and Tachikawa 2010

[56] Dimofte, Gaiotto, and Gukov 2010

[57] Witten 2009

[58] Witten 2012

[59] Khovanov 2000

[60] Gaiotto, Moore, and Neitzke 2013

[61] Aharony et al. 2008

[62] Witten 1989

[63] Dine 2000

[64] Candelas et al. 1985

[65] Yau and Nadis 2010, p. ix

[66] Yau and Nadis 2010, pp. 147–150

[67] Woit 2006

[68] Yau and Nadis 2010, p. 149

[69] Yau and Nadis 2010, p. 150

16.6.3 Bibliography

- Aharony, Ofer; Bergman, Oren; Jafferis, Daniel Louis; Maldacena, Juan (2008). "$N=6$ superconformal Chern-Simons-matter theories, M2-branes and their gravity duals". *Journal of High Energy Physics* **2008** (10): 091. arXiv:0806.1218. Bibcode:2008JHEP...10..091A. doi:10.1088/1126-6708/2008/10/091.

- Alday, Luis; Gaiotto, Davide; Tachikawa, Yuji (2010). "Liouville correlation functions from four-dimensional gauge theories". *Letters in Mathematical Physics* **91** (2): 167–197. arXiv:0906.3219. Bibcode:2010LMaPh..91..167A. doi:10.1007/s11005-010-0369-5.

- Banks, Tom; Fischler, Willy; Schenker, Stephen; Susskind, Leonard (1997). "M theory as a matrix model: A conjecture". *Physical Review D* **55** (8): 5112. arXiv:hep-th/9610043. Bibcode:1997PhRvD..55.5112B. doi:10.1103/physrevd.55.5112.

- Becker, Katrin; Becker, Melanie; Schwarz, John (2007). *String theory and M-theory: A modern introduction*. Cambridge University Press. ISBN 978-0-521-86069-7.

- Bergshoeff, Eric; Sezgin, Ergin; Townsend, Paul (1987). "Supermembranes and eleven-dimensional supergravity". *Physics Letters B* **189** (1): 75–78. Bibcode:1987PhLB..189...75B. doi:10.1016/0370-2693(87)91272-X.

- Candelas, Philip; Horowitz, Gary; Strominger, Andrew; Witten, Edward (1985). "Vacuum configurations for superstrings". *Nuclear Physics B* **258**: 46–74. Bibcode:1985NuPhB.258...46C. doi:10.1016/0550-3213(85)90602-9.

- Connes, Alain (1994). *Noncommutative Geometry*. Academic Press. ISBN 978-0-12-185860-5.

- Connes, Alain; Douglas, Michael; Schwarz, Albert (1998). "Noncommutative geometry and matrix theory". *Journal of High Energy Physics*. 19981 (2): 003. arXiv:hep-th/9711162. Bibcode:1998JHEP...02..003C. doi:10.1088/1126-6708/1998/02/003.

- Cremmer, Eugene; Julia, Bernard; Scherk, Joel (1978). "Supergravity theory in eleven dimensions". *Physics Letters B* **76** (4): 409–412. Bibcode:1978PhLB...76..409C. doi:10.1016/0370-2693(78)90894-8.

- Dimofte, Tudor; Gaiotto, Davide; Gukov, Sergei (2010). "Gauge theories labelled by three-manifolds". *Communications in Mathematical Physics* **325** (2): 367–419. Bibcode:2014CMaPh.325..367D. doi:10.1007/s00220-013-1863-2.

- Dine, Michael (2000). "TASI Lectures on M Theory Phenomenology". arXiv:hep-th/0003175.

- Dirac, Paul (1962). "An extensible model of the electron". *Proceedings of the Royal Society of London. A. Mathematical and Physical Sciences* **268** (1332): 57–67. Bibcode:1962RSPSA.268...57D. doi:10.1098/rspa.1962.0124.

- Duff, Michael (1996). "M-theory (the theory formerly known as strings)". *International Journal of Modern Physics A* **11** (32): 6523–41. arXiv:hep-th/9608117. Bibcode:1996IJMPA..11.5623D. doi:10.1142/S0217751X96002583.

- Duff, Michael (1998). "The theory formerly known as strings". *Scientific American* **278** (2): 64–9. doi:10.1038/scientificamerican0298-64.

- Duff, Michael; Howe, Paul; Inami, Takeo; Stelle, Kellogg (1987). "Superstrings in $D=10$ from supermembranes in $D=11$". *Nuclear Physics B* **191** (1): 70–74. Bibcode:1987PhLB..191...70D. doi:10.1016/0370-2693(87)91323-2.

- Gaiotto, Davide; Moore, Gregory; Neitzke, Andrew (2013). "Wall-crossing, Hitchin systems, and the WKB approximation". *Advances in Mathematics* **2341**: 239–403. arXiv:0907.3987. doi:10.1016/j.aim.2012.09.027.

- Greene, Brian (2000). *The Elegant Universe: Superstrings, Hidden Dimensions, and the Quest for the Ultimate Theory*. Random House. ISBN 978-0-9650888-0-0.

- Griffiths, David (2004). *Introduction to Quantum Mechanics*. Pearson Prentice Hall. ISBN 978-0-13-111892-8.

- Hořava, Petr; Witten, Edward (1996a). "Heterotic and Type I string dynamics from eleven dimensions". *Nuclear Physics B* **460** (3): 506–524. arXiv:hep-th/9510209. Bibcode:1996NuPhB.460..506H. doi:10.1016/0550-3213(95)00621-4.

- Hořava, Petr; Witten, Edward (1996b). "Eleven dimensional supergravity on a manifold with boundary". *Nuclear Physics B* **475** (1): 94–114. arXiv:hep-th/9603142. Bibcode:1996NuPhB.475...94H. doi:10.1016/0550-3213(96)00308-2.

- Hull, Chris; Townsend, Paul (1995). "Unity of superstring dualities". *Nuclear Physics B* **4381** (1): 109–137. arXiv:hep-th/9410167. Bibcode:1995NuPhB.438..109H. doi:10.1016/0550-3213(94)00559-W.

- Khovanov, Mikhail (2000). "A categorification of the Jones polynomial". *Duke Mathematical Journal* **1011** (3): 359–426. doi:10.1215/S0012-7094-00-10131-7.

- Klebanov, Igor; Maldacena, Juan (2009). "Solving Quantum Field Theories via Curved Spacetimes" (PDF). *Physics Today* **62**: 28. Bibcode:2009PhT....62a..28K. doi:10.1063/1.3074260. Retrieved May 2013.

- Maldacena, Juan (1998). "The Large N limit of superconformal field theories and supergravity". *Advances in Theoretical and Mathematical Physics* **2**: 231–252. arXiv:hep-th/9711200. Bibcode:1998AdTMP...2..231M. doi:10.1063/1.59653.

- Maldacena, Juan (2005). "The Illusion of Gravity" (PDF). *Scientific American* **293** (5): 56–63. Bibcode:2005SciAm.293e..56M. doi:10.1038/scientificamerican1105-56. PMID 16318027. Retrieved July 2013.

- Montonen, Claus; Olive, David (1977). "Magnetic monopoles as gauge particles?". *Physics Letters B* **72** (1): 117–120. Bibcode:1977PhLB...72..117M. doi:10.1016/0370-2693(77)90076-4.

- Moore, Gregory (2005). "What is ... a Brane?" (PDF). *Notices of the AMS* **52**: 214. Retrieved June 2013.

- Moore, Gregory (2012). "Lecture Notes for Felix Klein Lectures" (PDF). Retrieved 14 August 2013.

- Nahm, Walter (1978). "Supersymmetries and their representations". *Nuclear Physics B* **135** (1): 149–166. Bibcode:1978NuPhB.135..149N. doi:10.1016/0550-3213(78)90218-3.

- Nekrasov, Nikita; Schwarz, Albert (1998). "Instantons on noncommutative R^4 and (2,0) superconformal six dimensional theory". *Communications in Mathematical Physics* **198** (3): 689–703. arXiv:hep-th/9802068. Bibcode:1998CMaPh.198..689N. doi:10.1007/s002200050490.

- Peskin, Michael; Schroeder, Daniel (1995). *An Introduction to Quantum Field Theory*. Westview Press. ISBN 978-0-201-50397-5.

- Randall, Lisa; Sundrum, Raman (1999). "An alternative to compactification". *Physical Review Letters* **83** (23): 4690. arXiv:hep-th/9906064. Bibcode:1999PhRvL..83.4690R. doi:10.1103/PhysRevLett.83.4690.

- Seiberg, Nathan; Witten, Edward (1999). "String Theory and Noncommutative Geometry". *Journal of High Energy Physics* **1999** (9): 032. arXiv:hep-th/9908142. Bibcode:1999JHEP...09..032S. doi:10.1088/1126-6708/1999/09/032.

- Sen, Ashoke (1993). "Electric-magnetic duality in string theory". *Nuclear Physics B* **404** (1): 109–126. arXiv:hep-th/9207053. Bibcode:1993NuPhB.404..109S. doi:10.1016/0550-3213(93)90475-5.

- Sen, Ashoke (1994a). "Strong-weak coupling duality in four-dimensional string theory". *International Journal of Modern Physics A* **9** (21): 3707–3750. arXiv:hep-th/9402002. Bibcode:1994IJMPA...9.3707S. doi:10.1142/S0217751X94001497.

- Sen, Ashoke (1994b). "Dyon-monopole bound states, self-dual harmonic forms on the multi-monopole moduli space, and $SL(2,\mathbf{Z})$ invariance in string theory". *Physics Letters B* **329** (2): 217–221. arXiv:hep-th/9402032. Bibcode:1994PhLB..329..217S. doi:10.1016/0370-2693(94)90763-3.

- Strominger, Andrew (1990). "Heterotic solitons". *Nuclear Physics B* **343** (1): 167–184. Bibcode:1990NuPhB.343..167S. doi:10.1016/0550-3213(90)90599-9.

- van Nieuwenhuizen, Peter (1981). "Supergravity". *Physics Reports* **68** (4): 189–398. Bibcode:1981PhR....68..189V. doi:10.1016/0370-1573(81)90157-5.

- Wald, Robert (1984). *General Relativity*. University of Chicago Press. ISBN 978-0-226-87033-5.

- Witten, Edward (1989). "Quantum Field Theory and the Jones Polynomial". *Communications in Mathematical Physics* **121** (3): 351–399. Bibcode:1989CMaPh.121..351W. doi:10.1007/BF01217730. MR 0990772.

- Witten, Edward (1995). "String theory dynamics in various dimensions". *Nuclear Physics B* **443** (1): 85–126. arXiv:hep-th/9503124. Bibcode:1995NuPhB.443...85W. doi:10.1016/0550-3213(95)00158-O.

- Witten, Edward (2009). "Geometric Langlands from six dimensions". arXiv:0905.2720 [hep-th].

- Witten, Edward (2012). "Fivebranes and knots". *Quantum Topology* **3** (1): 1–137. doi:10.4171/QT/26.

- Woit, Peter (2006). *Not Even Wrong: The Failure of String Theory and the Search for Unity in Physical Law*. Basic Books. p. 105. ISBN 0-465-09275-6.

- Yau, Shing-Tung; Nadis, Steve (2010). *The Shape of Inner Space: String Theory and the Geometry of the Universe's Hidden Dimensions*. Basic Books. ISBN 978-0-465-02023-2.

- Zee, Anthony (2010). *Quantum Field Theory in a Nutshell* (2nd ed.). Princeton University Press. ISBN 978-0-691-14034-6.

- Zwiebach, Barton (2009). *A First Course in String Theory*. Cambridge University Press. ISBN 978-0-521-88032-9.

16.7 External links

- The Elegant Universe—A three-hour miniseries with Brian Greene on the series *Nova* (original PBS broadcast dates: October 28, 8–10 p.m. and November 4, 8–9 p.m., 2003). Various images, texts, videos and animations explaining string theory and M-theory.

- Superstringtheory.com—The "Official String Theory Web Site", created by Patricia Schwarz. References on string theory and M-theory for the layperson and expert.

- Not Even Wrong—Peter Woit's blog on physics in general, and string theory in particular.

Chapter 17

Ekpyrotic universe

The **ekpyrotic (ĕk′pī-rŏt′ĭk) universe**, or **ekpyrotic scenario**, is a cosmological model of the early universe that explains the origin of the large-scale structure of the cosmos. The model has also been incorporated in the **cyclic universe** theory (or **ekpyrotic cyclic universe** theory), which proposes a complete cosmological history, both the past and future.

The original ekpyrotic model was introduced by Justin Khoury, Burt Ovrut, Paul Steinhardt and Neil Turok in 2001. Steinhardt created the name based on the early word ekpyrosis (ancient Greek ἐκπύρωσις ,ekpurōsis). [1] Ekpyrosis means conflagration in Greek; it refers to an ancient Stoic cosmological model in which the universe is caught in an eternal cycle of fiery birth, cooling and rebirth. [2]

The name is well-suited to the theory, which addresses the fundamental question that remains unanswered by the big bang inflationary model: what happened before the big bang? The explanation, according to the ekpyrotic theory, is that the big bang was actually a big bounce, a transition from a previous epoch of contraction to the present epoch of expansion. The key events that shaped our universe occurred before the bounce, and, in a cyclic version, the universe bounces at regular intervals. [3]

The original ekpyrotic models relied on string theory, branes and extra dimensions, but most contemporary ekyprotic and cyclic models use the same physical ingredients as inflationary models (quantum fields evolving in ordinary space-time). The theory has shown impressive success in accurately describing what we know so far about our universe. It predicts a uniform, flat universe with patterns of hot spots and colds spots now visible in the cosmic microwave background (CMB), and has been confirmed by the WMAP and Planck satellite experiments. Discovery of the CMB was originally considered a landmark test of the big bang, but proponents of the ekpyrotic and cyclic theories have shown that the CMB is also consistent with a big bounce, as posited by the ekpyrotic and cyclic theories. The search for primordial gravitational waves in the CMB (which produce patterns of polarized light known as B-modes) may eventually help scientists distinguish between the rival theories, since the ekpyrotic and cyclic models predict that no B-mode patterns should be observed.

A key advantage of ekpyrotic and cyclic models is that they do not produce a multiverse. This is important, because, when the effects of quantum fluctuations are properly included in the big bang inflationary model, they prevent the universe from achieving the uniformity and flatness that the cosmologists are trying to explain. Instead, inflated quantum fluctuations cause the universe to break up into patches with every conceivable combination of physical properties. Instead of making clear predictions, the big bang inflationary theory allows any outcome, so that the properties we observe may be viewed as random chance, resulting from the particular patch of the multiverse in which the Earth resides. Most regions of the multiverse would have very different properties.

Nobel laureate Steven Weinberg has suggested that if the multiverse is true, "the hope of finding a rational explanation for the precise values of quark masses and other constants of the standard model that we observe in our big bang is doomed, for their values would be an accident of the particular part of the multiverse in which we live." [4]

The idea that the properties of our universe are an accident and come from a theory that allows a multiverse of other possibilities is hard to reconcile with fact that the universe is extraordinarily simple (uniform and flat) on large scales and that elementary particles appear to be described by simple symmetries and interactions. Also, the accidental concept cannot be falsified by experiment since any future experiments can be viewed as yet other accidental aspects.

In ekpyrotic and cyclic models, smoothing and flattening occurs during a period of slow contraction, so quantum fluctuations are not inflated and cannot produce a multiverse. As a result, the ekpyrotic and cyclic models predict simple physical properties that are consistent with current experimental evidence without producing a multiverse. This makes these models valuable and, at the same time, vulnerable since they can be falsified by experiments.

17.1 See also

- Cosmic inflation
- Cyclic model

17.2 Notes and references

[1] [http://en.wiktionary.org/wiki/ekpyrotic "Wikitionary"] Check |url= scheme (help).

[2] 'The dissolution of the universe into fire'. In Stoic philosophy, *ekpyrosis*, all-engulfing cosmic fire, represents the contractive phase of eternally-recurring destruction and recreation. On "ekpyrosis" see generally Michael Lapidge, 'Stoic Cosmology,' in John M. Rist, *The Stoics,* Cambridge University Press, 1978, pp. 161–186, pp. 180–184

[3] Steinhardt, P. J.; Turok, N. (2002-04-25). "A Cyclic Model of the Universe". *Science* **296** (5572): 1436–1439. arXiv:hep-th/0111030v2. Bibcode:2002Sci...296.1436S. doi:10.1126/science.1070462. PMID 11976408.

[4] Weinberg, Steven (November 20, 2007). "Physics: What we do and don't know". *The New York Review of Books.*

17.3 Further reading

- P. J. Steinhardt and N. Turok, Endless Universe: Beyond the Big Bang, (Doubleday, 2007)

- A Brief Introduction to the Ekpyrotic Universe Steinhardt, Paul J., Department of Physics, Princeton University

- Greene, Brian, *The Elegant Universe: Superstrings, Hidden Dimensions, and the Quest for the Ultimate Theory*, Vintage (2000).

- Kallosh, Renata, Kofman, Lev and Linde, Andrei, Pyrotechnic Universe (the first paper to point out problems with the theory)

- Whitehouse, David, "Before the Big Bang". BBC News. April 10, 2001.

- Discover Magazine, Before the Big Bang February 2004 issue.

- Parallel Universes, (BBC Two 9.00pm Thursday February 14, 2002)

- 'Brane-Storm' Challenges Part of Big Bang Theory

Chapter 18

String theory landscape

The **string theory landscape** refers to the huge number of possible false vacua in string theory.[1] The large number of theoretically allowed configurations has prompted suggestions that certain physical mysteries, particularly relating to the fine-tuning of constants like the cosmological constant or the Higgs boson mass, may be explained not by a physical mechanism but by assuming that many different vacua are physically realized.[2] The *anthropic landscape* thus refers to the collection of those portions of the landscape that are suitable for supporting intelligent life, an application of the anthropic principle that selects a subset of the otherwise possible configurations.

In string theory the number of false vacua is thought to be somewhere between 10^{10} to 10^{100}.[1] The large number of possibilities arises from different choices of Calabi–Yau manifolds and different values of generalized magnetic fluxes over different homology cycles. If one assumes that there is no structure in the space of vacua, the problem of finding one with a sufficiently small cosmological constant is NP complete,[3] being a version of the subset sum problem.

18.1 Anthropic principle

Main article: Anthropic principle

The idea of the string theory landscape has been used to propose a concrete implementation of the anthropic principle, the idea that fundamental constants may have the values they have not for fundamental physical reasons, but rather because such values are necessary for life (and hence intelligent observers to measure the constants). In 1987, Steven Weinberg proposed that the observed value of the cosmological constant was so small because it is impossible for life to occur in a universe with a much larger cosmological constant.[4] In order to implement this idea in a concrete physical theory, it is necessary to postulate a multiverse in which fundamental physical parameters can take different values. This has been realized in the context of eternal inflation.

18.2 Bayesian probability

Main article: Bayesian probability

Some physicists, starting with Weinberg, have proposed that Bayesian probability can be used to compute probability distributions for fundamental physical parameters, where the probability $P(x)$ of observing some fundamental parameters x is given by,

$$P(x) = P_{\text{prior}}(x) \times P_{\text{selection}}(x),$$

where P_{prior} is the prior probability, from fundamental theory, of the parameters x and $P_{\text{selection}}$ is the anthropic selection function, determined by the number of "observers" that would occur in the universe with parameters x. These probabilistic arguments are the most controversial aspect of the landscape. Technical criticisms of these proposals have pointed out that:

- The function P_{prior} is completely unknown in string theory and may be impossible to define or interpret in any sensible probabilistic way.

- The function $P_{\text{selection}}$ is completely unknown, since so little is known about the origin of life. Simplified criteria (such as the number of galaxies) must be used as a proxy for the number of observers. Moreover, it may never be possible to compute it for parameters radically different from those of the observable universe.

(Interpreting probability in a context where it is only possible to draw one sample from a distribution is problematic in frequentist probability but not in Bayesian probability, which is not defined in terms of the frequency of repeated events.)

Various physicists have tried to address these objections, and the ideas remain extremely controversial both within and outside the string theory community. These ideas have been reviewed by Carroll.[5]

18.3 Simplified approaches

Tegmark *et al.* have recently considered these objections and proposed a simplified anthropic scenario for axion dark matter in which they argue that the first two of these problems do not apply.[6]

Vilenkin and collaborators have proposed a consistent way to define the probabilities for a given vacuum.[7]

A problem with many of the simplified approaches people have tried is that they "predict" a cosmological constant that is too large by a factor of 10–1000 (depending on one's assumptions) and hence suggest that the cosmic acceleration should be much more rapid than is observed.[8][9][10]

18.4 Criticism

Although few dispute the idea that string theory appears to have an unimaginably large number of metastable vacua, the existence - meaning and scientific relevance of the anthropic landscape - remain highly controversial. Prominent proponents of the idea include Andrei Linde, Sir Martin Rees and especially Leonard Susskind, who advocate it as a solution to the cosmological-constant problem. Opponents, such as David Gross, suggest that the idea is inherently unscientific, unfalsifiable or premature. A famous debate on the anthropic landscape of string theory is the Smolin–Susskind debate on the merits of the landscape.

The term "landscape" comes from evolutionary biology (see *Fitness landscape*) and was first applied to cosmology by Lee Smolin in his book.[11] It was first used in the context of string theory by Susskind.

There are several popular books about the anthropic principle in cosmology.[12] Two popular physics blogs are opposed to this use of the anthropic principle.[13]

18.5 See also

- Extra dimensions
- Compactification

18.6 References

[1] The most commonly quoted number is of the order 10^{500}. See M. Douglas, "The statistics of string / M theory vacua", *JHEP* **0305**, 46 (2003). arXiv:hep-th/0303194; S. Ashok and M. Douglas, "Counting flux vacua", *JHEP* **0401**, 060 (2004).

[2] L. Susskind, "The anthropic landscape of string theory", arXiv:hep-th/0302219.

[3] Frederik Denef; Douglas, Michael R. (2006). "Computational complexity of the landscape". *Annals of Physics* **322** (5): 1096–1142. arXiv:hep-th/0602072. Bibcode:2007AnPhy.322.1096D. doi:10.1016/j.aop.2006.07.013.

[4] S. Weinberg, "Anthropic bound on the cosmological constant", *Phys. Rev. Lett.* **59**, 2607 (1987).

[5] S. M. Carroll, "Is our universe natural?", arXiv:hep-th/0512148.

[6] M. Tegmark, A. Aguirre, M. Rees and F. Wilczek, "Dimensionless constants, cosmology and other dark matters", arXiv:astro-ph/0511774. F. Wilczek, "Enlightenment, knowledge, ignorance, temptation", arXiv:hep-ph/0512187. See also the discussion at .

[7] See, *e.g.* Alexander Vilenkin (2006). "A measure of the multiverse". *Journal of Physics A: Mathematical and Theoretical* **40** (25): 6777–6785. arXiv:hep-th/0609193. Bibcode:2007JPhA...40.6777V. doi:10.1088/1751-8113/40/25/S22.

[8] Abraham Loeb (2006). "An observational test for the anthropic origin of the cosmological constant". *JCAP* **0605**: 009. (subscription required (help)).

[9] Jaume Garriga & Alexander Vilenkin (2006). "Anthropic prediction for Lambda and the Q catastrophe". *Prog. Theor.Phys. Suppl.* **163**: 245–57. arXiv:hep-th/0508005. Bibcode:2006PThPS.163..245G. doi:10.1143/PTPS.163.245. (subscription required (help)).

[10] Delia Schwartz-Perlov & Alexander Vilenkin (2006). "Probabilities in the Bousso-Polchinski multiverse". *JCAP* **0606**: 010. (subscription required (help)).

[11] L. Smolin, "Did the universe evolve?", *Classical and Quantum Gravity* **9**, 173–191 (1992). L. Smolin, *The Life of the Cosmos* (Oxford, 1997)

[12] L. Susskind, *The cosmic landscape: string theory and the illusion of intelligent design* (Little, Brown, 2005). M. J. Rees, *Just six numbers: the deep forces that shape the universe* (Basic Books, 2001). R. Bousso and J. Polchinski, "The string theory landscape", *Sci. Am.* **291**, 60–69 (2004).

[13] Lubos Motl's blog criticized the anthropic principle and Peter Woit's blog frequently attacks the anthropic string landscape.

18.7 External links

- String theory landscape on arxiv.org
- On the computation of non-perturbative effective potentials in the string theory landscape, Mirjam Cvetič, Iñaki García-Etxebarria, James Halverson

Chapter 19

Holographic principle

The **holographic principle** is a property of string theories and a supposed property of quantum gravity that states that the description of a volume of space can be thought of as encoded on a boundary to the region—preferably a light-like boundary like a gravitational horizon. First proposed by Gerard 't Hooft, it was given a precise string-theory interpretation by Leonard Susskind[1] who combined his ideas with previous ones of 't Hooft and Charles Thorn.[1][2] As pointed out by Raphael Bousso,[3] Thorn observed in 1978 that string theory admits a lower-dimensional description in which gravity emerges from it in what would now be called a holographic way.

In a larger sense, the theory suggests that the entire universe can be seen as a two-dimensional information on the cosmological horizon, such that the three dimensions we observe are an effective description only at macroscopic scales and at low energies. Cosmological holography has not been made mathematically precise, partly because the particle horizon has a non-zero area and grows with time.[4][5]

The holographic principle was inspired by black hole thermodynamics, which conjectures that the maximal entropy in any region scales with the radius *squared*, and not cubed as might be expected. In the case of a black hole, the insight was that the informational content of all the objects that have fallen into the hole might be entirely contained in surface fluctuations of the event horizon. The holographic principle resolves the black hole information paradox within the framework of string theory.[6] However, there exist classical solutions to the Einstein equations that allow values of the entropy larger than those allowed by an area law, hence in principle larger than those of a black hole. These are the so-called "Wheeler's bags of gold". The existence of such solutions conflicts with the holographic interpretation, and their effects in a quantum theory of gravity including the holographic principle are not yet fully understood.[7]

19.1 Black hole entropy

Main article: Black hole thermodynamics

An object with relatively high entropy is microscopically random, like a hot gas. A known configuration of classical fields has zero entropy: there is nothing random about electric and magnetic fields, or gravitational waves. Since black holes are exact solutions of Einstein's equations, they were thought not to have any entropy either.

But Jacob Bekenstein noted that this leads to a violation of the second law of thermodynamics. If one throws a hot gas with entropy into a black hole, once it crosses the event horizon, the entropy would disappear. The random properties of the gas would no longer be seen once the black hole had absorbed the gas and settled down. One way of salvaging the second law is if black holes are in fact random objects, with an enormous entropy whose increase is greater than the entropy carried by the gas.

Bekenstein assumed that black holes are maximum entropy objects—that they have more entropy than anything else in the same volume. In a sphere of radius R, the entropy in a relativistic gas increases as the energy increases. The only known limit is gravitational; when there is too much energy the gas collapses into a black hole. Bekenstein used this to put an upper bound on the entropy in a region of space, and the bound was proportional to the area of the region. He concluded that the black hole entropy is directly proportional to the area of the event horizon.[8]

Stephen Hawking had shown earlier that the total horizon area of a collection of black holes always increases with time. The horizon is a boundary defined by light-like geodesics; it is those light rays that are just barely unable to escape. If neighboring geodesics start moving toward each other they eventually collide, at which point their extension is inside the black hole. So the geodesics are always moving apart, and the number of geodesics which generate the boundary, the area of the horizon, always increases. Hawking's result was called the second law of black hole thermodynamics, by analogy with the law of entropy increase, but at first, he did not take the analogy too seriously.

Hawking knew that if the horizon area were an actual entropy, black holes would have to radiate. When heat is added to a thermal system, the change in entropy is the increase in mass-energy divided by temperature:

$$dS = \frac{dM}{T}.$$

If black holes have a finite entropy, they should also have a finite temperature. In particular, they would come to equilibrium with a thermal gas of photons. This means that black holes would not only absorb photons, but they would also have to emit them in the right amount to maintain detailed balance.

Time independent solutions to field equations do not emit radiation, because a time independent background conserves energy. Based on this principle, Hawking set out to show that black holes do not radiate. But, to his surprise, a careful analysis convinced him that they do, and in just the right way to come to equilibrium with a gas at a finite temperature. Hawking's calculation fixed the constant of proportionality at 1/4; the entropy of a black hole is one quarter its horizon area in Planck units.[9]

The entropy is proportional to the logarithm of the number of microstates, the ways a system can be configured microscopically while leaving the macroscopic description unchanged. Black hole entropy is deeply puzzling — it says that the logarithm of the number of states of a black hole is proportional to the area of the horizon, not the volume in the interior.[10]

Later, Raphael Bousso came up with a covariant version of the bound based upon null sheets.

19.2 Black hole information paradox

Main article: Black hole information paradox

Hawking's calculation suggested that the radiation which black holes emit is not related in any way to the matter that they absorb. The outgoing light rays start exactly at the edge of the black hole and spend a long time near the horizon, while the infalling matter only reaches the horizon much later. The infalling and outgoing mass/energy only interact when they cross. It is implausible that the outgoing state would be completely determined by some tiny residual scattering.

Hawking interpreted this to mean that when black holes absorb some photons in a pure state described by a wave function, they re-emit new photons in a thermal mixed state described by a density matrix. This would mean that quantum mechanics would have to be modified, because in quantum mechanics, states which are superpositions with probability amplitudes never become states which are probabilistic mixtures of different possibilities.[note 1]

Troubled by this paradox, Gerard 't Hooft analyzed the emission of Hawking radiation in more detail. He noted that when Hawking radiation escapes, there is a way in which incoming particles can modify the outgoing particles. Their gravitational field would deform the horizon of the black hole, and the deformed horizon could produce different outgoing particles than the undeformed horizon. When a particle falls into a black hole, it is boosted relative to an outside observer, and its gravitational field assumes a universal form. 't Hooft showed that this field makes a logarithmic tent-pole shaped bump on the horizon of a black hole, and like a shadow, the bump is an alternate description of the particle's location and mass. For a four-dimensional spherical uncharged black hole, the deformation of the horizon is similar to the type of deformation which describes the emission and absorption of particles on a string-theory world sheet. Since the deformations on the surface are the only imprint of the incoming particle, and since these deformations would have to completely determine the outgoing particles, 't Hooft believed that the correct description of the black hole would be by some form of string theory.

This idea was made more precise by Leonard Susskind, who had also been developing holography, largely independently. Susskind argued that the oscillation of the horizon of a black hole is a complete description[note 2] of both the infalling and outgoing matter, because the world-sheet theory of string theory was just such a holographic description. While short strings have zero entropy, he could identify long highly excited string states with ordinary black holes. This was a deep advance because it revealed that strings have a classical interpretation in terms of black holes.

This work showed that the black hole information paradox is resolved when quantum gravity is described in an unusual string-theoretic way assuming the string-theoretical description is complete, unambiguous and non-redundant.[12] The space-time in quantum gravity would emerge as an effective description of the theory of oscillations of a lower-dimensional black-hole horizon, and suggest that any black hole with appropriate properties, not just strings, would serve as a basis for a description of string theory.

In 1995, Susskind, along with collaborators Tom Banks, Willy Fischler, and Stephen Shenker, presented a formulation of the new M-theory using a holographic description in terms of charged point black holes, the D0 branes of type IIA string theory. The Matrix theory they proposed was first suggested as a description of two branes in 11-dimensional supergravity by Bernard de Wit, Jens Hoppe, and Hermann Nicolai. The later authors reinterpreted the same matrix models as a description of the dynamics of point black holes in particular limits. Holography allowed them to conclude that the dynamics of these black holes give a complete non-perturbative formulation of M-theory. In 1997, Juan Maldacena gave the first holographic descriptions of a higher-dimensional object, the 3+1-dimensional type IIB membrane, which resolved a long-standing problem of finding a string description which describes a gauge theory. These developments simultaneously explained how string theory is related to some forms of supersymmetric quantum field theories.

19.3 Limit on information density

Entropy, if considered as information (see information entropy), is measured in bits. The total quantity of bits is related to the total degrees of freedom of matter/energy.

For a given energy in a given volume, there is an upper limit to the density of information (the Bekenstein bound) about the whereabouts of all the particles which compose matter in that volume, suggesting that matter itself cannot be subdivided infinitely many times and there must be an ultimate level of fundamental particles. As the degrees of freedom of a particle are the product of all the degrees of freedom of its sub-particles, were a particle to have infinite subdivisions into lower-level particles, then the degrees of freedom of the original particle must be infinite, violating the maximal limit of entropy density. The holographic principle thus implies that the subdivisions must stop at some level, and that the fundamental particle is a bit (1 or 0) of information.

The most rigorous realization of the holographic principle is the AdS/CFT correspondence by Juan Maldacena. However, J.D. Brown and Marc Henneaux had rigorously proved already in 1986, that the asymptotic symmetry of 2+1 dimensional gravity gives rise to a Virasoro algebra, whose corresponding quantum theory is a 2-dimensional conformal field theory.[13]

19.4 High-level summary

The physical universe is widely seen to be composed of "matter" and "energy". In his 2003 article published in Scientific American magazine, Jacob Bekenstein summarized a current trend started by John Archibald Wheeler, which suggests scientists may *"regard the physical world as made of information, with energy and matter as incidentals."*

Bekenstein asks "Could we, as William Blake memorably penned, 'see a world in a grain of sand,' or is that idea no more than 'poetic license,'"[14] referring to the holographic principle.

19.4.1 Unexpected connection

Bekenstein's topical overview "A Tale of Two Entropies"[15] describes potentially profound implications of Wheeler's trend, in part by noting a previously unexpected connection between the world of information theory and classical physics. This connection was first described shortly after the seminal 1948 papers of American applied mathematician Claude E. Shannon introduced today's most widely used measure of information content, now known as Shannon entropy. As an objective measure of the quantity of information, Shannon entropy has been enormously useful, as the design of all modern communications and data storage devices, from cellular phones to modems to hard disk drives and DVDs, rely on Shannon entropy.

In thermodynamics (the branch of physics dealing with heat), entropy is popularly described as a measure of the "disorder" in a physical system of matter and energy. In 1877 Austrian physicist Ludwig Boltzmann described it more precisely in terms of the *number of distinct microscopic states* that the particles composing a macroscopic "chunk" of matter could be in while still *looking* like the same macroscopic "chunk". As an example, for the air in a room, its thermodynamic entropy would equal the logarithm of the count of all the ways that the individual gas molecules could be distributed in the room, and all the ways they could be moving.

19.4.2 Energy, matter, and information equivalence

Shannon's efforts to find a way to quantify the information contained in, for example, an e-mail message, led him unexpectedly to a formula with the same form as Boltzmann's. In an article in the August 2003 issue of Scientific American titled "Information in the Holographic Universe", Bekenstein summarizes that *"Thermodynamic entropy and Shannon entropy are conceptually equivalent: the number of arrangements that are counted by Boltzmann entropy reflects the amount of Shannon information one would need to implement any particular arrangement..."* of matter and energy. The only salient difference between the thermodynamic entropy of physics and Shannon's entropy of information is in the units of measure; the former is expressed in units of energy divided by temperature, the latter in *essentially dimensionless* "bits" of information.

The holographic principle states that the entropy of *ordinary mass* (not just black holes) is also proportional to surface area and not volume; that volume itself is illusory and the universe is really a hologram which is isomorphic to the information "inscribed" on the surface of its boundary.[10]

19.5 Experimental tests

The Fermilab physicist Craig Hogan claims that the holographic principle would imply quantum fluctuations in spatial position[16] that would lead to apparent background noise or "holographic noise" measurable at gravitational wave detectors, in particular GEO 600.[17] However these claims have not been widely accepted, or cited, among quantum gravity researchers and appear to be in direct conflict with string theory calculations.[18]

Analyses in 2011 of measurements of gamma ray burst GRB 041219A in 2004 by the INTEGRAL space observatory launched in 2002 by the European Space Agency shows that Craig Hogan's noise is absent down to a scale of 10^{-48} meters, as opposed to scale of 10^{-35} meters predicted by Hogan, and the scale of 10^{-16} meters found in measurements of the GEO 600 instrument.[19] Research continues at Fermilab under Hogan as of 2013.[20]

Jacob Bekenstein also claims to have found a way to test the holographic principle with a tabletop photon experiment.[21]

19.6 Tests of Maldacena's conjecture

Main article: Maldacena conjecture

Hyakutake et al. in 2013/4 published two papers[22] that bring computational evidence that Maldacena's conjecture is true. One paper computes the internal energy of a black hole, the position of its event horizon, its entropy and other properties based on the predictions of string theory and the effects of virtual particles. The other paper calculates the internal energy of the corresponding lower-dimensional cosmos with no gravity. The two simulations match. The papers are not an actual proof of Maldacena's conjecture for all cases but a demonstration that the conjecture works for a particular theoretical case and a verification of the AdS/CFT correspondence for a particular situation.[23]

19.7 See also

- Bekenstein bound
- Beyond black holes
- Bousso's holographic bound
- Brane cosmology
- Entropic gravity
- Implicate and explicate order according to David Bohm
- Margolus–Levitin theorem
- Physical cosmology
- Quantum foam
- Simulated reality

19.8 Notes

[1] except in the case of measurements, which the black hole should not be performing

[2] "Complete description" means all the *primary* qualities. For example, John Locke (and before him Robert Boyle) determined these to be *size, shape, motion, number,* and *solidity.* Such *secondary quality* information as *color, aroma, taste* and *sound*,[11] or internal quantum state is not information that is implied to be preserved in the surface fluctuations of the event horizon. (See however "path integral quantization")

19.9 References

General

- Bousso, Raphael (2002). "The holographic principle". *Reviews of Modern Physics* **74** (3): 825–874. arXiv:hep-th/0203101. Bibcode:2002RvMP...74..825B. doi:10.1103/RevModPhys.74.825.

- 't Hooft, Gerard (1993). "Dimensional Reduction in Quantum Gravity". arXiv:gr-qc/9310026.. 't Hooft's original paper.

Citations

[1] Susskind, Leonard (1995). "The World as a Hologram". *Journal of Mathematical Physics* **36** (11): 6377–6396. arXiv:hep-th/9409089. Bibcode:1995JMP....36.6377S. doi:10.1063/1.531249.

[2] Thorn, Charles B. (27–31 May 1991). *Reformulating string theory with the 1/N expansion*. International A.D. Sakharov Conference on Physics. Moscow. pp. 447–54. arXiv:hep-th/9405069. ISBN 978-1-56072-073-7.

[3] Bousso, Raphael (2002). "The Holographic Principle". *Reviews of Modern Physics* **74** (3): 825–874. arXiv:hep-th/0203101. Bibcode:2002RvMP...74..825B. doi:10.1103/RevModPhys.74.825.

[4] Lloyd, Seth (2002-05-24). "Computational Capacity of the Universe". *Physical Review Letters* **88** (23): 237901. arXiv:quant-ph/0110141. Bibcode:2002PhRvL..88w7901L. doi:10.1103/PhysRevLett.88.237901. PMID 12059399.

[5] Davies, Paul. "Multiverse Cosmological Models and the Anthropic Principle". *CTNS*. Retrieved 2008-03-14.

[6] Susskind, L. (2008). *The Black Hole War – My Battle with Stephen Hawking to Make the World Safe for Quantum Mechanics*. Little, Brown and Company.

[7] Marolf, Donald (April 2009). "Black Holes, AdS, and CFTs". *General Relativity and Gravitation* **41** (4): 903–17. arXiv:0810.4886. Bibcode:2009GReGr..41..903M. doi:10.1007/s10714-008-0749-7.

[8] Bekenstein, Jacob D. (January 1981). "Universal upper bound on the entropy-to-energy ratio for bounded systems". *Physical Review D* **23** (215): 287–298. Bibcode:1981PhRvD..23..287B. doi:10.1103/PhysRevD.23.287.

[9] Majumdar, Parthasarathi (1998). "Black Hole Entropy and Quantum Gravity" **73**. p. 147. arXiv:gr-qc/9807045. Bibcode:1999InJPB..73..147M.

[10] Bekenstein, Jacob D. (August 2003). "Information in the Holographic Universe — Theoretical results about black holes suggest that the universe could be like a gigantic hologram". *Scientific American*: p. 59.

[11] Dennett, Daniel (1991). *Consciousness Explained*. New York: Back Bay Books. p. 371. ISBN 0-316-18066-1.

[12] Susskind, L. (February 2003). "The Anthropic landscape of string theory". arXiv:hep-th/0302219.

[13] Brown, J. D. & Henneaux, M. (1986). "Central charges in the canonical realization of asymptotic symmetries: an example from three-dimensional gravity". *Communications in Mathematical Physics* **104** (2): 207–226. Bibcode:1986CMaPh.104..207B. doi:10.1007/BF01211590..

[14] Information in the Holographic Universe

[15] http://webcache.googleusercontent.com/search?q=cache:E360V697cvgJ:ref-sciam.livejournal.com/1190.html&hl=en&gl=us&strip=1

[16] Hogan, Craig J. (2008). "Measurement of quantum fluctuations in geometry". *Physical Review D* **77** (10): 104031. arXiv:0712.3419. Bibcode:2008PhRvD..77j4031H. doi:10.1103/PhysRevD.77.104031..

[17] Chown, Marcus (15 January 2009). "Our world may be a giant hologram". *NewScientist*. Retrieved 2010-04-19.

[18] "Consequently, he ends up with inequalities of the type... Except that one may look at the actual equations of Matrix theory and see that none of these commutators is nonzero... The last displayed inequality above obviously can't be a consequence of quantum gravity because it doesn't depend on G at all! However, in the G→0 limit, one must reproduce non-gravitational physics in the flat Euclidean background spacetime. Hogan's rules don't have the right limit so they can't be right." – Lubos Motl, Hogan's holographic noise doesn't exist, Feb 7, 2012

[19] "Integral challenges physics beyond Einstein". European Space Agency. 30 June 2011. Retrieved 3 February 2013.

[20] "Frequently Asked Questions for the Holometer at Fermilab". 6 July 2013. Retrieved 14 February 2014.

[21] Cowen, Ron (22 November 2012). "Single photon could detect quantum-scale black holes". *Nature*. Retrieved 3 February 2013.

[22] Cowen, Ron (10 December 2013). "Simulations back up theory that Universe is a hologram". *Nature News*. doi:10.1038/nature.2013.14328. Hyakutake, Yoshifumi (March 2014). "Quantum near-horizon geometry of a black 0-brane". *Progress of Theoretical and Experimental Physics* **2014** (3): 033B04. arXiv:1311.7526. Bibcode:2014PTEP.2014c3B04H. doi:10.1093/ptep/ptu028. Hanada, Masanori; Hyakutake, Yoshifumi; Ishiki, Goro; Nishimura, Jun (23 May 2014). "Holographic description of a quantum black hole on a computer". *Science* **344** (6186): 882–5. arXiv:1311.5607. Bibcode:2014Sci...344..882H. doi:10.1126/science.1250122.

[23] Yirka, Bob (December 13, 2013). "New work gives credence to theory of universe as a hologram". Phys.org.

19.10 External links

- UC Berkeley's Raphael Bousso gives an introductory lecture on the holographic principle - Video.
- *Scientific American* article on holographic principle by Jacob Bekenstein

Chapter 20

Simulated reality

Simulated reality is the hypothesis that reality could be simulated—for example by computer simulation—to a degree indistinguishable from "true" reality. It could contain conscious minds which may or may not be fully aware that they are living inside a simulation.

This is quite different from the current, technologically achievable concept of virtual reality. Virtual reality is easily distinguished from the experience of actuality; participants are never in doubt about the nature of what they experience. Simulated reality, by contrast, would be hard or impossible to separate from "true" reality.

There has been much debate over this topic, ranging from philosophical discourse to practical applications in computing.

20.1 Types of simulation

20.1.1 Brain-computer interface

Main articles: brain-computer interface and brain in a vat

In brain-computer interface simulations, each participant enters from outside, directly connecting their brain to the simulation computer. The computer transmits sensory data to the participant, reads and responds to their desires and actions in return; in this manner they interact with the simulated world and receive feedback from it. The participant may be induced by any number of possible means to forget, temporarily or otherwise, that they are inside a virtual realm (e.g. "passing through the veil", a term borrowed from Christian tradition, which describes the passage of a soul from an earthly body to an afterlife). While inside the simulation, the participant's consciousness is represented by an avatar, which can look very different from the participant's actual appearance.

20.1.2 Virtual people

Main article: Artificial consciousness

In a virtual-people simulation, every inhabitant is a native of the simulated world. They do not have a "real" body in the external reality of the physical world. Instead, each is a fully simulated entity, possessing an appropriate level of consciousness that is implemented using the simulation's own logic (i.e. using its own physics). As such, they could be downloaded from one simulation to another, or even archived and resurrected at a later time. It is also possible that a simulated entity could be moved out of the simulation entirely by means of mind transfer into a synthetic body.

20.2 Arguments

20.2.1 Simulation argument

The simulation hypothesis was first published by Hans Moravec.[1][2][3] Later, the philosopher Nick Bostrom developed an expanded argument examining the probability of our reality being a simulacrum.[4] His argument states that at least one of the following statements is very likely to be true:

> 1. Human civilization is unlikely to reach a level of technological maturity capable of producing simulated realities, or such simulations are physically impossible to construct.

> 2. A comparable civilization reaching aforementioned technological status will likely not produce a significant number of simulated realities (one that might push the probable existence of digital entities beyond the probable number of "real" entities in a Universe) for any of a number of reasons, such as, diversion of computational processing power for other tasks, ethical considera-

tions of holding entities captive in simulated realities, etc.

3. Any entities with our general set of experiences are almost certainly living in a simulation.

In greater detail, Bostrom is attempting to prove a tripartite disjunction, that at least one of these propositions must be true. His argument rests on the premise that given sufficiently advanced technology, it is possible to represent the populated surface of the Earth without recourse to digital physics; that the qualia experienced by a simulated consciousness is comparable or equivalent to that of a naturally occurring human consciousness; and that one or more levels of simulation within simulations would be feasible given only a modest expenditure of computational resources in the real world.

If one assumes first that humans will not be destroyed nor destroy themselves before developing such a technology, and, next, that human descendants will have no overriding legal restrictions or moral compunctions against simulating biospheres or their own historical biosphere, then it would be unreasonable to count ourselves among the small minority of genuine organisms who, sooner or later, will be vastly outnumbered by artificial simulations.

Epistemologically, it is not impossible to tell whether we are living in a simulation. For example, Bostrom suggests that a window could *popup* saying: "You are living in a simulation. Click here for more information." However, imperfections in a simulated environment might be difficult for the native inhabitants to identify, and for purposes of authenticity, even the simulated memory of a blatant revelation might be purged programmatically. Nonetheless, should any evidence come to light, either for or against the skeptical hypothesis, it would radically alter the aforementioned probability.

The simulation argument also has implications for existential risks. If we are living in a simulation, then it's possible that our simulation could get shut down. Many futurists have speculated about how we can avoid this outcome. Ray Kurzweil argues in *The Singularity is Near* that we should be interesting to our simulators, and that bringing about the Singularity is probably the most interesting event that could happen. The philosopher Phil Torres has argued that the simulation argument itself leads to the conclusion that, if we run simulations in the future, then there almost certainly exists a stack of nested simulations, with ours located towards the bottom. Since annihilation is inherited downwards, any terminal event in a simulation "above" ours would be a terminal event for us. If there are many simulations above us, then the risk of an existential catastrophe could be significant.[5]

20.2.2 Relativity of reality

As to the question of whether we are living in a simulated reality or a 'real' one, the answer may be 'indistinguishable', in principle. In a commemorative article dedicated to the 'The World Year of Physics 2005', physicist Bin-Guang Ma proposed the theory of 'Relativity of reality'.[6] The notion appears in ancient philosophy: Zhuangzi's 'Butterfly Dream', and analytical psychology.[7] Without special knowledge of a reference world, one cannot say with absolute skeptical certainty one is experiencing "reality".

20.2.3 Computationalism

Main articles: Computationalism and Mathematical universe hypothesis

Computationalism is a philosophy of mind theory stating that cognition is a form of computation. It is relevant to the Simulation hypothesis in that it illustrates how a simulation could contain conscious subjects, as required by a "virtual people" simulation. For example, it is well known that physical systems can be simulated to some degree of accuracy. If computationalism is correct, and if there is no problem in generating artificial consciousness or cognition, it would establish the theoretical possibility of a simulated reality. However, the relationship between cognition and phenomenal qualia of consciousness is disputed. It is possible that consciousness requires a vital substrate that a computer cannot provide, and that simulated people, while behaving appropriately, would be philosophical zombies. This would undermine Nick Bostrom's simulation argument; we cannot be a simulate consciousness, if consciousness, as we know it, cannot be simulated. However, the skeptical hypothesis remains intact, we could still be envatted brains, existing as conscious beings within a simulated environment, even if consciousness cannot be simulated.

Some theorists[8][9] have argued that if the "consciousness-is-computation" version of computationalism and mathematical realism (or radical mathematical Platonism)[10] are true then consciousnesses is computation, which in principle is platform independent, and thus admits of simulation. This argument states that a "Platonic realm" or ultimate ensemble would contain every algorithm, including those which implement consciousness. Hans Moravec has explored the simulation hypothesis and has argued for a kind of mathematical Platonism according to which every object (including e.g. a stone) can be regarded as implementing every possible computation.[1]

20.2.4 Dreaming

Further information: Dream argument

A dream could be considered a type of simulation capable of fooling someone who is asleep. As a result, the "dream hypothesis" cannot be ruled out, although it has been argued that common sense and considerations of simplicity rule against it.[11] One of the first philosophers to question the distinction between reality and dreams was Zhuangzi, a Chinese philosopher from the 4th century BC. He phrased the problem as the well-known "Butterfly Dream," which went as follows:

> Once Zhuangzi dreamt he was a butterfly, a butterfly flitting and fluttering around, happy with himself and doing as he pleased. He didn't know he was Zhuangzi. Suddenly he woke up and there he was, solid and unmistakable Zhuangzi. But he didn't know if he was Zhuangzi who had dreamt he was a butterfly, or a butterfly dreaming he was Zhuangzi. Between Zhuangzi and a butterfly there must be *some* distinction! This is called the Transformation of Things. (2, tr. Burton Watson 1968:49)

The philosophical underpinnings of this argument are also brought up by Descartes, who was one of the first Western philosophers to do so. In *Meditations on First Philosophy*, he states "... there are no certain indications by which we may clearly distinguish wakefulness from sleep",[12] and goes on to conclude that "It is possible that I am dreaming right now and that all of my perceptions are false".[12]

Chalmers (2003) discusses the dream hypothesis, and notes that this comes in two distinct forms:

- that he is *currently* dreaming, in which case many of his beliefs about the world are incorrect;
- that he has *always* been dreaming, in which case the objects he perceives actually exist, albeit in his imagination.[13]

Both the dream argument and the simulation hypothesis can be regarded as skeptical hypotheses; however in raising these doubts, just as Descartes noted that his own thinking led him to be convinced of his own existence, the existence of the argument itself is testament to the possibility of its own truth.

Another state of mind in which some argue an individual's perceptions have no physical basis in the real world is called psychosis though psychosis may have a physical basis in the real world and explanations vary.

20.2.5 Computability of physics

Further information: Computational universe theory and The Emperor's New Mind

A decisive refutation of any claim that our reality is computer-simulated would be the discovery of some uncomputable physics, because if reality is doing something that no computer can do, it cannot be a computer simulation. (*Computability* generally means computability by a Turing machine. Hypercomputation (super-Turing computation) introduces other possibilities which will be dealt with separately.) In fact, known physics is held to be (Turing) computable,[14] but the statement "physics is computable" needs to be qualified in various ways. Before symbolic computation, a number, thinking particularly of a real number, one with an infinite number of digits, was said to be computable if a Turing machine will continue to spit out digits endlessly, never reaching a "final digit".[15] This runs counter, however, to the idea of simulating physics in real time (or any plausible kind of time). Known physical laws (including those of quantum mechanics) are very much infused with real numbers and continua, and the universe seems to be able to decide their values on a moment-by-moment basis. As Richard Feynman put it:[16]

> "It always bothers me that, according to the laws as we understand them today, it takes a computing machine an infinite number of logical operations to figure out what goes on in no matter how tiny a region of space, and no matter how tiny a region of time. How can all that be going on in that tiny space? Why should it take an infinite amount of logic to figure out what one tiny piece of space/time is going to do? So I have often made the hypotheses that ultimately physics will not require a mathematical statement, that in the end the machinery will be revealed, and the laws will turn out to be simple, like the chequer board with all its apparent complexities".

The objection could be made that the simulation does not have to run in "real time".[17] It misses an important point, though: the shortfall is not linear; rather it is a matter of performing an infinite number of computational steps in a finite time.[18]

Note that these objections all relate to the idea of reality being *exactly* simulated. Ordinary computer simulations as used by physicists are always approximations.

These objections do not apply if the hypothetical simulation is being run on a hypercomputer, a hypothetical machine more powerful than a Turing machine.[19] Unfortunately, there is no way of working out if computers running a simulation are capable of doing things that computers in the simulation cannot do. The laws of physics inside a simulation and those outside it do not have to be the same, and simulations of different physical laws have been constructed.[20] The problem now is that there is no evidence that can conceivably be produced to show that the universe is *not* any kind of computer, making the simulation hypothesis unfalsifiable and therefore scientifically unacceptable, at least by Popperian standards.[21]

All conventional computers, however, are less than hyper-computational, and the simulated reality hypothesis is usually expressed in terms of conventional computers, i.e. Turing machines.

Roger Penrose, an English mathematical physicist, presents the argument that human consciousness is non-algorithmic, and thus is not capable of being modeled by a conventional Turing machine-type of digital computer. Penrose hypothesizes that quantum mechanics plays an essential role in the understanding of human consciousness. He sees the collapse of the quantum wavefunction as playing an important role in brain function. (See consciousness causes collapse).

20.2.6 CantGoTu environments

Further information: The Fabric of Reality and Computational universality

In his book *The Fabric of Reality*, David Deutsch discusses how the limits to computability imposed by Gödel's Incompleteness Theorem affect the Virtual Reality rendering process.[22][23] In order to do this, Deutsch invents the notion of a CantGoTu environment (named after Cantor, Gödel, and Turing), using Cantor's diagonal argument to construct an 'impossible' Virtual Reality which a physical VR generator would not be able to generate. The way that this works is to imagine that all VR environments renderable by such a generator can be enumerated, and that we label them VR1, VR2, etc. Slicing time up into discrete chunks we can create an environment which is unlike VR1 in the first timeslice, unlike VR2 in the second timeslice and so on. This environment is not in the list, and so it cannot be generated by the VR generator. Deutsch then goes on to discuss a universal VR generator, which as a physical device would not be able to render all possible environments, but would be able to render those environments which can be rendered by all other physical VR generators. He argues that 'an environment which can be rendered' corresponds to a set of mathematical questions whose answers can be calculated, and discusses various forms of the Turing Principle, which in its initial form refers to the fact that it is possible to build a universal computer which can be programmed to execute any computation that any other machine can do. Attempts to capture the process of virtual reality rendering provides us with a version which states: "It is possible to build a virtual-reality generator, whose repertoire includes every physically possible environment". In other words, a single, buildable physical object can mimic all the behaviours and responses of any other physically possible process or object. This, it is claimed, is what makes reality comprehensible.

Later on in the book, Deutsch goes on to argue for a very strong version of the Turing principle, namely: "It is possible to build a virtual reality generator whose repertoire includes every *physically possible* environment." However, in order to include *every physically possible environment*, the computer would have to be able to include a recursive simulation of the environment containing *itself*. Even so, a computer running a simulation need not have to run every possible physical moment to be plausible to its inhabitants.

20.2.7 Nested simulations

The existence of simulated reality is unprovable in any concrete sense: any "evidence" that is directly observed could be another simulation itself. In other words, there is an infinite regress problem with the argument. Even if we are a simulated reality, there is no way to be sure the beings running the simulation are not themselves a simulation, and the operators of *that* simulation are not a simulation.[24]

"Recursive simulation involves a simulation, or an entity in the simulation, creating another instance of the same simulation, running it and using its results" (Pooch and Sullivan 2000).[25]

20.2.8 Peer-to-Peer Explanation of Quantum Phenomena

In two recent articles, the philosopher Marcus Arvan has argued that a new version of the simulation hypothesis, the Peer-to-Peer Simulation Hypothesis, provides a unified explanation of a wide variety of quantum phenomena. According to Arvan, peer-to-peer networking (networking involving no central "dedicated server") inherently gives rise to (i) Quantum superposition, (ii) Quantum indeterminacy, (iii) The quantum measurement problem, (iv) Wave-particle duality, (iv) Quantum wave-function "collapse", (v) Quantum entanglement, (vi) a minimum space-time distance (e.g. the Planck length), and (vii) The relativity of time to observers.[26][27]

20.3 In fiction

Main article: Simulated reality in fiction

Simulated reality is a theme that pre-dates science fiction. In Medieval and Renaissance religious theatre, the concept of the "world as theater" is frequent. Simulated reality in fiction has been explored by many authors, game designers, and film directors.

20.4 See also

- Artificial life
- Artificial reality
- Augmented reality
- Artificial society
- Boltzmann brain
- Computational sociology
- Consensus reality
- Cyberpsychology
- Digital philosophy
- Digital physics
- The Experience Machine
- Holodeck (*Star Trek: The Next Generation*)
- Hyperreality
- Holographic Universe
- Infosphere
- Interactive online characters
- Margolus–Levitin theorem
- Maya (illusion)
- Mind uploading
- Molecular modeling
- Metaverse
- Tipler's "Omega point"
- Omnidirectional treadmill
- Philosophy of information
- Pseudorealism
- *Ready Player One*
- Reality in Buddhism
- *Simulacra and Simulation*
- Simulacrum
- Simulated reality in fiction
- Simulation hypothesis
- Social simulation
- Theory of knowledge
- Virtual economy
- Virtual Reality Addiction
- Virtual worlds
- Zeno's paradoxes

20.4.1 Major contributing thinkers

- Jean Baudrillard
- Nick Bostrom and his simulation argument
- René Descartes (1596–1650) and his Evil Demon, sometimes also called his 'Evil Genius'[28]
- Philip K. Dick for "We Can Remember It for You Wholesale"
- George Berkeley (1685–1753) and his "immaterialism" (later referred to as subjective idealism by others)
- Stanislaw Lem who presented the idea e.g. in "Further reminiscences of Ijon Tichy" (chapter I) part of The Star Diaries
- Plato (424/423 BC – 348/347 BC) and his Allegory of the Cave
- Zeno of Elea
- Zhuangzi (around the 4th century BCE) and his Chinese Butterfly Dream

20.5 Bibliography

- Copleston, Frederick (1993) [1946]. "XIX Theory of Knowledge". *A History of Philosophy, Volume I: Greece and Rome*. New York: Image Books (Doubleday). p. 160. ISBN 0-385-46843-1.

- Copleston, Frederick (1994) [1960]. "II Descartes (I)". *A History of Philosophy, Volume IV: Modern Philosophy*. New York: Image Books (Doubleday). p. 86. ISBN 0-385-47041-X.

- Deutsch, David (1997). *The Fabric of Reality*. London: Penguin Science (Allen Lane). ISBN 0-14-014690-3.

- Lloyd, Seth (2006). *Programming the Universe: A Quantum Computer Scientist Takes On the Cosmos*. Knopf. ISBN 978-1-4000-4092-6.

- Tipler, Frank (1994). *The Physics of Immortality*. Doubleday. ISBN 0-385-46799-0.

- Lem, Stanislaw (1964). *Summa Technologiae*. ISBN 3-518-37178-9.

20.6 References

[1] Moravec, Hans, Simulation, Consciousness, Existence

[2] Moravec, Hans, Platt, Charles Superhumanism

[3] Moravec, Hans Pigs in Cyberspace

[4] Are You Living in a Computer Simulation? by Nick Bostrom. July 2002. Accessed *21 December 2006*

[5] Why Running Simulations May Mean the End is Near

[6] About Mechanics of Virtual Reality, by Bin-Guang Ma, (2005)

[7] Warburg, B. (1942). The Relativity of Reality. Reflections on the Limitations of Thought and the Genesis of the Need of Causality: by René Laforgue. Translated by Anne Jouard. New York: Nervous and Mental Disease Monographs, 1940. 92 pp.. Psychoanal Q., 11:562.

[8] Bruno Marchal

[9] Russel Standish

[10] Hut, P.; Alford, M.; Tegmark, M. (2006). "On Math, Matter and Mind". *Foundations of Physics* **36**: 765–94. arXiv:physics/0510188. Bibcode:2006FoPh...36..765H. doi:10.1007/s10701-006-9048-x.

[11] "There is no logical impossibility in the supposition that the whole of life is a dream, in which we ourselves create all the objects that come before us. But although this is not logically impossible, there is no reason whatever to suppose that it is true; and it is, in fact, a less simple hypothesis, viewed as a means of accounting for the facts of our own life, than the common-sense hypothesis that there really are objects independent of us, whose action on us causes our sensations." Bertrand Russell, *The Problems of Philosophy*

[12] René Descartes, Meditations on the First Philosophy, from Descartes, The Philosophical Works of Descartes, trans. Elizabeth S. Haldane and G.R.T. Ross (Cambridge: Cambridge University Press, 1911 – reprinted with corrections 1931), Volume I, 145-46.

[13] Chalmers, J., The Matrix as Metaphysics, Department of Philosophy, University of Arizona

[14] *PHYSICS, PHILOSOPHY AND QUANTUM TECHNOLOGY*

[15] Alan Turing, *On computable numbers, with an application to the Entscheidungsproblem*, Proceedings of the London Mathematical Society, Series 2, 42 (1936), pp. 230–265. (online version). Computable numbers (and Turing machines) were introduced in this paper; the definition of computable numbers uses infinite decimal sequences.

[16] Feynman, R. The Character of Physical Law, page 57.

[17] Subjective time

[18] "But ordinary computing systems, such as Turing Machines (TM), can only take a finite number of states. Even if we combine the internal states of a TM with the content of the machine's tape to increase the number of possible states, the total number of states that a TM can be in is only countably infinite. Moreover, TMs can only follow a countable number of state space trajectories. The same point applies to any ordinary computing system of the kinds used in scientific modelling. So ordinary computational descriptions do not have a cardinality of states and state space trajectories that is sufficient for them to map onto ordinary mathematical descriptions of natural systems. Thus, from the point of view of strict mathematical description, the thesis that everything is a computing system in this second sense cannot be supported"*Computational Modelling vs. Computational Explanation: Is Everything a Turing Machine, and Does It Matter to the Philosophy of Mind?*

[19] Ord, Toby (2002). "Hypercomputation: computing more than the Turing machine". arXiv:math/0209332. Bibcode:2002math......9332O.

[20] "The Cosmology Machine takes data from billions of observations about the behaviour of stars, gases and the mysterious dark matter throughout the universe and then calculates, at ultra high speeds, how galaxies and solar systems evolved. By testing different theories of cosmic evolution it can simulate virtual universes to test which ideas come closest to explaining the real universe."*Cosmology Machine creates the Universe*

[21] Popper, K. *Science as Falsification*

[22] Deutsch, David (1998). *The Fabric of Reality: The Science of Parallel Universes—and Its Implications*. Penguin Group US. pp. 105–107. ISBN 9781101550632. My question about the ultimate limits of virtual reality can he stated like this: what constraints, if any, do the laws of physics impose on the repertoires of virtual-reality generators? ... Known physics provides no way other than free fall, even in principle, of removing an object's weight. ... Stated generally, the problem is this. To override the normal functioning of the sense organs, we must send them images resembling those that would be produced by the environment being simulated. We must also intercept and suppress the images produced by the user's actual environment. But these image manipulations are physical operations, and can be performed only by processes available in the real physical world. Light and sound can be physically absorbed and replaced fairly easily. But as I have said, that is not true of gravity: the laws of physics do not happen to permit it.

[23] Note: from the perspective of its inhabitants, a Virtual Reality is a Simulated Reality:
Deutsch, David (1998). *The Fabric of Reality: The Science of Parallel Universes—and Its Implications*. Penguin Group US. pp. 58,179. ISBN 9781101550632. Reality might consist of one person, presumably you, dreaming a lifetimes experiences. Or it might consist of just you and me. Or just the planet Earth and its inhabitants... [Living] processeses and virtual-reality renderings are, superficial differences aside, the same...

[24] Bostrom, Nick (2009). "The Simulation Argument: Some Explanations" (PDF). If each first-level ancestor-simulation run by the non-Sims requires more resources (because they contain within themselves additional second-level ancestor-simulations run by the Sims), the non-Sims might well respond by producing fewer first-level ancestor-simulations. Conversely, the cheaper it is for the non-Sims to run a simulation, the more simulations they may run. It is therefore unclear whether the total number of ancestor-simulations would be greater if Sims run ancestor-simulations than if they do not.

[25] Pooch, U.W.; Sullivan, F.J. (2000). "Recursive simulation to aid models of decisionmaking". *Simulation Conference* (Winter ed.) **1**. doi:10.1109/WSC.2000.899898. ISBN 0-7803-6579-8.

[26] Arvan, Marcus (2014). "A Unified Explanation of Quantum Phenomena? The Case for the Peer-to-Peer Simulation Hypothesis as an Interdisciplinary Research Program". *Philosophical Forum* (4 ed.) **45**. doi:10.1111/phil.12043.

[27] Arvan, Marcus (2013). "A New Theory of Free Will". *Philosophical Forum* (1 ed.) **44**. doi:10.1111/phil.12000.

[28] Chalmers, David (2005). "The Matrix as Metaphysics". In C. Grau. *Philosophers Explore the Matrix*. Oxford University Press. pp. 157–158. ISBN 9780195181067. LCCN 2004059977. Evil Genius Hypothesis: I have a disembodied mind, and an evil genius is feeding me sensory inputs to give the appearance of an external world. This is René Descartes's classical skeptical hypothesis... Dream Hypothesis: I am now and have always been dreaming. Descartes raised the question: how do you know that you are not currently dreaming? Morpheus raises a similar question: 'Have you ever had a dream, Neo, that you were so sure was real. What if you were unable to wake from that dream? How would you know the difference between the dream world and the real world?'... I think this case is analogous to the Evil Genius Hypothesis: it's just that the role of the "evil genius" is played by a part of my own cognitive system! If my dream-generating system simulates all of space-time, we have something like the original Matrix Hypothesis. p.22

20.7 External links

- *Anthropic-principle.com"* Website maintained by Nick Bostrom with a collection of papers on SR and related topics.

- *The Big Brother Universe*

- *Computer Universes and an Algorithmic Theory of Everything* by Jürgen Schmidhuber

- *The Computational Requirements for the Matrix* discussion on Slashdot.

- *Computationalism: The Very Idea*, an overview of computationalism by David Davenport.

- *The Cutting Edge of Haptics*, an article in MIT's Technology review on touch illusion technology by Duncan Graham-Rowe.

- *God Is the Machine* Wired article by Kevin Kelly.

- "That Mysterious Flow". *Davies, Paul, Sept. (2002) Scientific American* **287** (3): 40–45.

- *Simulated Universe* Paper by Brent Silby provides objections to the Simulation Argument (also published in Philosophy Now: Issue 75, 2009).

- *The Simulation Argument* Website by Nick Bostrom, Director, Future of Humanity Institute, Oxford University. Includes his original paper.

- *Simulation, Consciousness, Existence* by Hans Moravec.

- Simulism, a wiki devoted to the possibility that our reality is a simulation.

20.7. EXTERNAL LINKS

- *Superhumanism, an interview of Hans Moravec.* by Charles Platt.
- *What We Still Don't Know*, Channel 4 documentary by British Astronomer Royal Sir Martin Rees.
- *Zombies* – Philosophical zombie article by Robert Kirk in the Stanford Encyclopedia of Philosophy.

Chapter 21

Black-hole cosmology

A **black-hole cosmology** (also called **Schwarzschild cosmology** or **black-hole cosmological model**) is a cosmological model in which the observable universe is the interior of a black hole. Such models were originally proposed by theoretical physicist Raj Pathria,[1] and concurrently by mathematician I. J. Good.[2]

Any such model requires that the Hubble radius of the observable universe is equal to its Schwarzschild radius, that is, the product of its mass and the Schwarzschild proportionality constant. This is indeed known to be nearly the case; however, most cosmologists consider this close match a coincidence.[3]

In the version as originally proposed by Pathria and Good, and studied more recently by, among others, Nikodem Popławski, [4] the observable universe is the interior of a black hole existing as one of possibly many inside a larger universe, or multiverse.

According to general relativity, the gravitational collapse of a sufficiently compact mass forms a singular Schwarzschild black hole. In the Einstein-Cartan-Sciama-Kibble theory of gravity, however, it forms a regular Einstein-Rosen bridge, or wormhole. Schwarzschild wormholes and Schwarzschild black holes are different, mathematical solutions of general relativity and the Einstein–Cartan theory. Yet for distant observers, the exteriors of both solutions with the same mass are indistinguishable. The Einstein–Cartan theory extends general relativity by removing a constraint of the symmetry of the affine connection and regarding its antisymmetric part, the torsion tensor, as a dynamical variable. Torsion naturally accounts for the quantum-mechanical, intrinsic angular momentum (spin) of matter. The minimal coupling between torsion and Dirac spinors generates a repulsive spin-spin interaction which is significant in fermionic matter at extremely high densities. Such an interaction prevents the formation of a gravitational singularity. Instead, the collapsing matter reaches an enormous but finite density and rebounds, forming the other side of an Einstein-Rosen bridge, which grows as a new universe.[5] Accordingly, the Big Bang was a nonsingular Big Bounce at which the universe had a finite, minimum scale factor.[6]

21.1 References

[1] Pathria, R. K. (1972). "The Universe as a Black Hole". *Nature* **240** (5379): 298–299. Bibcode:1972Natur.240..298P. doi:10.1038/240298a0.

[2] Good, I. J. (July 1972). "Chinese universes". *Physics Today* **25** (7): 15. Bibcode:1972PhT....25g..15G. doi:10.1063/1.3070923.

[3] Landsberg, P. T. (1984). "Mass Scales and the Cosmological Coincidences". *Annalen der Physik* **496** (2): 88–92. Bibcode:1984AnP...496...88L. doi:10.1002/andp.19844960203.

[4] Popławski, N. J. (2010). "Radial motion into an Einstein-Rosen bridge". *Physics Letters B* **687** (2–3): 110–113. arXiv:0902.1994. Bibcode:2010PhLB..687..110P. doi:10.1016/j.physletb.2010.03.029.

[5] Popławski, N. J. (2010). "Cosmology with torsion: An alternative to cosmic inflation". *Physics Letters B* **694** (3): 181–185. arXiv:1007.0587. Bibcode:2010PhLB..694..181P. doi:10.1016/j.physletb.2010.09.056.

[6] Popławski, N. (2012). "Nonsingular, big-bounce cosmology from spinor-torsion coupling". *Physical Review D* **85** (10): 107502. arXiv:1111.4595. Bibcode:2012PhRvD..85j7502P. doi:10.1103/PhysRevD.85.107502.

Chapter 22

Anthropic principle

The **anthropic principle** (from Greek *anthropos*, meaning "human") is the philosophical consideration that observations of the universe must be compatible with the conscious and sapient life that observes it. Some proponents of the anthropic principle reason that it explains why the universe has the age and the fundamental physical constants necessary to accommodate conscious life. As a result, they believe it is unremarkable that the universe's fundamental constants happen to fall within the narrow range thought to be compatible with life.[1][2]

The strong anthropic principle (SAP) as explained by John D. Barrow and Frank Tipler states that this is all the case because the universe is compelled to eventually have conscious and sapient life emerge within it. Some critics of the SAP argue in favor of a weak anthropic principle (WAP) similar to the one defined by Brandon Carter, which states that the universe's ostensible fine tuning is the result of selection bias: i.e., only in a universe capable of eventually supporting life will there be living beings capable of observing and reflecting upon fine tuning. Most often such arguments draw upon some notion of the multiverse for there to be a statistical population of universes to select from and from which selection bias (our observance of *only* this universe, compatible with life) could occur.

22.1 Definition and basis

The principle was formulated as a response to a series of observations that the laws of nature and parameters of the universe take on values that are consistent with conditions for life as we know it rather than a set of values that would not be consistent with life on Earth. The anthropic principle states that this is a necessity, because if life were impossible, no living entity would be there to observe it, and thus would not be known. That is, it must be possible to observe *some* universe, and hence, the laws and constants of any such universe must accommodate that possibility.

The term *anthropic* in "anthropic principle" has been argued[3] to be a misnomer.[4] While singling out our kind of carbon-based life, none of the finely tuned phenomena require human life or some kind of carbon chauvinism.[5][6] Any form of life or any form of heavy atom, stone, star or galaxy would do; nothing specifically human or anthropic is involved.

The anthropic principle has given rise to some confusion and controversy, partly because the phrase has been applied to several distinct ideas. All versions of the principle have been accused of discouraging the search for a deeper physical understanding of the universe. The anthropic principle is often criticized for lacking falsifiability and therefore critics of the anthropic principle may point out that the anthropic principle is a non-scientific concept, even though the weak anthropic principle, *"conditions that are observed in the universe must allow the observer to exist"*,[7] is "easy" to support in mathematics and philosophy, i.e. it is a tautology or truism. However, building a substantive argument based on a tautological foundation is problematic. Stronger variants of the anthropic principle are not tautologies and thus make claims considered controversial by some and that are contingent upon empirical verification.[8][9]

22.2 Anthropic coincidences

Main article: Fine-tuned Universe

In 1961, Robert Dicke noted that the age of the universe, as seen by living observers, cannot be random.[10] Instead, biological factors constrain the universe to be more or less in a "golden age," neither too young nor too old.[11] If the universe were one tenth as old as its present age, there would not have been sufficient time to build up appreciable levels of metallicity (levels of elements besides hydrogen and helium) especially carbon, by nucleosynthesis. Small rocky planets did not yet exist. If the universe were 10 times older than it actually is, most stars would be too old to remain on the main sequence and would have turned into white dwarfs, aside from the dimmest red dwarfs, and stable planetary systems would have already come to an end. Thus, Dicke explained the coincidence between large dimensionless numbers constructed from the constants of physics and the age of the universe, a coincidence which had inspired Dirac's varying-G theory.

Dicke later reasoned that the density of matter in the universe must be almost exactly the critical density needed to prevent the Big Crunch (the "Dicke coincidences" argument). The most recent measurements may suggest that the observed density of baryonic matter, and some theoretical predictions of the amount of dark matter account for about 30% of this critical density, with the rest contributed by a cosmological constant. Steven Weinberg[12] gave an anthropic explanation for this fact: he noted that the cosmological constant has a remarkably low value, some 120 orders of magnitude smaller than the value particle physics predicts (this has been described as the "worst prediction in physics").[13] However, if the cosmological constant were only one order of magnitude larger than its observed value, the universe would suffer catastrophic inflation, which would preclude the formation of stars, and hence life.

The observed values of the dimensionless physical constants (such as the fine-structure constant) governing the four fundamental interactions are balanced as if fine-tuned to permit the formation of commonly found matter and subsequently the emergence of life.[14] A slight increase in the strong interaction would bind the dineutron and the diproton, and nuclear fusion would have converted all hydrogen in the early universe to helium. Water, as well as sufficiently long-lived stable stars, both essential for the emergence of life as we know it, would not exist. More generally, small changes in the relative strengths of the four fundamental interactions can greatly affect the universe's age, structure, and capacity for life.

22.3 Origin

The phrase "anthropic principle" first appeared in Brandon Carter's contribution to a 1973 Kraków symposium honouring Copernicus's 500th birthday. Carter, a theoretical astrophysicist, articulated the Anthropic Principle in reaction to the Copernican Principle, which states that humans do not occupy a privileged position in the Universe. As Carter said: "Although our situation is not necessarily *central*, it is inevitably privileged to some extent."[15] Specifically, Carter disagreed with using the Copernican principle to justify the Perfect Cosmological Principle, which states that all large regions *and times* in the universe must be statistically identical. The latter principle underlay the steady-state theory, which had recently been falsified by the 1965 discovery of the cosmic microwave background radiation. This discovery was unequivocal evidence that the universe has changed radically over time (for example, via the Big Bang).

Carter defined two forms of the anthropic principle, a "weak" one which referred only to anthropic selection of privileged spacetime locations in the universe, and a more controversial "strong" form which addressed the values of the fundamental constants of physics.

Roger Penrose explained the weak form as follows:

> The argument can be used to explain why the conditions happen to be just right for the existence of (intelligent) life on the Earth at the present time. For if they were not just right, then we should not have found ourselves to be here now, but somewhere else, at some other appropriate time. This principle was used very effectively by Brandon Carter and Robert Dicke to resolve an issue that had puzzled physicists for a good many years. The issue concerned various striking numerical relations that are observed to hold between the physical constants (the gravitational constant, the mass of the proton, the age of the universe, etc.). A puzzling aspect of this was that some of the relations hold only at the present epoch in the Earth's history, so we appear, coincidentally, to be living at a very special time (give or take a few million years!). This was later explained, by Carter and Dicke, by the fact that this epoch coincided with the lifetime of what are called main-sequence stars, such as the Sun. At any other epoch, so the argument ran, there would be no intelligent life around in order to measure the physical constants in question — so the coincidence had to hold, simply because there would be intelligent life around only at the particular time that the coincidence did hold!
>
> — *The Emperor's New Mind*, Chapter 10

One reason this is plausible is that there are many other places and times in which we can imagine finding ourselves. But when applying the strong principle, we only have one universe, with one set of fundamental parameters, so what exactly is the point being made? Carter offers two possibilities: First, we can use our own existence to make "predictions" about the parameters. But second, "as a last resort", we can convert these predictions into *explanations* by assuming that there *is* more than one universe, in fact a large and possibly infinite collection of universes, something that is now called the multiverse ("world ensemble" was Carter's term), in which the parameters (and perhaps the laws of physics) vary across universes. The strong principle then becomes an example of a selection effect, exactly analogous to the weak principle. Postulating a multiverse is certainly a radical step, but taking it could provide at least a partial answer to a question which had seemed to be out of the reach of normal science: "why do the fundamental laws of physics take the particular form we observe and not another?"

Since Carter's 1973 paper, the term "anthropic principle" has been extended to cover a number of ideas which differ in important ways from those he espoused. Particular confusion was caused in 1986 by the book *The Anthropic Cosmological Principle* by John D. Barrow and Frank Tipler,[16] published that year which distinguished between "weak" and "strong" anthropic principle in a way very different from Carter's, as discussed in the next section.

Carter was not the first to invoke some form of the anthropic principle. In fact, the evolutionary biologist Alfred Russel Wallace anticipated the anthropic principle as long ago as 1904: "Such a vast and complex universe as that which we know exists around us, may have been absolutely required ... in order to produce a world that should be precisely adapted in every detail for the orderly development of life culminating in man."[17] In 1957, Robert Dicke wrote: "The age of the Universe 'now' is not random but conditioned by biological factors ... [changes in the values of the fundamental constants of physics] would preclude the existence of man to consider the problem."[18]

22.4 Variants

Weak anthropic principle (WAP) (Carter): "we must be prepared to take account of the fact that our location in the universe is *necessarily* privileged to the extent of being compatible with our existence as observers." Note that for Carter, "location" refers to our location in time as well as space.

Strong anthropic principle (SAP) (Carter): "the universe (and hence the fundamental parameters on which it depends) must be such as to admit the creation of observers within it at some stage. To paraphrase Descartes, *cogito ergo mundus talis est.*"
The Latin tag ("I think, therefore the world is such [as it is]") makes it clear that "must" indicates a deduction from the fact of our existence; the statement is thus a truism.

In their 1986 book, *The Anthropic Cosmological Principle*, John Barrow and Frank Tipler depart from Carter and define the WAP and SAP as follows:[19][20]

Weak anthropic principle (WAP) (Barrow and Tipler): "The observed values of all physical and cosmological quantities are not equally probable but they take on values restricted by the requirement that there exist sites where carbon-based life can evolve and by the requirements that the universe be old enough for it to have already done so."[21]

Unlike Carter they restrict the principle to carbon-based life, rather than just "observers." A more important difference is that they apply the WAP to the fundamental physical constants, such as the fine structure constant, the number of spacetime dimensions, and the cosmological constant — topics that fall under Carter's SAP.

Strong anthropic principle (SAP) (Barrow and Tipler): "The Universe must have those properties which allow life to develop within it at some stage in its history."[22]
This looks very similar to Carter's SAP, but unlike the case with Carter's SAP, the "must" is an imperative, as shown by the following three possible elaborations of the SAP, each proposed by Barrow and Tipler:[23]

- "There exists one possible Universe 'designed' with the goal of generating and sustaining 'observers'."

  This can be seen as simply the classic design argument restated in the garb of contemporary cosmology. It implies that the purpose of the universe is to give rise to intelligent life, with the laws of nature and their fundamental physical constants set to ensure that life as we know it will emerge and evolve.

- "Observers are necessary to bring the Universe into being."

  Barrow and Tipler believe that this is a valid conclusion from quantum mechanics, as John Archibald Wheeler has suggested, especially via his idea that information is the fundamental reality, see It from bit, and his **Participatory Anthropic Principle (PAP)** which is an interpretation of quantum mechanics associated with the ideas of John von Neumann and Eugene Wigner.

- "An ensemble of other different universes is necessary for the existence of our Universe."

  By contrast, Carter merely says that an ensemble of universes is necessary for the SAP to count as an explanation.

Modified anthropic principle (MAP) (Schmidhuber): The 'problem' of existence is only relevant to a species capable of formulating the question. Prior to *Homo sapiens* intellectual evolution to the point where the nature of the observed universe - and humans' place within same - spawned deep inquiry into its origins, the 'problem' simply did not exist.[24]

The philosophers John Leslie[25] and Nick Bostrom[26] reject the Barrow and Tipler SAP as a fundamental misreading of Carter. For Bostrom, Carter's anthropic principle just warns us to make allowance for **anthropic bias**, that is, the bias created by anthropic selection effects (which Bostrom calls "observation" selection effects) — the necessity for observers to exist in order to get a result. He writes:

> Many 'anthropic principles' are simply confused. Some, especially those drawing inspiration from Brandon Carter's seminal papers, are sound, but... they are too weak to do any real scientific work. In particular, I argue that existing methodology does not permit any observational consequences to be derived from contemporary cosmological theories, though these theories quite plainly can be and are being tested empirically by astronomers. What is needed to bridge this methodological gap is a more adequate formulation of how observation selection effects are to be taken into account.
> — *Anthropic Bias*, Introduction[27]

Strong self-sampling assumption (SSSA) (Bostrom): "Each observer-moment should reason as if it were randomly selected from the class of all observer-moments in its reference class."

Analysing an observer's experience into a sequence of "observer-moments" helps avoid certain paradoxes; but the main ambiguity is the selection of the appropriate "reference class": for Carter's WAP this might correspond to all real or potential observer-moments in our universe; for the SAP, to all in the multiverse. Bostrom's mathematical development shows that choosing either too broad or too narrow a reference class leads to counter-intuitive results, but he is not able to prescribe an ideal choice.

According to Jürgen Schmidhuber, the anthropic principle essentially just says that the conditional probability of finding yourself in a universe compatible with your existence is always 1. It does not allow for any additional nontrivial predictions such as "gravity won't change tomorrow." To gain more predictive power, additional assumptions on the prior distribution of alternative universes are necessary.[24][28]

Playwright and novelist Michael Frayn describes a form of the Strong Anthropic Principle in his 2006 book *The Human Touch*, which explores what he characterises as "the central oddity of the Universe":

> It's this simple paradox. The Universe is very old and very large. Humankind, by comparison, is only a tiny disturbance in one small corner of it - and a very recent one. Yet the Universe is only very large and very old because we are here to say it is... And yet, of course, we all know perfectly well that it is what it is whether we are here or not.[29]

22.5 Character of anthropic reasoning

Carter chose to focus on a tautological aspect of his ideas, which has resulted in much confusion. In fact, anthropic reasoning interests scientists because of something that is only implicit in the above formal definitions, namely that we should give serious consideration to there being other universes with different values of the "fundamental parameters" — that is, the dimensionless physical constants and initial conditions for the Big Bang. Carter and others have argued that life as we know it would not be possible in most such universes. In other words, the universe we are in is fine tuned to permit life. Collins & Hawking (1973) characterized Carter's then-unpublished big idea as the postulate that "there is not one universe but a whole infinite ensemble of universes with all possible initial conditions".[30] If this is granted, the anthropic principle provides a plausible explanation for the fine tuning of our universe: the "typical" universe is not fine-tuned, but given enough universes, a small fraction thereof will be capable of supporting intelligent life. Ours must be one of these, and so the observed fine tuning should be no cause for wonder.

Although philosophers have discussed related concepts for centuries, in the early 1970s the only genuine physical theory yielding a multiverse of sorts was the many-worlds interpretation of quantum mechanics. This would allow variation in initial conditions, but not in the truly fundamental constants. Since that time a number of mechanisms for producing a multiverse have been suggested: see the review by Max Tegmark.[31] An important development in the 1980s was the combination of inflation theory with the hypothesis that some parameters are determined by symmetry breaking in the early universe, which allows parameters previously thought of as "fundamental constants" to vary over very large distances, thus eroding the distinction between Carter's weak and strong principles. At the beginning of the 21st century, the string landscape emerged as a mechanism for varying essentially all the constants, including the number of spatial dimensions.[32]

The anthropic idea that fundamental parameters are selected from a multitude of different possibilities (each actual in some universe or other) contrasts with the traditional hope of physicists for a theory of everything having no free parameters: as Einstein said, "What really interests me is whether God had any choice in the creation of the world." In 2002, proponents of the leading candidate for a "theory of everything", string theory, proclaimed "the end of the anthropic principle"[33] since there would be no free parameters to select. Ironically, string theory now seems to offer no hope of predicting fundamental parameters, and now some who advocate it invoke the anthropic principle as well (see below).

The modern form of a design argument is put forth by Intelligent design. Proponents of intelligent design often cite the fine-tuning observations that (in part) preceded the formulation of the anthropic principle by Carter as a proof of an intelligent designer. Opponents of intelligent design are not limited to those who hypothesize that other universes exist; they may also argue, anti-anthropically, that the universe is less fine-tuned than often claimed, or that accepting fine tuning as a brute fact is less astonishing than the idea of an intelligent creator. Furthermore, even accepting fine tuning, Sober (2005)[34] and Ikeda and Jefferys,[35][36] argue that the Anthropic Principle as conventionally stated actually undermines intelligent design; see fine-tuned universe.

Paul Davies's book *The Goldilocks Enigma* (2006) reviews the current state of the fine tuning debate in detail, and concludes by enumerating the following responses to that debate:

1. The absurd universe: Our universe just happens to be the way it is.

2. The unique universe: There is a deep underlying unity in physics which necessitates the Universe being the way it is. Some Theory of Everything will explain why the various features of the Universe must have exactly the values that we see.

3. The multiverse: Multiple universes exist, having all possible combinations of characteristics, and we inevitably find ourselves within a universe that allows us to exist.

4. Intelligent Design: A creator designed the Universe with the purpose of supporting complexity and the emergence of intelligence.

5. The life principle: There is an underlying principle that constrains the Universe to evolve towards life and mind.

6. The self-explaining universe: A closed explanatory or causal loop: "perhaps only universes with a capacity for consciousness can exist." This is Wheeler's Participatory Anthropic Principle (PAP).

7. The fake universe: We live inside a virtual reality simulation.

Omitted here is Lee Smolin's model of cosmological natural selection, also known as "fecund universes," which proposes that universes have "offspring" which are more plentiful if they resemble our universe. Also see Gardner (2005).[37]

Clearly each of these hypotheses resolve some aspects of the puzzle, while leaving others unanswered. Followers of Carter would admit only option 3 as an anthropic explanation, whereas 3 through 6 are covered by different versions of Barrow and Tipler's SAP (which would also include 7 if it is considered a variant of 4, as in Tipler 1994).

The anthropic principle, at least as Carter conceived it, can be applied on scales much smaller than the whole universe. For example, Carter (1983)[38] inverted the usual line of reasoning and pointed out that when interpreting the evolutionary record, one must take into account cosmological and astrophysical considerations. With this in mind, Carter concluded that given the best estimates of the age of the universe, the evolutionary chain culminating in *Homo sapiens* probably admits only one or two low probability links. Antonio Feoli and Salvatore Rampone dispute this conclusion, arguing instead that the estimated size of our universe and the number of planets in it allows for a higher bound, so that there is no need to invoke intelligent design to explain evolution.[39]

22.6 Observational evidence

No possible observational evidence bears on Carter's WAP, as it is merely advice to the scientist and asserts nothing debatable. The obvious test of Barrow's SAP, which says that the universe is "required" to support life, is to find evidence of life in universes other than ours. Any other universe is, by most definitions, unobservable (otherwise it would be included in *our* portion of *this* universe). Thus, in principle Barrow's SAP cannot be falsified by observing a universe in which an observer cannot exist.

Philosopher John Leslie[40] states that the Carter SAP (with multiverse) predicts the following:

- Physical theory will evolve so as to strengthen the hypothesis that early phase transitions occur probabilistically rather than deterministically, in which case there will be no deep physical reason for the values of fundamental constants;

- Various theories for generating multiple universes will prove robust;
- Evidence that the universe is fine tuned will continue to accumulate;
- No life with a non-carbon chemistry will be discovered;
- Mathematical studies of galaxy formation will confirm that it is sensitive to the rate of expansion of the universe.

Hogan[41] has emphasised that it would be very strange if all fundamental constants were strictly determined, since this would leave us with no ready explanation for apparent fine tuning. In fact we might have to resort to something akin to Barrow and Tipler's SAP: there would be no option for such a universe *not* to support life.

Probabilistic predictions of parameter values can be made given:

1. a particular multiverse with a "measure", i.e. a well defined "density of universes" (so, for parameter X, one can calculate the prior probability $P(X_0)\, dX$ that X is in the range $X_0 < X < X_0 + dX$), and
2. an estimate of the number of observers in each universe, $N(X)$ (e.g., this might be taken as proportional to the number of stars in the universe).

The probability of observing value X is then proportional to $N(X)\, P(X)$. A generic feature of an analysis of this nature is that the expected values of the fundamental physical constants should not be "over-tuned," i.e. if there is some perfectly tuned predicted value (e.g. zero), the observed value need be no closer to that predicted value than what is required to make life possible. The small but finite value of the cosmological constant can be regarded as a successful prediction in this sense.

One thing that would *not* count as evidence for the Anthropic Principle is evidence that the Earth or the solar system occupied a privileged position in the universe, in violation of the Copernican principle (for possible counterevidence to this principle, see Copernican principle), unless there was some reason to think that that position was a necessary condition for our existence as observers.

22.7 Applications of the principle

22.7.1 The nucleosynthesis of carbon-12

Fred Hoyle may have invoked anthropic reasoning to predict an astrophysical phenomenon. He is said to have reasoned from the prevalence on Earth of life forms whose chemistry was based on carbon-12 atoms, that there must be an undiscovered resonance in the carbon-12 nucleus facilitating its synthesis in stellar interiors via the triple-alpha process. He then calculated the energy of this undiscovered resonance to be 7.6 million electron-volts.[42][43] Willie Fowler's research group soon found this resonance, and its measured energy was close to Hoyle's prediction.

However, a recently released paper argues that Hoyle did not use anthropic reasoning to make this prediction.[44]

22.7.2 Cosmic inflation

Main article: Cosmic inflation

Don Page criticized the entire theory of cosmic inflation as follows.[45] He emphasized that initial conditions which made possible a thermodynamic arrow of time in a universe with a Big Bang origin, must include the assumption that at the initial singularity, the entropy of the universe was low and therefore extremely improbable. Paul Davies rebutted this criticism by invoking an inflationary version of the anthropic principle.[46] While Davies accepted the premise that the initial state of the visible universe (which filled a microscopic amount of space before inflating) had to possess a very low entropy value — due to random quantum fluctuations — to account for the observed thermodynamic arrow of time, he deemed this fact an advantage for the theory. That the tiny patch of space from which our observable universe grew had to be extremely orderly, to allow the post-inflation universe to have an arrow of time, makes it unnecessary to adopt any "ad hoc" hypotheses about the initial entropy state, hypotheses other Big Bang theories require.

22.7.3 String theory

Main article: String theory landscape

String theory predicts a large number of possible universes, called the "backgrounds" or "vacua." The set of these vacua is often called the "multiverse" or "anthropic landscape" or "string landscape." Leonard Susskind has argued that the existence of a large number of vacua puts anthropic reasoning on firm ground: only universes whose properties are such as to allow observers to exist are observed, while a possibly much larger set of universes lacking such properties go unnoticed.

Steven Weinberg[47] believes the Anthropic Principle may be appropriated by cosmologists committed to nontheism, and refers to that Principle as a "turning point" in modern science because applying it to the string landscape "...may explain how the constants of nature that we observe can take values suitable for life without being fine-tuned by a benevolent creator." Others, most notably David Gross but also Lubos Motl, Peter Woit, and Lee Smolin, argue that this is not predictive. Max Tegmark,[48] Mario Livio, and Martin Rees[49] argue that only some aspects of a physical theory need be observable and/or testable for the theory to be accepted, and that many well-accepted theories are far from completely testable at present.

Jürgen Schmidhuber (2000–2002) points out that Ray Solomonoff's theory of universal inductive inference and its extensions already provide a framework for maximizing our confidence in any theory, given a limited sequence of physical observations, and some prior distribution on the set of possible explanations of the universe.

22.7.4 Spacetime

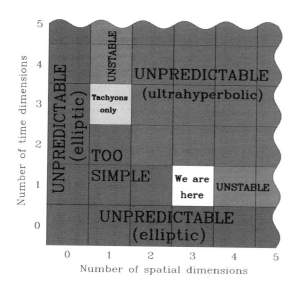

Properties of n+m-*dimensional spacetimes*

Main article: Spacetime

There are two kinds of dimensions, spatial (bidirectional) and temporal (unidirectional). Let the number of spatial dimensions be N and the number of temporal dimensions be T. That $N = 3$ and $T = 1$, setting aside the compactified dimensions invoked by string theory and undetectable to date, can be explained by appealing to the physical consequences of letting N differ from 3 and T differ from 1. The argument is often of an anthropic character and possibly the first of its kind, albeit before the complete concept came into vogue. Immanuel Kant argued that 3-dimensional space was a consequence of the inverse square law of universal gravitation. While Kant's argument is historically important, John D. Barrow says that it "...gets the punch-line back to front: it is the three-dimensionality of space that explains why we see inverse-square force laws in Nature, not vice-versa." (Barrow 2002: 204). This is because the law of gravitation (or any other inverse-square law) follows from the concept of flux and the proportional relationship of flux density and the strength of field. If $N = 3$, then 3-dimensional solid objects have surface areas proportional to the square of their size in any selected spatial dimension. In particular, a sphere of radius r has area of $4\pi r^2$. More generally, in a space of N dimensions, the strength of the gravitational attraction between two bodies separated by a distance of r would be inversely proportional to r^{N-1}.

In 1920, Paul Ehrenfest showed that if there is only one time dimension and greater than three spatial dimensions, the orbit of a planet about its Sun cannot remain stable. The same is true of a star's orbit around the center of its galaxy.[50] Ehrenfest also showed that if there are an even number of spatial dimensions, then the different parts of a wave impulse will travel at different speeds. If there are $5 + 2k$ spatial dimensions, where k is a whole number, then wave impulses become distorted. In 1922, Hermann Weyl showed that Maxwell's theory of electromagnetism works only with three dimensions of space and one of time.[51] Finally, Tangherlini showed in 1963 that when there are more than three spatial dimensions, electron orbitals around nuclei cannot be stable; electrons would either fall into the nucleus or disperse.[52]

Max Tegmark expands on the preceding argument in the following anthropic manner.[53] If T differs from 1, the behavior of physical systems could not be predicted reliably from knowledge of the relevant partial differential equations. In such a universe, intelligent life capable of manipulating technology could not emerge. Moreover, if $T > 1$, Tegmark maintains that protons and electrons would be unstable and could decay into particles having greater mass than themselves. (This is not a problem if the particles have a sufficiently low temperature).

22.8 The Anthropic Cosmological Principle

A thorough extant study of the anthropic principle is the book *The Anthropic Cosmological Principle* by John D. Barrow, a cosmologist, and Frank J. Tipler, a cosmologist and mathematical physicist. This book sets out in detail the many known anthropic coincidences and constraints, including many found by its authors. While the book is primarily a work of theoretical astrophysics, it also touches on quantum physics, chemistry, and earth science. An entire chapter argues that *Homo sapiens* is, with high probability, the only intelligent species in the Milky Way.

The book begins with an extensive review of many topics in the history of ideas the authors deem relevant to the anthropic principle, because the authors believe that principle has important antecedents in the notions of teleology and intelligent design. They discuss the writings of Fichte, Hegel, Bergson, and Alfred North Whitehead, and the Omega Point cosmology of Teilhard de Chardin. Barrow and Tipler carefully distinguish teleological reasoning from *eutaxiological* reasoning; the former asserts that order must have a consequent purpose; the latter asserts more modestly that order must have a planned cause. They attribute this important but nearly always overlooked distinction to an obscure 1883 book by L. E. Hicks.[54]

Seeing little sense in a principle requiring intelligent life to emerge while remaining indifferent to the possibility of its eventual extinction, Barrow and Tipler propose the final anthropic principle (FAP): Intelligent information-processing must come into existence in the universe, and, once it comes into existence, it will never die out.[55]

Barrow and Tipler submit that the FAP is both a valid physical statement and "closely connected with moral values." FAP places strong constraints on the structure of the universe, constraints developed further in Tipler's *The Physics of Immortality*.[56] One such constraint is that the universe must end in a big crunch, which seems unlikely in view of the tentative conclusions drawn since 1998 about dark energy, based on observations of very distant supernovas.

In his review[57] of Barrow and Tipler, Martin Gardner ridiculed the FAP by quoting the last two sentences of their book as defining a Completely Ridiculous Anthropic Principle (CRAP):

> At the instant the Omega Point is reached, life will have gained control of *all* matter and forces not only in a single universe, but in all universes whose existence is logically possible; life will have spread into *all* spatial regions in all universes which could logically exist, and will have stored an infinite amount of information, including *all* bits of knowledge which it is logically possible to know. And this is the end.[58]

22.9 Criticisms

Carter has frequently regretted his own choice of the word "anthropic," because it conveys the misleading impression that the principle involves humans specifically, rather than intelligent observers in general.[59] Others[60] have criticised the word "principle" as being too grandiose to describe straightforward applications of selection effects.

A common criticism of Carter's SAP is that it is an easy deus ex machina which discourages searches for physical explanations. To quote Penrose again: "it tends to be invoked by theorists whenever they do not have a good enough theory to explain the observed facts."[61]

Carter's SAP and Barrow and Tipler's WAP have been dismissed as truisms or trivial tautologies, that is, statements true solely by virtue of their logical form (the conclusion is identical to the premise) and not because a substantive claim is made and supported by observation of reality. As such, they are criticized as an elaborate way of saying "if things were different, they would be different," which is a valid statement, but does not make a claim of some factual alternative over another.

Critics of the Barrow and Tipler SAP claim that it is neither testable nor falsifiable, and thus is not a scientific statement but rather a philosophical one. The same criticism has been leveled against the hypothesis of a multiverse, although some argue that it does make falsifiable predictions. A modified version of this criticism is that we understand so little about the emergence of life, especially intelligent life, that it is effectively impossible to calculate the number of observers in each universe. Also, the prior distribution of universes as a function of the fundamental constants is easily modified to get any desired result.[62]

Many criticisms focus on versions of the strong anthropic principle, such as Barrow and Tipler's *anthropic cosmological principle*, which are teleological notions that tend to describe the existence of life as a *necessary prerequisite* for the observable constants of physics. Similarly, Stephen Jay Gould,[63][64] Michael Shermer,[65] and others claim that the stronger versions of the anthropic principle seem to reverse known causes and effects. Gould compared the claim that the universe is fine-tuned for the benefit of our kind of life to saying that sausages were made long and narrow so that they could fit into modern hotdog buns, or saying that ships had been invented to house barnacles. These critics cite the vast physical, fossil, genetic, and other biological evidence consistent with life having been fine-tuned through natural selection to adapt to the physical and geophysical environment in which life exists. Life appears to have adapted to the universe, and not vice versa.

Some applications of the anthropic principle have been criticized as an argument by lack of imagination, for tacitly assuming that carbon compounds and water are the only possible chemistry of life (sometimes called "carbon chauvinism", see also alternative biochemistry).[66] The range of fundamental physical constants consistent with the evolution of carbon-based life may also be wider than those who advocate a fine tuned universe have argued.[67] For instance, Harnik et al.[68] propose a weakless universe in which the weak nuclear force is eliminated. They show that this has no significant effect on the other fundamental interactions, provided some adjustments are made in how those interactions work. However, if some of the fine-tuned details of our universe were violated, that would rule out complex structures of any kind — stars, planets, galaxies, etc.

Lee Smolin has offered a theory designed to improve on the lack of imagination that anthropic principles have been accused of. He puts forth his fecund universes theory, which assumes universes have "offspring" through the creation of black holes whose offspring universes have values of physical constants that depend on those of the mother universe.[69]

Some versions of the anthropic principle are only interesting if the range of physical constants that allow certain kinds of life are unlikely in a landscape of possible universes. But Lee Smolin assumes that conditions for carbon based life are similar to conditions for black hole creation, which would change the a priori distribution of universes such that universes containing life would be likely. In Smolin vs. Susskind: The Anthropic Principle[70] the string theorist Leonard Susskind disagrees about some assumptions in Lee Smolin's theory, while Smolin defends his theory.

The philosophers of cosmology John Earman,[71] Ernan McMullin,[72] and Jesús Mosterín contend that "in its weak version, the anthropic principle is a mere tautology, which does not allow us to explain anything or to predict anything that we did not already know. In its strong version, it is a gratuitous speculation".[73] A further criticism by Mosterín concerns the flawed "anthropic" inference from the assumption of an infinity of worlds to the existence of one like ours:

> The suggestion that an infinity of objects characterized by certain numbers or properties implies the existence among them of objects with any combination of those numbers or characteristics [...] is mistaken. An infinity does not imply at all that any arrangement is present or repeated. [...] The assumption that all possible worlds are realized in an infinite universe is equivalent to the assertion that any infinite set of numbers contains all numbers (or at least all Gödel numbers of the [defining] sequences), which is obviously false.

22.10 See also

- Big Bounce
- Boltzmann brain
- Doomsday argument
- Goldilocks principle
- The Great Filter
- Infinite monkey theorem
- Inverse gambler's fallacy
- Mediocrity principle
- Metaphysical naturalism
- Neocatastrophism
- Puddle thinking
- Quark mass and congeniality to life
- Rare Earth hypothesis
- Triple-alpha process
- Teleology

22.11 Footnotes

[1] Anthropic Principle

[2] James Schombert, Department of Physics at University of Oregon

[3] Mosterín J., (2005), *Antropic Explanations in Cosmology*, in Hajek, Valdés & Westerstahl (eds.), *Proceedings of the 12th International Congress of Logic, Methodology and Philosophy of Science*; http://philsci-archive.pitt.edu/1658/"

[4] "anthropic" means "of or pertaining to mankind or humans"

[5] The Anthropic Principle, Victor J. Stenger

[6] Anthropic Bias, Nick Bostrom, p.6

[7] Merriam-Webster Online Dictionary

[8] The Strong Anthropic Principle and the Final Anthropic Principle

[9] On Knowing, Sagan from Pale Blue Dot

[10] Dicke, R. H. (1961). "Dirac's Cosmology and Mach's Principle". *Nature* 192 (4801): 440–441. Bibcode:1961Natur.192..440D. doi:10.1038/192440a0.

[11] Davies, P. (2006). *The Goldilocks Enigma*. Allen Lane. ISBN 0-7139-9883-0.

[12] Weinberg, S. (1987). "Anthropic bound on the cosmological constant". *Physical Review Letters* 59 (22): 2607–2610. Bibcode:1987PhRvL..59.2607W. doi:10.1103/PhysRevLett.59.2607. PMID 10035596.

[13] New Scientist Space Blog: Physicists debate the nature of space-time - New Scientist

[14] How Many Fundamental Constants Are There? John Baez, mathematical physicist. U. C. Riverside, April 22, 2011

[15] Carter, B. (1974). "Large Number Coincidences and the Anthropic Principle in Cosmology". *IAU Symposium 63: Confrontation of Cosmological Theories with Observational Data*. Dordrecht: Reidel. pp. 291–298.; republished in General Relativity and Gravitation (Nov. 2011), Vol. 43, Iss. 11, p. 3225-3233, with an introduction by George Ellis (available on Arxiv

[16] Barrow, John D.; Tipler, Frank J. (1988). *The Anthropic Cosmological Principle*. Oxford University Press. ISBN 978-0-19-282147-8. LCCN 87028148.

[17] Wallace, A. R. (1904). *Man's place in the universe: a study of the results of scientific research in relation to the unity or plurality of worlds, 4th ed.* London: George Bell & Sons. pp. 256–7.

[18] Dicke, R. H. (1957). "Gravitation without a Principle of Equivalence". *Reviews of Modern Physics* **29** (3): 363–376. Bibcode:1957RvMP...29..363D. doi:10.1103/RevModPhys.29.363.

[19] Barrow, John D. (1997). "Anthropic Definitions". *Quarterly Journal of the Royal Astronomical Society* **24**: 146–53. Bibcode:1983QJRAS..24..146B.

[20] Barrow & Tipler's definitions are quoted verbatim at *Genesis of Eden Diversity Encyclopedia*.

[21] Barrow and Tipler 1986: 16.

[22] Barrow and Tipler 1986: 21.

[23] Barrow and Tipler 1986: 22.

[24] Jürgen Schmidhuber, 2000, "Algorithmic theories of everything."

[25] Leslie, J. (1986). "Probabilistic Phase Transitions and the Anthropic Principle". *Origin and Early History of the Universe: LIEGE 26*. Knudsen. pp. 439–444.

[26] Bostrom, N. (2002). *Anthropic Bias: Observation Selection Effects in Science and Philosophy*. Routledge. ISBN 0-415-93858-9. 5 chapters available online.

[27] Bostrom, N. (2002), op. cit.

[28] Jürgen Schmidhuber, 2002, "The Speed Prior: A New Simplicity Measure Yielding Near-Optimal Computable Predictions." *Proc. 15th Annual Conference on Computational Learning Theory* (COLT 2002), Sydney, Australia, Lecture Notes in Artificial Intelligence. Springer: 216-28.

[29] Michael Frayn, *The Human Touch*. Faber & Faber ISBN 0-571-23217-5

[30] Collins C. B., Hawking, S. W. (1973). "Why is the universe isotropic?". *Astrophysical Journal* **180**: 317–334. Bibcode:1973ApJ...180..317C. doi:10.1086/151965.

[31] Tegmark, M. (1998). "Is 'the theory of everything' merely the ultimate ensemble theory?". *Annals of Physics* **270**: 1–51. arXiv:gr-qc/9704009. Bibcode:1998AnPhy.270....1T. doi:10.1006/aphy.1998.5855.

[32] Strictly speaking, the number of non-compact dimensions, see String theory.

[33] Kane, Gordon L., Perry, Malcolm J., and Zytkow, Anna N. (2002). "The Beginning of the End of the Anthropic Principle". *New Astronomy* **7**: 45–53. arXiv:astro-ph/0001197. Bibcode:2002NewA....7...45K. doi:10.1016/S1384-1076(01)00088-4.

[34] Sober, Elliott, 2005, "The Design Argument" in Mann, W. E., ed., *The Blackwell Guide to the Philosophy of Religion*. Blackwell Publishers. Archived January 4, 2015 at the Wayback Machine

[35] Ikeda, M. and Jefferys, W., "The Anthropic Principle Does Not Support Supernaturalism," in *The Improbability of God*, Michael Martin and Ricki Monnier, Editors, pp. 150-166. Amherst, N.Y.: Prometheus Press. ISBN 1-59102-381-5

[36] Ikeda, M. and Jefferys, W. (2006). Unpublished FAQ "The Anthropic Principle Does Not Support Supernaturalism."

[37] Gardner, James N., 2005, "The Physical Constants as Biosignature: An anthropic retrodiction of the Selfish Biocosm Hypothesis," *International Journal of Astrobiology*.

[38] Carter, B.; McCrea, W. H. (1983). "The anthropic principle and its implications for biological evolution". *Philosophical Transactions of the Royal Society* **A310** (1512): 347–363. Bibcode:1983RSPTA.310..347C. doi:10.1098/rsta.1983.0096.

[39] Feoli, A. and Rampone, S.; Rampone (1999). "Is the Strong Anthropic Principle too weak?". *Nuovo Cim.* **B114**: 281–289. arXiv:gr-qc/9812093. Bibcode:1999NCimB.114..281F.

[40] Leslie, J. (1986) op. cit.

[41] Hogan, Craig (2000). "Why is the universe just so?". *Reviews of Modern Physics* **72** (4): 1149–1161. arXiv:astro-ph/9909295. Bibcode:2000RvMP...72.1149H. doi:10.1103/RevModPhys.72.1149.

[42] University of Birmingham Life, Bent Chains and the Anthropic Principle Archived March 14, 2014 at the Wayback Machine

[43] *Rev. Mod. Phys.* 29 (1957) 547

[44] Kragh, Helge (2010) When is a prediction anthropic? Fred Hoyle and the 7.65 MeV carbon resonance. http://philsci-archive.pitt.edu/5332/

[45] Page, D.N. (1983). "Inflation does not explain time asymmetry". *Nature* **304** (5921): 39. Bibcode:1983Natur.304...39P. doi:10.1038/304039a0.

[46] Davies, P.C.W. (1984). "Inflation to the universe and time asymmetry". *Nature* **312** (5994): 524. Bibcode:1984Natur.312..524D. doi:10.1038/312524a0.

[47] Weinberg, S. (2007). "Living in the multiverse". In B. Carr (ed). *Universe or multiverse?*. Cambridge University Press. ISBN 0-521-84841-5. preprint

[48] Tegmark (1998) op. cit.

[49] Livio, M. and Rees, M. J. (2003). "Anthropic reasoning". *Science* **309** (5737): 1022–3. Bibcode:2005Sci...309.1022L. doi:10.1126/science.1111446. PMID 16099967.

[50] Ehrenfest, Paul (1920). "How do the fundamental laws of physics make manifest that Space has 3 dimensions?". *Annalen der Physik* **61** (5): 440. Bibcode:1920AnP...366..440E. doi:10.1002/andp.19203660503.. Also see Ehrenfest, P. (1917) "In what way does it become manifest in the fundamental laws of physics that space has three dimensions?" *Proceedings of the Amsterdam Academy* 20: 200.

[51] Weyl, H. (1922) *Space, time, and matter*. Dover reprint: 284.

[52] Tangherlini, F. R. (1963). "Atoms in Higher Dimensions". *Nuovo Cimento* **14** (27): 636.

[53] Tegmark, Max (April 1997). "On the dimensionality of spacetime" (PDF). *Classical and Quantum Gravity* **14** (4): L69–L75. arXiv:gr-qc/9702052. Bibcode:1997CQGra..14L..69T. doi:10.1088/0264-9381/14/4/002. Retrieved 2006-12-16.

[54] Hicks, L. E. (1883). *A Critique of Design Arguments*. New York: Scribner's.

[55] Barrow and Tipler 1986: 23

[56] Tipler, F. J. (1994). *The Physics of Immortality*. DoubleDay. ISBN 0-385-46798-2.

[57] Gardner, M., "WAP, SAP, PAP, and FAP," *The New York Review of Books 23*, No. 8 (May 8, 1986): 22-25.

[58] Barrow and Tipler 1986: 677

[59] e.g. Carter (2004) op. cit.

[60] e.g. message from Martin Rees presented at the Kavli-CERCA conference (see video in External links)

[61] Penrose, R. (1989). *The Emperor's New Mind*. Oxford University Press. ISBN 0-19-851973-7. Chapter 10.

[62] Starkman, G. D., Trotta, R. (2006). "Why Anthropic Reasoning Cannot Predict Λ". *Physical Review Letters* **97** (20): 201301. arXiv:astro-ph/0607227. Bibcode:2006PhRvL..97t1301S. doi:10.1103/PhysRevLett.97.201301. PMID 17155671. See also this news story.

[63] Gould, Stephen Jay (1998). "Clear Thinking in the Sciences". *Lectures at Harvard University*.

[64] Gould, Stephen Jay (2002). *Why People Believe Weird Things: Pseudoscience, Superstition, and Other Confusions of Our Time*. ISBN 0-7167-3090-1.

[65] Shermer, Michael (2007). *Why Darwin Matters*. ISBN 0-8050-8121-6.

[66] e.g. Carr, B. J., Rees, M. J. (1979). "The anthropic principle and the structure of the physical world". *Nature* **278** (5705): 605–612. Bibcode:1979Natur.278..605C. doi:10.1038/278605a0.

[67] Stenger, Victor J (2000). *Timeless Reality: Symmetry, Simplicity, and Multiple Universes*. Prometheus Books. ISBN 1-57392-859-3.

[68] Harnik, R., Kribs, G., Perez, G. (2006). "A Universe without Weak interactions". *Physical Review* **D74** (3): 035006. arXiv:hep-ph/0604027. Bibcode:2006PhRvD..74c5006H. doi:10.1103/PhysRevD.74.035006.

[69] Lee Smolin (2001). Tyson, Neil deGrasse and Soter, Steve, ed. *Cosmic Horizons: Astronomy at the Cutting Edge*. The New Press. pp. 148–152. ISBN 978-1-56584-602-9.

[70] Smolin vs. Susskind: The Anthropic Principle

[71] Earman John (1987). "The SAP also rises: A critical examination of the anthropic principle". *American Philosophical Quarterly* **24**: 307–317.

[72] McMullin, Ernan. (1994). "Fine-tuning the Universe?" In M. Shale & G. Shields (ed.), *Science, Technology, and Religious Ideas*, Lanham: University Press of America.

[73] Mosterín, Jesús. (2005). Op. cit.

22.12 References

- Barrow, John D.; Tipler, Frank J. (1988). *The Anthropic Cosmological Principle*. Oxford University Press. ISBN 978-0-19-282147-8. LCCN 87028148.

- Cirkovic, M. M. (2002). "On the First Anthropic Argument in Astrobiology". *Earth, Moon, and Planets* **91** (4): 243–254. doi:10.1023/A:1026266630823.

- Cirkovic, M. M. (2004). "The Anthropic Principle and the Duration of the Cosmological Past". *Astronomical and Astrophysical*

- *Transactions* **23** (6): 567–597. arXiv:astro-ph/0505005. Bibcode:2004A&AT...23..567C. doi:10.1080/10556790412331335327.

- Conway Morris, Simon (2003). *Life's Solution: Inevitable Humans in a Lonely Universe.* Cambridge University Press.

- Craig, William Lane (1987). "Critical review of *The Anthropic Cosmological Principle*". *International Philosophical Quarterly* **27**: 437–47. doi:10.5840/ipq198727433.

- Hawking, Stephen W. (1988). *A Brief History of Time.* New York: Bantam Books. p. 174. ISBN 0-553-34614-8.

- Stenger, Victor J. (1999), "Anthropic design," *The Skeptical Inquirer 23* (August 31, 1999): 40-43

- Mosterín, Jesús (2005). "Anthropic Explanations in Cosmology." In P. Háyek, L. Valdés and D. Westerstahl (ed.), *Logic, Methodology and Philosophy of Science, Proceedings of the 12th International Congress of the LMPS.* London: King's College Publications, pp. 441–473. ISBN 1-904987-21-4.

- Taylor, Stuart Ross (1998). *Destiny or Chance: Our Solar System and Its Place in the Cosmos.* Cambridge University Press. ISBN 0-521-78521-9.

- Tegmark, Max (1997). "On the dimensionality of spacetime". *Classical and Quantum Gravity* **14** (4): L69–L75. arXiv:gr-qc/9702052. Bibcode:1997CQGra..14L..69T. doi:10.1088/0264-9381/14/4/002. A simple anthropic argument for why there are 3 spatial and 1 temporal dimensions.

- Tipler, F. J. (2003). "Intelligent Life in Cosmology". *International Journal of Astrobiology* **2** (2): 141–48. arXiv:0704.0058. Bibcode:2003IJAsB...2..141T. doi:10.1017/S1473550403001526.

- Walker, M. A., and Cirkovic, M. M. (2006). "Anthropic Reasoning, Naturalism and the Contemporary Design Argument". *International Studies in the Philosophy of Science* **20** (3): 285–307. doi:10.1080/02698590600960945. Shows that some of the common criticisms of AP based on its relationship with numerology or the theological Design Argument are wrong.

- Ward, P. D., and Brownlee, D. (2000). *Rare Earth: Why Complex Life is Uncommon in the Universe.* Springer Verlag. ISBN 0-387-98701-0.

- Vilenkin, Alex (2006). *Many Worlds in One: The Search for Other Universes.* Hill and Wang. ISBN 978-0-8090-9523-0.

22.13 External links

- Nick Bostrom: web site devoted to the Anthropic Principle.

- Gijsbers, Victor. (2000). Theistic Anthropic Principle Refuted Positive Atheism Magazine.

- Chown, Marcus, Anything Goes, *New Scientist*, 6 June 1998. On Max Tegmark's work.

- Stephen Hawking, Steven Weinberg, Alexander Vilenkin, David Gross and Lawrence Krauss: Debate on Anthropic Reasoning Kavli-CERCA Conference Video Archive.

- Sober, Elliott R. 2009, "Absence of Evidence and Evidence of Absence -- Evidential Transitivity in Connection with Fossils, Fishing, Fine-Tuning, and Firing Squads." Philosophical Studies, 2009, 143: 63-90.

- "Anthropic Coincidence"—the anthropic controversy as a segue to Lee Smolin's theory of cosmological natural selection.

- Leonard Susskind and Lee Smolin debate the Anthropic Principle.

- debate among scientists on arxiv.org.

- Evolutionary Probability and Fine Tuning

- Benevolent Design and the Anthropic Principle at MathPages

- Critical review of "The Privileged Planet"

- The Anthropic Principle - a review.

- Berger, Daniel, 2002, "An impertinent résumé of the Anthropic Cosmological Principle." A critique of Barrow & Tipler.

- Jürgen Schmidhuber: Papers on algorithmic theories of everything and the Anthropic Principle's lack of predictive power.

- Paul Davies: Cosmic Jackpot Interview about the Anthropic Principle (starts at 40 min), 15 May 2007.

Chapter 23

Wilkinson Microwave Anisotropy Probe

"WMAP" redirects here. WMAP may also refer to either radio station WXNC or WGSP-FM.

The **Wilkinson Microwave Anisotropy Probe (WMAP)**, originally known as the **Microwave Anisotropy Probe (MAP)** was a spacecraft operating from 2001 to 2010 which measured differences across the sky in the temperature of the cosmic microwave background (CMB) – the radiant heat remaining from the Big Bang.[3][4] Headed by Professor Charles L. Bennett of Johns Hopkins University, the mission was developed in a joint partnership between the NASA Goddard Space Flight Center and Princeton University.[5] The WMAP spacecraft was launched on June 30, 2001 from Florida. The WMAP mission succeeded the COBE space mission and was the second medium-class (MIDEX) spacecraft in the NASA Explorers program. In 2003, MAP was renamed WMAP in honor of cosmologist David Todd Wilkinson (1935–2002),[5] who had been a member of the mission's science team. After 9 years of operations, WMAP was switched off in 2010, following the launch of the more advanced Planck spacecraft by ESA in 2009.

WMAP's measurements played a key role in establishing the current Standard Model of Cosmology: the Lambda-CDM model. The WMAP data are very well fit by a universe that is dominated by dark energy in the form of a cosmological constant. Other cosmological data are also consistent, and together tightly constrain the Model. In the Lambda-CDM model of the universe, the age of the universe is 13.772 ± 0.059 billion years. The WMAP mission's determination of the age of the universe to better than 1% precision was recognized by the Guinness Book of World Records.[6] The current expansion rate of the universe is (see Hubble constant) of 69.32 ± 0.80 km·s^{-1}·Mpc^{-1}. The content of the universe presently consists of $4.628\% \pm 0.093\%$ ordinary baryonic matter; $24.02\% + 0.88\%$
-0.87% Cold dark matter (CDM) that neither emits nor absorbs light; and $71.35\% + 0.95\%$
-0.96% of dark energy in the form of a cosmological constant that accelerates the expansion of the universe.[7] Less than 1% of the current contents of the universe is in neutrinos, but WMAP's measurements have found, for the first time in 2008, that the data prefers the existence of a cosmic neutrino background[8] with an effective number of neutrino species of 3.26 ± 0.35. The contents point to a Euclidean flat geometry, with curvature (Ω_k) of $-0.0027 + 0.0039$
-0.0038. The WMAP measurements also support the cosmic inflation paradigm in several ways, including the flatness measurement.

The mission has won various awards: according to *Science* magazine, the WMAP was the *Breakthrough of the Year for 2003*.[9] This mission's results papers were first and second in the "Super Hot Papers in Science Since 2003" list.[10] Of the all-time most referenced papers in physics and astronomy in the INSPIRE-HEP database, only three have been published since 2000, and all three are WMAP publications. Bennett, Lyman A. Page, Jr., and David N. Spergel, the latter both of Princeton University, shared the 2010 Shaw Prize in astronomy for their work on WMAP.[11] Bennett and the WMAP science team were awarded the 2012 Gruber Prize in cosmology.

As of October 2010, the WMAP spacecraft is derelict in a heliocentric graveyard orbit after 9 years of operations.[12] All WMAP data are released to the public and have been subject to careful scrutiny. The final official data release was the nine-year release in 2012.[13][14]

Some aspects of the data are statistically unusual for the Standard Model of Cosmology. For example, the largest angular-scale measurement, the quadrupole moment, is somewhat smaller than the Model would predict, but this discrepancy is not highly significant.[15] A large cold spot and other features of the data are more statistically significant, and research continues into these.

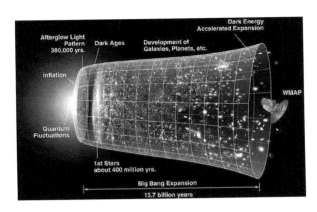

The universe's timeline, from inflation to the WMAP.

23.1 Objectives

The WMAP objective is to measure the temperature differences in the Cosmic Microwave Background (CMB) radiation. The anisotropies then are used to measure the universe's geometry, content, and evolution; and to test the Big Bang model, and the cosmic inflation theory.[16] For that, the mission is creating a full-sky map of the CMB, with a 13 arcminute resolution via multi-frequency observation. The map requires the fewest systematic errors, no correlated pixel noise, and accurate calibration, to ensure angular-scale accuracy greater than its resolution.[16] The map contains 3,145,728 pixels, and uses the HEALPix scheme to pixelize the sphere.[17] The telescope also measures the CMB's E-mode polarization,[16] and foreground polarization;[8] its life is 27 months; 3 to reach the L_2 position, 2 years of observation.[16]

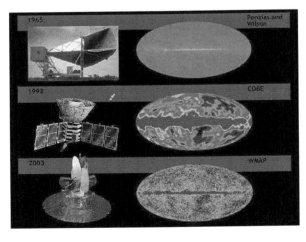

A comparison of the sensitivity of WMAP with COBE and Penzias and Wilson's telescope. Simulated data.

23.2 Development

The MAP mission was proposed to NASA in 1995, selected for definition study in 1996, and approved for development in 1997.[18][19]

The WMAP was preceded by two missions to observe the CMB; (i) the Soviet RELIKT-1 that reported the upper-limit measurements of CMB anisotropies, and (ii) the U.S. COBE satellite that first reported large-scale CMB fluctuations. The WMAP is 45 times more sensitive, with 33 times the angular resolution of its COBE satellite predecessor.[20] The successor European Planck mission (operational 2009-2013) had a higher resolution and higher sensitivity than WMAP and observed in 9 frequency bands rather than WMAP's 5, allowing improved astrophysical foreground models.

23.3 Spacecraft

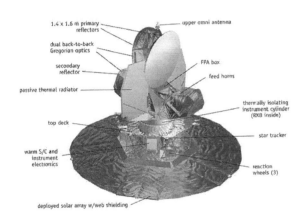

WMAP spacecraft diagram

The telescope's primary reflecting mirrors are a pair of Gregorian 1.4m x 1.6m dishes (facing opposite directions), that focus the signal onto a pair of 0.9m x 1.0m secondary reflecting mirrors. They are shaped for optimal performance: a carbon fibre shell upon a Korex core, thinly-coated with aluminium and silicon oxide. The secondary reflectors transmit the signals to the corrugated feedhorns that sit on a focal plane array box beneath the primary reflectors.[16]

The receivers are polarization-sensitive differential radiometers measuring the difference between two telescope beams. The signal is amplified with HEMT low-noise amplifiers, built by the National Radio Astronomy Observatory. There are 20 feeds, 10 in each direction, from which a radiometer collects a signal; the measure is

23.4 Launch, trajectory, and orbit

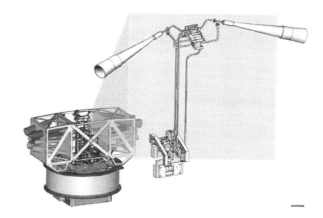

Illustration of WMAP's receivers

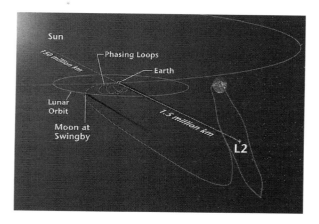

The WMAP's trajectory and orbit.

the difference in the sky signal from opposite directions. The directional separation azimuth is 180 degrees; the total angle is 141 degrees.[16] To avoid collecting Milky Way galaxy foreground signals, the WMAP uses five discrete radio frequency bands, from 23 GHz to 94 GHz.[16]

The WMAP's base is a 5.0m-diameter solar panel array that keeps the instruments in shadow during CMB observations, (by keeping the craft constantly angled at 22 degrees, relative to the Sun). Upon the array sit a bottom deck (supporting the warm components) and a top deck. The telescope's cold components: the focal-plane array and the mirrors, are separated from the warm components with a cylindrical, 33 cm-long thermal isolation shell atop the deck.[16]

Passive thermal radiators cool the WMAP to ca. 90 degrees K; they are connected to the low-noise amplifiers. The telescope consumes 419 W of power. The available telescope heaters are emergency-survival heaters, and there is a transmitter heater, used to warm them when off. The WMAP spacecraft's temperature is monitored with platinum resistance thermometers.[16]

The WMAP's calibration is effected with the CMB dipole and measurements of Jupiter; the beam patterns are measured against Jupiter. The telescope's data are relayed daily via a 2 GHz transponder providing a 667kbit/s downlink to a 70m Deep Space Network telescope. The spacecraft has two transponders, one a redundant back-up; they are minimally active – ca. 40 minutes daily – to minimize radio frequency interference. The telescope's position is maintained, in its three axes, with three reaction wheels, gyroscopes, two star trackers and sun sensors, and is steered with eight hydrazine thrusters.[16]

WMAP launches from Kennedy Space Center, June 30, 2001.

The WMAP spacecraft arrived at the Kennedy Space Center on April 20, 2001. After being tested for two months, it was launched via Delta II 7425 rocket on June 30, 2001.[18][20] It began operating on its internal power five minutes before its launching, and so continued operating until the solar panel array deployed. The WMAP was activated and monitored while it cooled. On July 2, it began working, first with in-flight testing (from launching until August 17), then began constant, formal work.[20] Afterwards, it effected three Earth-Moon phase loops, measuring its sidelobes, then flew by the Moon on July 30, en route to the Sun-Earth L_2 Lagrangian point, arriving there on October 1, 2001, becoming, thereby, the first CMB observation mission permanently posted there.[18]

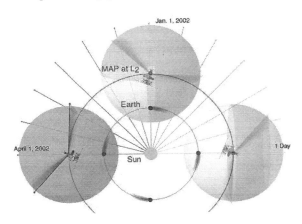

WMAP's orbit and sky scan strategy

The spacecraft's location at Lagrange 2, (1.5 million kilometers from Earth) minimizes the amount of contaminating solar, terrestrial, and lunar emissions registered, and thermally stabilizes it. To view the entire sky, without looking to the Sun, the WMAP traces a path around L_2 in a Lissajous orbit ca. 1.0 degree to 10 degrees,[16] with a 6-month period.[18] The telescope rotates once every 2 minutes, 9 seconds" (0.464 rpm) and precesses at the rate of 1 revolution per hour.[16] WMAP measures the entire sky every six months, and completed its first, full-sky observation in April 2002.[19]

23.5 Foreground radiation subtraction

The WMAP observes in five frequencies, permitting the measurement and subtraction of foreground contamination (from the Milky Way and extra-galactic sources) of the CMB. The main emission mechanisms are synchrotron radiation and free-free emission (dominating the lower frequencies), and astrophysical dust emissions (dominating the higher frequencies). The spectral properties of these emissions contribute different amounts to the five frequencies, thus permitting their identification and subtraction.[16]

Foreground contamination is removed in several ways. First, subtract extant emission maps from the WMAP's measurements; second, use the components' known spectral values to identify them; third, simultaneously fit the position and spectra data of the foreground emission, using extra data sets. Foreground contamination also is reduced by using only the full-sky map portions with the least foreground contamination, whilst masking the remaining map portions.[16]

23.6 Measurements and discoveries

23.6.1 One-year data release

1 year WMAP image of background cosmic radiation (2003).

On February 11, 2003, NASA published the First-year's worth of WMAP data. The latest calculated age and composition of the early universe were presented. In addition, an image of the early universe, that "contains such stunning detail, that it may be one of the most important scientific results of recent years" was presented. The newly released data surpass previous CMB measurements.[5]

Based upon the Lambda-CDM model, the WMAP team produced cosmological parameters from the WMAP's first-year results. Three sets are given below; the first and second sets are WMAP data; the difference is the addition of spectral indices, predictions of some inflationary models. The third data set combines the WMAP constraints with those from other CMB experiments (ACBAR and CBI), and constraints from the 2dF Galaxy Redshift Survey and Lyman alpha forest measurements. Note that there are degenerations among the parameters, the most significant is between n_s and τ ; the errors given are at 68% confidence.[21]

Using the best-fit data and theoretical models, the WMAP team determined the times of important universal events, including the redshift of reionization, 17±4; the redshift of decoupling, 1089±1 (and the universe's age at decoupling, 379+8 −7 kyr); and the redshift of matter/radiation equality, 3233+194 −210. They determined the thickness of the surface of last scattering to be 195±2 in redshift, or 118+3 −2 kyr. They determined the current density of baryons,

23.6. MEASUREMENTS AND DISCOVERIES

$(2.5\pm0.1)\times10^{-7}$ cm^{-1}, and the ratio of baryons to photons, $6.1^{+0.3}_{-0.2}\times10^{-10}$. The WMAP's detection of an early reionization excluded warm dark matter.[21]

The team also examined Milky Way emissions at the WMAP frequencies, producing a 208-point source catalogue. Also, they observed the Sunyaev-Zel'dovich effect at 2.5 σ the strongest source is the Coma cluster.[17]

23.6.2 Three-year data release

5-year WMAP image of background cosmic radiation (2008).

3-year WMAP image of background cosmic radiation (2006).

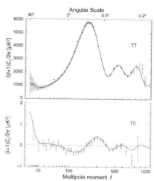

The five-year total-intensity and polarization spectra from WMAP

The Three-year WMAP data were released on March 17, 2006. The data included temperature and polarization measurements of the CMB, which provided further confirmation of the standard flat Lambda-CDM model and new evidence in support of inflation.

The 3-year WMAP data alone shows that the universe must have dark matter. Results were computed both only using WMAP data, and also with a mix of parameter constraints from other instruments, including other CMB experiments (ACBAR, CBI and BOOMERANG), SDSS, the 2dF Galaxy Redshift Survey, the Supernova Legacy Survey and constraints on the Hubble constant from the Hubble Space Telescope.[22]

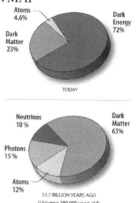

Matter/energy content in the current universe (top) and at the time of photon decoupling in the recombination epoch 380,000 years after the Big Bang (bottom)

[a] ^ Optical depth to reionization improved due to polarization measurements.[23]

[b] ^ < 0.30 when combined with SDSS data. No indication of non-gaussianity.[22]

23.6.3 Five-year data release

The Five-year WMAP data were released on February 28, 2008. The data included new evidence for the cosmic neutrino background, evidence that it took over half a billion years for the first stars to reionize the universe, and new constraints on cosmic inflation.[24]

The improvement in the results came from both having an extra 2 years of measurements (the data set runs between midnight on August 10, 2001 to midnight of August 9, 2006), as well as using improved data processing techniques and a better characterization of the instrument, most notably of the beam shapes. They also make use of the 33 GHz observations for estimating cosmological parameters; previously only the 41 GHz and 61 GHz channels had been used. Finally, improved masks were used to remove foregrounds.[8]

Improvements to the spectra were in the 3rd acoustic peak, and the polarization spectra.[8]

The measurements put constraints on the content of the universe at the time that the CMB was emitted; at the time 10% of the universe was made up of neutrinos, 12% of atoms, 15% of photons and 63% dark matter. The contribution of dark energy at the time was negligible.[24] It also constrained the content of the present-day universe; 4.6% atoms, 23% dark matter and 72% dark energy.[8]

The WMAP five-year data was combined with measurements from Type Ia supernova (SNe) and Baryon acoustic oscillations (BAO).[8]

The elliptical shape of the WMAP skymap is the result of a Mollweide projection.[25]

The data puts a limits on the value of the tensor-to-scalar ratio, $r < 0.22$ (95% certainty), which determines the level at which gravitational waves affect the polarization of the CMB, and also puts limits on the amount of primordial non-gaussianity. Improved constraints were put on the redshift of reionization, which is 10.9 ± 1.4, the redshift of decoupling, 1090.88 ± 0.72 (as well as age of universe at decoupling, $376.971 + 3.162 - 3.167$ kyr) and the redshift of matter/radiation equality, $3253 + 89 - 87$.[8]

The extragalactic source catalogue was expanded to include 390 sources, and variability was detected in the emission from Mars and Saturn.[8]

23.6.4 Seven-year data release

7-year WMAP image of background cosmic radiation (2010).

The Seven-year WMAP data were released on January 26, 2010. As part of this release, claims for inconsistencies with the standard model were investigated.[26] Most were shown not to be statistically significant, and likely due to *a posteriori* selection (where one sees a weird deviation, but fails to consider properly how hard one has been looking; a deviation with 1:1000 likelihood will typically be found if one tries one thousand times). For the deviations that do remain, there are no alternative cosmological ideas (for instance, there seem to be correlations with the ecliptic pole). It seems most likely these are due to other effects, with the report mentioning uncertainties in the precise beam shape and other possible small remaining instrumental and analysis issues.

The other confirmation of major significance is of the total amount of matter/energy in the Universe in the form of Dark Energy – 72.8% (within 1.6%) as non 'particle' background, and Dark Matter – 22.7% (within 1.4%) of non baryonic (sub atomic) 'particle' energy. This leaves matter, or baryonic particles (atoms) at only 4.56% (within 0.16%).

23.6.5 Nine-year data release

9-year WMAP image of background cosmic radiation (2012).

On December 20, 2012, the Nine-year WMAP data and related images were released. 13.772 ± 0.059 billion-year-old temperature fluctuations and a temperature range of ± 200 micro-Kelvin are shown in the image. In addition, the study found that "95-percent" of the early universe is composed of dark matter and energy, the curvature of space is less than 0.4 percent of "flat" and the universe emerged from the cosmic Dark Ages "about 400 million years" after the Big Bang.[13][14][30]

23.7 Main result

The main result of the mission is contained in the various oval maps of the CMB spectrum over the years. These oval images present the temperature distribution gained by the WMAP team from the observations by the telescope of the mission. Measured is the temperature obtained from

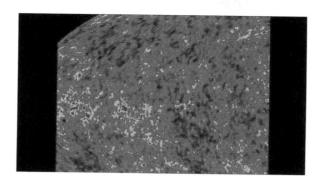

Interviews with Dr. Charles Bennett and Dr. Lyman Page about WMAP.

a Planck's law interpretation of the microwave background. The oval map covers the whole sky. The results describe the state of the universe only some hundred-thousand years after the Big Bang, which happened roughly 13.8 billion years before our time. The microwave background is very homogeneous in temperature (the relative variations from the mean, which presently is still 2.7 kelvins, are only of the order of 5×10^{-5}. The temperature variations corresponding to the local directions are presented through different colours (the "red" directions are hotter, the "blue" directions cooler than the average).

23.8 Follow-on missions and future measurements

The original timeline for WMAP gave it two years of observations; these were completed by September 2003. Mission extensions were granted in 2002, 2004, 2006, and 2008 giving the spacecraft a total of 9 observing years, which ended August 2010[18] and in October 2010 the spacecraft was moved to a heliocentric "graveyard" orbit[12] outside L2, in which it orbits the Sun 14 times every 15 years.

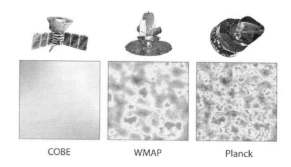

Comparison of CMB results from COBE, WMAP and Planck – March 21, 2013.

The Planck spacecraft, launched on May 14, 2009, also measures the CMB and aims to refine the measurements made by WMAP, both in total intensity and polarization. Various ground- and balloon-based instruments have also made CMB contributions, and others are being constructed to do so. Many are aimed at searching for the B-mode polarization expected from the simplest models of inflation, including EBEX, Spider, BICEP2, Keck, QUIET, CLASS, SPTpol and others.

On 21 March 2013, the European-led research team behind the Planck cosmology probe released the mission's all-sky map of the cosmic microwave background.[31][32] The map suggests the universe is slightly older than thought. According to the map, subtle fluctuations in temperature were imprinted on the deep sky when the cosmos was about 370,000 years old. The imprint reflects ripples that arose as early, in the existence of the universe, as the first nonillionth (10^{-30}) of a second. Apparently, these ripples gave rise to the present vast cosmic web of galaxy clusters and dark matter. Based on the 2013 data, the universe contains 4.9% ordinary matter, 26.8% dark matter and 68.3% dark energy. On 5 February 2015, new data was released by the Planck mission, according to which the age of the universe is 13.799 ± 0.021 billion years old and the Hubble constant was measured to be 67.74 ± 0.46 (km/s)/Mpc.[33]

23.9 See also

- European Space Agency Planck (spacecraft)
- Illustris project
- List of cosmic microwave background experiments
- List of cosmological computation software

23.10 Further reading

- Wilkinson Microwave Anisotrophy Probe Charles L. Bennett Scholarpedia, 2(10):4731. doi:10.4249/scholarpedia.4731

23.11 References

23.11.1 Footnotes

[1] Liz Citrin. "WMAP The Wilkinson Microwave Anisotropy Probe". Retrieved July 8, 2015.

[2] "WMAP News: Events Timeline". NASA. December 27, 2010. Retrieved July 8, 2015.

[3] "Wilkinson Microwave Anisotropy Probe: Overview". *Legacy Archive for Background Data Analysis (LAMBDA)*. Greenbelt, Maryland: NASA's High Energy Astrophysics Science Archive Research Center (HEASARC). August 4, 2009. Retrieved September 24, 2009. The WMAP (Wilkinson Microwave Anisotropy Probe) mission is designed to determine the geometry, content, and evolution of the universe via a 13 arcminute FWHM resolution full sky map of the temperature anisotropy of the cosmic microwave background radiation.

[4] "Tests of Big Bang: The CMB". *Universe 101: Our Universe*. NASA. July 2009. Retrieved September 24, 2009. Only with very sensitive instruments, such as COBE and WMAP, can cosmologists detect fluctuations in the cosmic microwave background temperature. By studying these fluctuations, cosmologists can learn about the origin of galaxies and large scale structures of galaxies and they can measure the basic parameters of the Big Bang theory.

[5] "New image of infant universe reveals era of first stars, age of cosmos, and more". NASA / WMAP team. February 11, 2003. Archived from the original on February 27, 2008. Retrieved April 27, 2008.

[6] *Guinness World Records 2010: Thousands of new records in The Book of the Decade!*. p. 7. ISBN 978-0553593372.

[7] J. Beringer et al. (Particle Data Group), Phys. Rev. D86, 010001 (2012) and 2013 partial update for the 2014 edition.

[8] Hinshaw et al. (2009)

[9] Seife (2003)

[10] ""Super Hot" Papers in Science". in-cites. October 2005. Retrieved April 26, 2008.

[11] "Announcement of the Shaw Laureates 2010". Archived from the original on June 4, 2010.

[12] O'Neill, Ian (2010-10-07). "MISSION COMPLETE! WMAP FIRES ITS THRUSTERS FOR THE LAST TIME". Discovery News. Retrieved 2013-01-27.

[13] Gannon, Megan (December 21, 2012). "New 'Baby Picture' of Universe Unveiled". Space.com. Retrieved December 21, 2012.

[14] Bennett, C.L.; Larson, L.; Weiland, J.L.; Jarosk, N.; et al. (2013). "Nine-Year Wilkinson Microwave Anisotropy Probe (WMAP) Observations: Final Maps and Results". *The Astrophysical Journal Supplement* **208** (2): 20. arXiv:1212.5225. Bibcode:2013ApJS..208...20B. doi:10.1088/0067-0049/208/2/20.

[15] O'Dwyer et al., 2004. The Astrophysical Journal, Volume 617, Issue 2, pp. L99-L102.

[16] Bennett et al. (2003a)

[17] Bennett et al. (2003b)

[18] "WMAP News: Facts". NASA. April 22, 2008. Retrieved April 27, 2008.

[19] "WMAP News: Events". NASA. April 17, 2008. Retrieved April 27, 2008.

[20] Limon et al. (2008)

[21] Spergel et al. (2003)

[22] Spergel et al. (2007)

[23] Hinshaw et al. (2007)

[24] "WMAP Press Release — WMAP reveals neutrinos, end of dark ages, first second of universe". NASA / WMAP team. March 7, 2008. Retrieved April 27, 2008.

[25] WMAP 1-year Paper Figures, Bennett, et al.

[26] Seven-Year Wilkinson Microwave Anisotropy Probe (WMAP) Observations: Are There Cosmic Microwave Background Anomalies?

[27] Table 8 on p. 39 of Jarosik, N.; et al. "Seven-Year Wilkinson Microwave Anisotropy Probe (WMAP) Observations: Sky Maps, Systematic Errors, and Basic Results" (PDF). WMAP Collaboration. nasa.gov. Retrieved December 4, 2010. (from NASA's WMAP Documents page)

[28] Percival, Will J.; et al. (February 2010). "Baryon Acoustic Oscillations in the Sloan Digital Sky Survey Data Release 7 Galaxy Sample". *Monthly Notices of the Royal Astronomical Society* **401** (4): 2148–2168. arXiv:0907.1660. Bibcode:2010MNRAS.401.2148P. doi:10.1111/j.1365-2966.2009.15812.x.

[29] Riess, Adam G.; et al. "A Redetermination of the Hubble Constant with the Hubble Space Telescope from a Differential Distance Ladder" (PDF). hubblesite.org. Retrieved December 4, 2010.

[30] Hinshaw et al., 2013

[31] Clavin, Whitney; Harrington, J.D. (21 March 2013). "Planck Mission Brings Universe Into Sharp Focus". NASA. Retrieved 21 March 2013.

[32] Staff (21 March 2013). "Mapping the Early Universe". *New York Times*. Retrieved 23 March 2013.

[33] Planck Collaboration (2015). "Planck 2015 results. XIII. Cosmological parameters (See Table 4 on page 31 of pfd).". arXiv:1502.01589.

23.11.2 Primary sources

- Bennett, C.; et al. (2003a). "The Microwave Anisotropy Probe (MAP) Mission". *Astrophysical Journal* **583** (1): 1–23. arXiv:astro-ph/0301158. Bibcode:2003ApJ...583....1B. doi:10.1086/345346.

- Bennett, C.; et al. (2003b). "First-Year Wilkinson Microwave Anisotropy Probe (WMAP) Observations: Foreground Emission". *Astrophysical Journal Supplement* **148** (1): 97–117. arXiv:astro-ph/0302208. Bibcode:2003ApJS..148...97B. doi:10.1086/377252.

- Hinshaw, G.; et al. (2007). "Three-Year Wilkinson Microwave Anisotropy Probe (WMAP1) Observations: Temperature Analysis". *Astrophysical Journal Supplement* **170** (2): 288–334. arXiv:astro-ph/0603451. Bibcode:2007ApJS..170..288H. doi:10.1086/513698.

- Hinshaw, G.; et al. (Feb 2009). WMAP Collaboration. "Five-Year Wilkinson Microwave Anisotropy Probe Observations: Data Processing, Sky Maps, and Basic Results". *The Astrophysical Journal Supplement* **180** (2): 225–245. arXiv:0803.0732. Bibcode:2009ApJS..180..225H. doi:10.1088/0067-0049/180/2/225.

- Limon, M.; et al. (March 20, 2008). "Wilkinson Microwave Anisotropy Probe (WMAP): Five–Year Explanatory Supplement" (PDF).

- Seife, Charles (2003). "Breakthrough of the Year: Illuminating the Dark Universe". *Science* **302** (5653): 2038–2039. doi:10.1126/science.302.5653.2038. PMID 14684787.

- Spergel, D. N.; et al. (2003). "First-Year Wilkinson Microwave Anisotropy Probe (WMAP) Observations: Determination of Cosmological Parameters". *Astrophysical Journal Supplement* **148** (1): 175–194. arXiv:astro-ph/0302209. Bibcode:2003ApJS..148..175S. doi:10.1086/377226.

- Sergel, D. N.; et al. (2007). "Three-Year Wilkinson Microwave Anisotropy Probe (WMAP) Observations: Implications for Cosmology". *Astrophysical Journal Supplement* **170** (2): 377–408. arXiv:astro-ph/0603449. Bibcode:2007ApJS..170..377S. doi:10.1086/513700.

- Komatsu; Dunkley; Nolta; Bennett; Gold; Hinshaw; Jarosik; Larson; et al. (2009). "Five-Year Wilkinson Microwave Anisotropy Probe (WMAP) Observations: Cosmological Interpretation". *The Astrophysical Journal Supplement Series* **180** (2): 330–376. arXiv:0803.0547. Bibcode:2009ApJS..180..330K. doi:10.1088/0067-0049/180/2/330.

- About WMAP and the Cosmic Microwave Background – Article at Space.com

- Big Bang glow hints at funnel-shaped Universe, NewScientist, April 15, 2004

- NASA March 16, 2006 WMAP inflation related press release

- Seife, Charles (2003). "With Its Ingredients MAPped, Universe's Recipe Beckons". *Science* **300** (5620): 730–731. doi:10.1126/science.300.5620.730. PMID 12730575.

23.12 External links

- Sizing up the universe

Chapter 24

Kolmogorov complexity

Not to be confused with descriptive complexity theory.

In algorithmic information theory (a subfield of computer science and mathematics), the **Kolmogorov complexity** (also known as **descriptive complexity**, **Kolmogorov–Chaitin complexity**, **algorithmic entropy**, or **program-size complexity**) of an object, such as a piece of text, is a measure of the computational resources needed to specify the object. It is named after Andrey Kolmogorov, who first published on the subject in 1963.[1][2]

For example, consider the following two strings of 32 lowercase letters and digits:

abababababababababababababababab
4c1j5b2p0cv4w1x8rx2y39umgw5q85s7

The first string has a short English-language description, namely "ab 16 times", which consists of **11** characters. The second one has no obvious simple description (using the same character set) other than writing down the string itself, which has **32** characters.

More formally, the complexity of a string is the length of the shortest possible description of the string in some fixed universal description language (the sensitivity of complexity relative to the choice of description language is discussed below). It can be shown that the Kolmogorov complexity of any string cannot be more than a few bytes larger than the length of the string itself. Strings, like the *abab* example above, whose Kolmogorov complexity is small relative to the string's size are not considered to be complex.

The notion of Kolmogorov complexity can be used to state and prove impossibility results akin to Cantor's diagonal argument, Gödel's incompleteness theorem, and Turing's halting problem.

24.1 Definition

The Kolmogorov complexity can be defined for any mathematical object, but for simplicity the scope of this article is restricted to strings. We must first specify a description language for strings. Such a description language can be based on any computer programming language, such as Lisp, Pascal, or Java virtual machine bytecode. If **P** is a program which outputs a string x, then **P** is a description of x. The length of the description is just the length of **P** as a character string, multiplied by the number of bits in a character (e.g. 7 for ASCII).

We could, alternatively, choose an encoding for Turing machines, where an *encoding* is a function which associates to each Turing Machine **M** a bitstring <**M**>. If **M** is a Turing Machine which, on input w, outputs string x, then the concatenated string <**M**> w is a description of x. For theoretical analysis, this approach is more suited for constructing detailed formal proofs and is generally preferred in the research literature. In this article, an informal approach is discussed.

Any string s has at least one description, namely the program:

function GenerateFixedString() **return** s

If a description of s, $d(s)$, is of minimal length (i.e. it uses the fewest bits), it is called a **minimal description** of s. Thus, the length of $d(s)$ (i.e. the number of bits in the description) is the **Kolmogorov complexity** of s, written $K(s)$. Symbolically,

$$K(s) = |d(s)|.$$

The length of the shortest description will depend on the choice of description language; but the effect of changing languages is bounded (a result called the *invariance theorem*).

24.2 Invariance theorem

24.2.1 Informal treatment

There are some description languages which are optimal, in the following sense: given any description of an object in a description language, I can use that description in my optimal description language with a constant overhead. The constant depends only on the languages involved, not on the description of the object, nor the object being described.

Here is an example of an optimal description language. A description will have two parts:

- The first part describes another description language.
- The second part is a description of the object in that language.

In more technical terms, the first part of a description is a computer program, with the second part being the input to that computer program which produces the object as output.

The invariance theorem follows: Given any description language L, the optimal description language is at least as efficient as L, with some constant overhead.

Proof: Any description D in L can be converted into a description in the optimal language by first describing L as a computer program P (part 1), and then using the original description D as input to that program (part 2). The total length of this new description D' is (approximately):

$$|D'| = |P| + |D|$$

The length of P is a constant that doesn't depend on D. So, there is at most a constant overhead, regardless of the object described. Therefore, the optimal language is universal up to this additive constant.

24.2.2 A more formal treatment

Theorem: If K_1 and K_2 are the complexity functions relative to Turing complete description languages L_1 and L_2, then there is a constant c – which depends only on the languages L_1 and L_2 chosen – such that

$$\forall s.\ -c \leq K_1(s) - K_2(s) \leq c.$$

Proof: By symmetry, it suffices to prove that there is some constant c such that for all strings s

$$K_1(s) \leq K_2(s) + c.$$

Now, suppose there is a program in the language L_1 which acts as an interpreter for L_2:

function InterpretLanguage(**string** p)

where p is a program in L_2. The interpreter is characterized by the following property:

> Running InterpretLanguage on input p returns the result of running p.

Thus, if **P** is a program in L_2 which is a minimal description of s, then InterpretLanguage(**P**) returns the string s. The length of this description of s is the sum of

1. The length of the program InterpretLanguage, which we can take to be the constant c.
2. The length of **P** which by definition is $K_2(s)$.

This proves the desired upper bound.

24.3 History and context

Algorithmic information theory is the area of computer science that studies Kolmogorov complexity and other complexity measures on strings (or other data structures).

The concept and theory of Kolmogorov Complexity is based on a crucial theorem first discovered by Ray Solomonoff, who published it in 1960, describing it in "A Preliminary Report on a General Theory of Inductive Inference"[3] as part of his invention of algorithmic probability. He gave a more complete description in his 1964 publications, "A Formal Theory of Inductive Inference," Part 1 and Part 2 in *Information and Control*.[4][5]

Andrey Kolmogorov later independently published this theorem in *Problems Inform. Transmission*[6] in 1965. Gregory Chaitin also presents this theorem in *J. ACM* – Chaitin's paper was submitted October 1966 and revised in December 1968, and cites both Solomonoff's and Kolmogorov's papers.[7]

The theorem says that, among algorithms that decode strings from their descriptions (codes), there exists an optimal one. This algorithm, for all strings, allows codes as short as allowed by any other algorithm up to an additive constant that depends on the algorithms, but not on the strings themselves. Solomonoff used this algorithm, and the code lengths it allows, to define a "universal probability" of a string on which inductive inference of the subsequent digits of the string can be based. Kolmogorov used this theorem to define several functions of strings, including complexity, randomness, and information.

When Kolmogorov became aware of Solomonoff's work, he acknowledged Solomonoff's priority.[8] For several years,

Solomonoff's work was better known in the Soviet Union than in the Western World. The general consensus in the scientific community, however, was to associate this type of complexity with Kolmogorov, who was concerned with randomness of a sequence, while Algorithmic Probability became associated with Solomonoff, who focused on prediction using his invention of the universal prior probability distribution. The broader area encompassing descriptional complexity and probability is often called Kolmogorov complexity. The computer scientist Ming Li considers this an example of the Matthew effect: "... to everyone who has more will be given ..."[9]

There are several other variants of Kolmogorov complexity or algorithmic information. The most widely used one is based on self-delimiting programs, and is mainly due to Leonid Levin (1974).

An axiomatic approach to Kolmogorov complexity based on Blum axioms (Blum 1967) was introduced by Mark Burgin in the paper presented for publication by Andrey Kolmogorov (Burgin 1982).

24.4 Basic results

In the following discussion, let $K(s)$ be the complexity of the string s.

It is not hard to see that the minimal description of a string cannot be too much larger than the string itself - the program GenerateFixedString above that outputs s is a fixed amount larger than s.

Theorem: There is a constant c such that

$$\forall s. \ K(s) \leq |s| + c.$$

24.4.1 Uncomputability of Kolmogorov complexity

Theorem: There exist strings of arbitrarily large Kolmogorov complexity. Formally: for each $n \in \mathbb{N}$, there is a string s with $K(s) \geq n$.[note 1]

Proof: Otherwise all infinitely many possible strings could be generated by the finitely many[note 2] programs with a complexity below n bits.

Theorem: K is not a computable function. In other words, there is no program which takes a string s as input and produces the integer $K(s)$ as output.

The following indirect **proof** uses a simple Pascal-like language to denote programs; for sake of proof simplicity assume its description (i.e. an interpreter) to have a length of 1400000 bits. Assume for contradiction there is a program

function KolmogorovComplexity(**string** s)

which takes as input a string s and returns $K(s)$; for sake of proof simplicity, assume its length to be 7000000000 bits. Now, consider the following program of length 1288 bits:

function GenerateComplexString() **for** i = 1 **to** infinity: **for each** string s **of** length exactly i **if** KolmogorovComplexity(s) >= 8000000000 **return** s

Using KolmogorovComplexity as a subroutine, the program tries every string, starting with the shortest, until it returns a string with Kolmogorov complexity at least 8000000000 bits,[note 3] i.e. a string that cannot be produced by any program shorter than 8000000000 bits. However, the overall length of the above program that produced s is only 7001401288 bits,[note 4] which is a contradiction. (If the code of KolmogorovComplexity is shorter, the contradiction remains. If it is longer, the constant used in GenerateComplexString can always be changed appropriately.)[note 5]

The above proof used a contradiction similar to that of the Berry paradox: "$_1$The $_2$smallest $_3$positive $_4$integer $_5$that $_6$cannot $_7$be $_8$defined $_9$in $_{10}$fewer $_{11}$than $_{12}$twenty $_{13}$English $_{14}$words". It is also possible to show the non-computability of K by reduction from the non-computability of the halting problem H, since K and H are Turing-equivalent.[10]

There is a corollary, humorously called the "full employment theorem" in the programming language community, stating that there is no perfect size-optimizing compiler.

24.4.2 Chain rule for Kolmogorov complexity

Main article: Chain rule for Kolmogorov complexity

The chain rule[11] for Kolmogorov complexity states that

$$K(X,Y) = K(X) + K(Y|X) + O(\log(K(X,Y))).$$

It states that the shortest program that reproduces X and Y is no more than a logarithmic term larger than a program to reproduce X and a program to reproduce Y given X. Using this statement, one can define an analogue of mutual information for Kolmogorov complexity.

24.5 Compression

It is straightforward to compute upper bounds for $K(s)$ – simply compress the string s with some method, implement the corresponding decompressor in the chosen language, concatenate the decompressor to the compressed string, and

24.6. CHAITIN'S INCOMPLETENESS THEOREM

measure the length of the resulting string – concretely, the size of a self-extracting archive in the given language.

A string s is compressible by a number c if it has a description whose length does not exceed $|s|-c$ bits. This is equivalent to saying that $K(s) \leq |s|-c$. Otherwise, s is incompressible by c. A string incompressible by 1 is said to be simply *incompressible* – by the pigeonhole principle, which applies because every compressed string maps to only one uncompressed string, incompressible strings must exist, since there are 2^n bit strings of length n, but only $2^n - 1$ shorter strings, that is, strings of length less than n, (i.e. with length $0, 1, ..., n-1$).[note 6]

For the same reason, most strings are complex in the sense that they cannot be significantly compressed – their $K(s)$ is not much smaller than $|s|$, the length of s in bits. To make this precise, fix a value of n. There are 2^n bitstrings of length n. The uniform probability distribution on the space of these bitstrings assigns exactly equal weight 2^{-n} to each string of length n.

Theorem: With the uniform probability distribution on the space of bitstrings of length n, the probability that a string is incompressible by c is at least $1 - 2^{-c+1} + 2^{-n}$.

To prove the theorem, note that the number of descriptions of length not exceeding $n-c$ is given by the geometric series:

$$1 + 2 + 2^2 + ... + 2^{n-c} = 2^{n-c+1} - 1.$$

There remain at least

$$2^n - 2^{n-c+1} + 1$$

bitstrings of length n that are incompressible by c. To determine the probability, divide by 2^n.

24.6 Chaitin's incompleteness theorem

We know that, in the set of all possible strings, most strings are complex in the sense that they cannot be described in any significantly "compressed" way. However, it turns out that the fact that a specific string is complex cannot be formally proven, if the complexity of the string is above a certain threshold. The precise formalization is as follows. First, fix a particular axiomatic system **S** for the natural numbers. The axiomatic system has to be powerful enough so that, to certain assertions **A** about complexity of strings, one can associate a formula **FA** in **S**. This association must have the following property:

If **FA** is provable from the axioms of **S**, then the corresponding assertion **A** must be true. This "formalization"

Kolmogorov complexity $K(s)$, and two computable lower bound functions prog1(s), prog2(s). The horizontal axis (logarithmic scale) enumerates all strings s, ordered by length; the vertical axis (linear scale) measures string length in bits. Most strings are incompressible, i.e. their Kolmogorov complexity exceeds their length by a constant amount. 17 compressible strings are shown in the picture, appearing as almost vertical slopes. Due to Chaitin's incompleteness theorem, the output of any program computing a lower bound of the Kolmogorov complexity cannot exceed some fixed limit, which is independent of the input string s.

can be achieved, either by an artificial encoding such as a Gödel numbering, or by a formalization which more clearly respects the intended interpretation of **S**.

Theorem: There exists a constant L (which only depends on the particular axiomatic system and the choice of description language) such that there does not exist a string s for which the statement

$$K(s) \geq L \text{ (as formalized in } \mathbf{S}\text{)}$$

can be proven within the axiomatic system **S**.

Note that, by the abundance of nearly incompressible strings, the vast majority of those statements must be true.

The proof of this result is modeled on a self-referential construction used in Berry's paradox. The proof is by contradiction. If the theorem were false, then

> **Assumption (X)**: For any integer n there exists a string s for which there is a proof in **S** of the formula "$K(s) \geq n$" (which we assume can be formalized in **S**).

We can find an effective enumeration of all the formal proofs in **S** by some procedure

function NthProof(**int** n)

which takes as input n and outputs some proof. This function enumerates all proofs. Some of these are proofs for formulas we do not care about here, since every possible proof in the language of **S** is produced for some n. Some of these are complexity formulas of the form $K(s) \geq n$ where s and n are constants in the language of **S**. There is a program

function NthProofProvesComplexityFormula(**int** n)

which determines whether the nth proof actually proves a

complexity formula $K(s) \geq L$. The strings s, and the integer L in turn, are computable by programs:

function StringNthProof(**int** n) **function** ComplexityLowerBoundNthProof(**int** n)

Consider the following program

function GenerateProvablyComplexString(**int** n) **for** i = 1 to infinity: **if** NthProofProvesComplexityFormula(i) **and** ComplexityLowerBoundNthProof(i) $\geq n$ **return** StringNthProof(i)

Given an n, this program tries every proof until it finds a string and a proof in the formal system **S** of the formula $K(s) \geq L$ for some $L \geq n$. The program terminates by our **Assumption (X)**. Now, this program has a length U. There is an integer n_0 such that $U + \log_2(n_0) + C < n_0$, where C is the overhead cost of

function GenerateProvablyParadoxicalString() **return** GenerateProvablyComplexString(n_0)

(note that n_0 is hard-coded into the above function, and the summand $\log_2(n_0)$ already allows for its encoding). The program GenerateProvablyParadoxicalString outputs a string s for which there exists an L such that $K(s) \geq L$ can be formally proved in **S** with $L \geq n_0$. In particular, $K(s) \geq n_0$ is true. However, s is also described by a program of length $U + \log_2(n_0) + C$, so its complexity is less than n_0. This contradiction proves **Assumption (X)** cannot hold.

Similar ideas are used to prove the properties of Chaitin's constant.

24.7 Minimum message length

The minimum message length principle of statistical and inductive inference and machine learning was developed by C.S. Wallace and D.M. Boulton in 1968. MML is Bayesian (i.e. it incorporates prior beliefs) and information-theoretic. It has the desirable properties of statistical invariance (i.e. the inference transforms with a re-parametrisation, such as from polar coordinates to Cartesian coordinates), statistical consistency (i.e. even for very hard problems, MML will converge to any underlying model) and efficiency (i.e. the MML model will converge to any true underlying model about as quickly as is possible). C.S. Wallace and D.L. Dowe (1999) showed a formal connection between MML and algorithmic information theory (or Kolmogorov complexity).

24.8 Kolmogorov randomness

Kolmogorov randomness – also called *algorithmic randomness* – defines a string (usually of bits) as being random if and only if it is shorter than any computer program that can produce that string. To make this precise, a universal computer (or universal Turing machine) must be specified, so that "program" means a program for this universal machine. A random string in this sense is "incompressible" in that it is impossible to "compress" the string into a program whose length is shorter than the length of the string itself. A counting argument is used to show that, for any universal computer, there is at least one algorithmically random string of each length. Whether any particular string is random, however, depends on the specific universal computer that is chosen.

This definition can be extended to define a notion of randomness for *infinite* sequences from a finite alphabet. These algorithmically random sequences can be defined in three equivalent ways. One way uses an effective analogue of measure theory; another uses effective martingales. The third way defines an infinite sequence to be random if the prefix-free Kolmogorov complexity of its initial segments grows quickly enough - there must be a constant c such that the complexity of an initial segment of length n is always at least $n-c$. This definition, unlike the definition of randomness for a finite string, is not affected by which universal machine is used to define prefix-free Kolmogorov complexity. [12]

24.9 Relation to entropy

For dynamical systems, entropy rate and algorithmic complexity of the trajectories are related by a theorem of Brudno, that the equality $K(x;T) = h(T)$ holds for almost all x.[13]

It can be shown[14] that for the output of Markov information sources, Kolmogorov complexity is related to the entropy of the information source. More precisely, the Kolmogorov complexity of the output of a Markov information source, normalized by the length of the output, converges almost surely (as the length of the output goes to infinity) to the entropy of the source.

24.10 Conditional versions

The conditional [Kolmogorov] complexity of two strings $K(x|y)$ is, roughly speaking, defined as the Kolmogorov complexity of x given y as an auxiliary input to the procedure.[15][16]

There is also a length-conditional complexity $K(x|l(x))$, which is the complexity of x given the length of x as known/input.[17]

24.11 See also

- Berry paradox
- Data compression
- Inductive inference
- Kolmogorov structure function
- Important publications in algorithmic information theory
- Solomonoff's theory of inductive inference
- Levenshtein distance
- Grammar induction

24.12 Notes

[1] However, an s with $K(s) = n$ needn't exist for every n. For example, if n isn't a multiple of 7 bits, no ASCII program can have a length of exactly n bits.

[2] There are $1 + 2 + 2^2 + 2^3 + \ldots + 2^n = 2^{n+1} - 1$ different program texts of length up to n bits; cf. geometric series. If program lengths are to be multiples of 7 bits, even fewer program texts exist.

[3] By the previous theorem, such a string exists, hence the for loop will eventually terminate.

[4] including the language interpreter and the subroutine code for KolmogorovComplexity

[5] If KolmogorovComplexity has length n bits, the constant m used in GenerateComplexString needs to be adapted to satisfy $n + 1400000 + 1218 + 7 \cdot \log_{10}(m) < m$, which is always possible since m grows faster than $\log_{10}(m)$.

[6] As there are $NL = 2^L$ strings of length L, the number of strings of lengths $L = 0, 1, \ldots, n-1$ is $N_0 + N_1 + \ldots + N_{n-1} = 2^0 + 2^1 + \ldots + 2^{n-1}$, which is a finite geometric series with sum $2^0 + 2^1 + \ldots + 2^{n-1} = 2^0 \times (1 - 2^n) / (1 - 2) = 2^n - 1$.

24.13 References

[1] Kolmogorov, Andrey (1963). "On Tables of Random Numbers". *Sankhyā Ser. A.* **25**: 369–375. MR 178484.

[2] Kolmogorov, Andrey (1998). "On Tables of Random Numbers". *Theoretical Computer Science* **207** (2): 387–395. doi:10.1016/S0304-3975(98)00075-9. MR 1643414.

[3] Solomonoff, Ray (February 4, 1960). "A Preliminary Report on a General Theory of Inductive Inference" (PDF). *Report V-131* (Cambridge, Ma.: Zator Co.). revision, Nov., 1960.

[4] Solomonoff, Ray (March 1964). "A Formal Theory of Inductive Inference Part I" (PDF). *Information and Control* **7** (1): 1–22. doi:10.1016/S0019-9958(64)90223-2.

[5] Solomonoff, Ray (June 1964). "A Formal Theory of Inductive Inference Part II" (PDF). *Information and Control* **7** (2): 224–254. doi:10.1016/S0019-9958(64)90131-7.

[6] Kolmogorov, A.N. (1965). "Three Approaches to the Quantitative Definition of Information". *Problems Inform. Transmission* **1** (1): 1–7.

[7] Chaitin, Gregory J. (1969). "On the Simplicity and Speed of Programs for Computing Infinite Sets of Natural Numbers" (PDF). *Journal of the ACM* **16** (3): 407–422. doi:10.1145/321526.321530.

[8] Kolmogorov, A. (1968). "Logical basis for information theory and probability theory". *IEEE Transactions on Information Theory* **14** (5): 662–664. doi:10.1109/TIT.1968.1054210.

[9] Li, Ming; Paul Vitanyi (1997-02-27). *An Introduction to Kolmogorov Complexity and Its Applications* (2nd ed.). Springer. p. 90. ISBN 0-387-94868-6.

[10] Stated without proof in: "*Course notes for Data Compression - Kolmogorov complexity*", 2005, P.B. Miltersen, p.7

[11] Zvonkin, A.; L. Levin (1970). "The complexity of finite objects and the development of the concepts of information and randomness by means of the theory of algorithms.". *Russian Mathematical Surveys* **25** (6). pp. 83–124.

[12] Martin-Löf, P. (1966). "The definition of random sequences" (PDF). *Information and Control* **9** (6): 602–619. doi:10.1016/s0019-9958(66)80018-9.

[13] Stefano Galatolo, Mathieu Hoyrup, Cristóbal Rojas (2010). "Effective symbolic dynamics, random points, statistical behavior, complexity and entropy" (PDF). *Information and Computation* **208**: 23–41. doi:10.1016/j.ic.2009.05.001.

[14] Alexei Kaltchenko (2004). "Algorithms for Estimating Information Distance with Application to Bioinformatics and Linguistics". *CoRR*. cs.CC/0404039.

[15] Jorma Rissanen (2007). *Information and Complexity in Statistical Modeling*. Springer Science & Business Media. p. 53. ISBN 978-0-387-68812-1.

[16] Ming Li; Paul M.B. Vitányi (2009). *An Introduction to Kolmogorov Complexity and Its Applications*. Springer Science & Business Media. pp. 105–106. ISBN 978-0-387-49820-1.

[17] Ming Li; Paul M.B. Vitányi (2009). *An Introduction to Kolmogorov Complexity and Its Applications*. Springer Science & Business Media. p. 119. ISBN 978-0-387-49820-1.

- Blum, M. (1967). "On the size of machines". *Information and Control* **11** (3): 257. doi:10.1016/S0019-9958(67)90546-3.

- Brudno, A. Entropy and the complexity of the trajectories of a dynamical system., Transactions of the Moscow Mathematical Society, 2:127{151, 1983.

- Burgin, M. (1982), "Generalized Kolmogorov complexity and duality in theory of computations", *Notices of the Russian Academy of Sciences*, v.25, No. 3, pp. 19–23.

- Cover, Thomas M. and Thomas, Joy A., *Elements of information theory*, 1st Edition. New York: Wiley-Interscience, 1991. ISBN 0-471-06259-6. 2nd Edition. New York: Wiley-Interscience, 2006. ISBN 0-471-24195-4.

- Lajos, Rónyai and Gábor, Ivanyos and Réka, Szabó, *Algoritmusok*. TypoTeX, 1999. ISBN 963-279-014-6

- Li, Ming and Vitányi, Paul (1997). *An Introduction to Kolmogorov Complexity and Its Applications*. Springer. ISBN 978-0387339986. First chapter on citeseer

- Yu Manin, *A Course in Mathematical Logic*, Springer-Verlag, 1977. ISBN 978-0-7204-2844-5

- Sipser, Michael, *Introduction to the Theory of Computation*, PWS Publishing Company, 1997. ISBN 0-534-95097-3.

- Wallace, C. S. and Dowe, D. L., Minimum Message Length and Kolmogorov Complexity, Computer Journal, Vol. 42, No. 4, 1999).

24.14 External links

- The Legacy of Andrei Nikolaevich Kolmogorov
- Chaitin's online publications
- Solomonoff's IDSIA page
- Generalizations of algorithmic information by J. Schmidhuber
- Ming Li and Paul Vitanyi, An Introduction to Kolmogorov Complexity and Its Applications, 2nd Edition, Springer Verlag, 1997.
- Tromp's lambda calculus computer model offers a concrete definition of K()
- Universal AI based on Kolmogorov Complexity ISBN 3-540-22139-5 by M. Hutter: ISBN 3-540-22139-5
- David Dowe's Minimum Message Length (MML) and Occam's razor pages.
- P. Grunwald, M. A. Pitt and I. J. Myung (ed.), Advances in Minimum Description Length: Theory and Applications, M.I.T. Press, April 2005, ISBN 0-262-07262-9.

Chapter 25

Possible world

"Possible worlds" redirects here. For other uses, see Possible Worlds.

In philosophy and logic, the concept of a **possible world** is used to express modal claims. The concept of possible worlds is common in contemporary philosophical discourse but has been disputed.

25.1 Possibility, necessity, and contingency

Further information: Modal logic § The ontology of possibility

Those theorists who use the concept of possible worlds consider the *actual* world to be one of the many possible worlds. For each distinct way the world could have been, there is said to be a distinct possible world; the actual world is the one we in fact live in. Among such theorists there is disagreement about the nature of possible worlds; their precise ontological status is disputed, and especially the difference, if any, in ontological status between the actual world and all the other possible worlds. One position on these matters is set forth in David Lewis's modal realism (see below). There is a close relation between propositions and possible worlds. We note that every proposition is either true or false at any given possible world; then the *modal status* of a proposition is understood in terms of the *worlds in which it is true* and *worlds in which it is false*. The following are among the assertions we may now usefully make:

- **True propositions** are those that are *true in the actual world* (for example: "Richard Nixon became president in 1969").

- **False propositions** are those that are *false in the actual world* (for example: "Ronald Reagan became president in 1969"). (Reagan did not run for president until 1976, and thus couldn't possibly have been elected.)

- **Possible propositions** are those that are *true in at least one possible world* (for example: "Hubert Humphrey became president in 1969"). (Humphrey did run for president in 1968, and thus could have been elected.) This includes propositions which are necessarily true, in the sense below.

- **Impossible propositions** (or *necessarily false propositions*) are those that are *true in no possible world* (for example: "Melissa and Toby are taller than each other at the same time").

- **Necessarily true propositions** (often simply called **necessary propositions**) are those that are *true in all possible worlds* (for example: "2 + 2 = 4"; "all bachelors are unmarried").[1]

- **Contingent propositions** are those that are *true in some possible worlds and false in others* (for example: "Richard Nixon became president in 1969" is *contingently true* and "Hubert Humphrey became president in 1969" is *contingently false*).

The idea of possible worlds is most commonly attributed to Gottfried Leibniz, who spoke of possible worlds as ideas in the mind of God and used the notion to argue that our actually created world must be "the best of all possible worlds". However, scholars have also found implicit traces of the idea in the works of Rene Descartes,[2] a major influence on Leibniz, Al-Ghazali (*The Incoherence of the Philosophers*), Averroes (*The Incoherence of the Incoherence*),[3] Fakhr al-Din al-Razi (*Matalib al-'Aliya*)[4] and John Duns Scotus.[3] The modern philosophical use of the notion was pioneered by David Lewis and Saul Kripke.

25.2 Formal semantics of modal logics

Main article: Modal logic § Semantics

A semantics for modal logic was first introduced in the late-1950s work of Saul Kripke and his colleagues. A statement in modal logic that is *possible* is said to be true in at least one possible world; a statement that is *necessary* is said to be true in all possible worlds.

25.3 From modal logic to philosophical tool

From this groundwork, the theory of possible worlds became a central part of many philosophical developments, from the 1960s onwards – including, most famously, the analysis of counterfactual conditionals in terms of "nearby possible worlds" developed by David Lewis and Robert Stalnaker. On this analysis, when we discuss what *would* happen *if* some set of conditions *were* the case, the truth of our claims is determined by what is true at the nearest possible world (or the *set* of nearest possible worlds) where the conditions obtain. (A possible world W_1 is said to be near to another possible world W_2 in respect of R to the degree that the same things happen in W_1 and W_2 in respect of R; the more different something happens in two possible worlds in a certain respect, the "further" they are from one another in that respect.) Consider this conditional sentence: "If George W. Bush hadn't become president of the U.S. in 2001, Al Gore would have." The sentence would be taken to express a claim that could be reformulated as follows: "In all nearest worlds to our actual world (nearest in relevant respects) where George W. Bush didn't become president of the U.S. in 2001, Al Gore became president of the U.S. then instead." And on this interpretation of the sentence, if there is or are some nearest worlds to the actual world (nearest in relevant respects) where George W. Bush didn't become president but Al Gore didn't either, then the claim expressed by this counterfactual would be false.

Today, possible worlds play a central role in many debates in philosophy, including especially debates over the Zombie Argument, and physicalism and supervenience in the philosophy of mind. Many debates in the philosophy of religion have been reawakened by the use of possible worlds. Intense debate has also emerged over the ontological status of possible worlds, provoked especially by David Lewis's defense of modal realism, the doctrine that talk about "possible worlds" is best explained in terms of innumerable, *really existing* worlds beyond the one we live in. The fundamental question here is: *given* that modal logic works, and that some possible-worlds semantics for modal logic is correct, *what has to be true* of the world, and just what *are* these possible worlds that we range over in our interpretation of modal statements? Lewis argued that what we range over are real, concrete *worlds* that exist just as unequivocally as our actual world exists, but that are distinguished from the actual world simply by standing in no spatial, temporal, or causal relations with the actual world. (On Lewis's account, the only "special" property that the *actual* world has is a relational one: that *we* are in it. This doctrine is called "the indexicality of actuality": "actual" is a merely indexical term, like "now" and "here".) Others, such as Robert Adams and William Lycan, reject Lewis's picture as metaphysically extravagant, and suggest in its place an interpretation of possible worlds as consistent, maximally complete sets of descriptions of or propositions about the world, so that a "possible world" is conceived of as a complete *description* of *a way the world could be* – rather than a *world that is that way*. (Lewis describes their position, and similar positions such as those advocated by Alvin Plantinga and Peter Forrest, as "*ersatz* modal realism", arguing that such theories try to get the benefits of possible worlds semantics for modal logic "on the cheap", but that they ultimately fail to provide an adequate explanation.) Saul Kripke, in *Naming and Necessity*, took explicit issue with Lewis's use of possible worlds semantics, and defended a *stipulative* account of possible worlds as purely *formal* (logical) entities rather than either really existent worlds or as some set of propositions or descriptions.

25.4 Possible-world theory in literary studies

Possible worlds theory in literary studies uses concepts from possible-world logic and applies them to worlds that are created by fictional texts, fictional universe. In particular, possible-world theory provides a useful vocabulary and conceptual framework with which to describe such worlds. However, a literary world is a specific type of possible world, quite distinct from the possible worlds in logic. This is because a literary text houses its own system of modality, consisting of actual worlds (actual events) and possible worlds (possible events). In fiction, the principle of simultaneity, it extends to cover the dimensional aspect, when it is contemplated that two or more physical objects, realities, perceptions and objects non-physical, can coexist in the same space-time. Thus, a literary universe is granted autonomy in much the same way as the actual universe.

Literary critics, such as Marie-Laure Ryan, Lubomír Doležel, and Thomas Pavel, have used possible-worlds theory to address notions of literary truth, the nature of fictionality, and the relationship between fictional worlds and reality. Taxonomies of fictional possibilities have also been proposed where the likelihood of a fictional world is assessed. Rein Raud has extended this approach onto "cultural" worlds, comparing possible worlds to the particular constructions of reality of different cultures.[5] Possible-world theory is also used within narratology to divide a specific text into its constituent worlds, possible and actual. In this approach, the modal structure of the fictional text is analysed in relation to its narrative and thematic concerns.

25.5 See also

- Standard translation, an embedding of modal logics into first-order logic which captures their possible world semantics
- N-universes
- Modal fictionalism
- Fictionalism

25.6 Notes

[1] See "A Priori and A Posteriori" (author: Jason S. Baehr), at *Internet Encyclopedia of Philosophy*: "A necessary proposition is one the truth value of which remains constant across all possible worlds. Thus a necessarily true proposition is one that is true in every possible world, and a necessarily false proposition is one that is false in every possible world. By contrast, the truth value of contingent propositions is not fixed across all possible worlds: for any contingent proposition, there is at least one possible world in which it is true and at least one possible world in which it is false." Accessed 7 July 2012.

[2] "Nor could we doubt that, if God had created many worlds, they would not be as true in all of them as in this one. Thus those who could examine sufficiently the consequences of these truths and of our rules, could be able to discover effects by their causes, and, to explain myself in the language of the schools, they could have a priori demonstrations of everything that could be produced in this new world." -The World, Chapter VII

[3] Taneli Kukkonen (2000), "Possible Worlds in the Tahâfut al-Falâsifa: Al-Ghazâlî on Creation and Contingency", *Journal of the History of Philosophy* **38** (4): 479–502, doi:10.1353/hph.2005.0033

[4] Adi Setia (2004), "Fakhr Al-Din Al-Razi on Physics and the Nature of the Physical World: A Preliminary Survey", *Islam & Science* **2**, retrieved 2010-03-02

[5] "Identity, Difference and Cultural Worlds", in Lang, V. & Kull, K. (eds) (2014) Estonian Approaches to Culture Theory. Approaches to Culture Theory 4, 164–179. University of Tartu Press, Tartu, https://www.academia.edu/9128313/Identity_Difference_and_Cultural_Worlds

25.7 Further reading

- D.M. Armstrong, *A World of States of Affairs* (1997. Cambridge: Cambridge University Press) ISBN 0-521-58948-7
- John Divers, *Possible Worlds* (2002. London: Routledge) ISBN 0-415-15556-8
- Paul Herrick, *The Many Worlds of Logic* (1999. Oxford: Oxford University Press) Chapters 23 and 24. ISBN 978-0-19-515503-7
- David Lewis, *On the Plurality of Worlds* (1986. Oxford & New York: Basil Blackwell) ISBN 0-631-13994-X
- Michael J. Loux [ed.] *The Possible and the Actual* (1979. Ithaca & London: Cornell University Press) ISBN 0-8014-9178-9
- G.W. Leibniz, *Theodicy* (2001. Wipf & Stock Publishers) ISBN 978-0-87548-437-2
- Brian Skyrms "Possible Worlds, Physics and Metaphysics" (1976. Philosophical Studies 30)

25.8 External links

- "Possible Worlds" - Stanford Encyclopedia of Philosophy
- "Possible worlds: what they are good for and what they are" — Alexander Pruss

Chapter 26

Modal realism

Modal realism is the view propounded by David Kellogg Lewis that all possible worlds are as real as the actual world. It is based on the following tenets: possible worlds exist; possible worlds are not different in kind from the actual world; possible worlds are irreducible entities; the term *actual* in *actual world* is indexical, i.e. any subject can declare their world to be the actual one, much as they label the place they are "here" and the time they are "now".

26.1 The term *possible world*

The term goes back to Leibniz's theory of possible worlds, used to analyse necessity, possibility, and similar modal notions. In short: the actual world is regarded as merely one among an infinite set of logically possible worlds, some "nearer" to the actual world and some more remote. A proposition is *necessary* if it is true in all possible worlds, and *possible* if it is true in at least one.

26.2 Main tenets of modal realism

At the heart of David Lewis's modal realism are six central doctrines about possible worlds:

1. Possible worlds exist – they are just as real as our world;
2. Possible worlds are the same sort of things as our world – they differ in content, not in kind;
3. Possible worlds cannot be reduced to something more basic – they are irreducible entities in their own right.
4. Actuality is indexical. When we distinguish our world from other possible worlds by claiming that it alone is actual, we mean only that it is *our* world.
5. Possible worlds are unified by the spatiotemporal interrelations of their parts; every world is spatiotemporally isolated from every other world.
6. Possible worlds are causally isolated from each other.

26.3 Reasons given by Lewis

Lewis backs modal realism for a variety of reasons. First, there doesn't seem to be a reason not to. Many abstract mathematical entities are held to exist simply because they are useful. For example, sets are useful, abstract mathematical constructs that were only conceived in the 19th century. Sets are now considered to be objects in their own right, and while this is a philosophically unintuitive idea, its usefulness in understanding the workings of mathematics makes belief in it worthwhile. The same should go for possible worlds. Since these constructs have helped us make sense of key philosophical concepts in epistemology, metaphysics, philosophy of mind, etc., their existence should be uncritically accepted on pragmatic grounds.

Lewis believes that the concept of alethic modality can be reduced to talk of real possible worlds. For example, to say "x is possible" is to say that there exists a possible world where x is true. To say "x is necessary" is to say that in all possible worlds x is true. The appeal to possible worlds provides a sort of economy with the least number of undefined primitives/axioms in our ontology.

Taking this latter point one step further, Lewis argues that modality cannot be made sense of *without* such a reduction. He maintains that we cannot determine that x is possible without a conception of what a real world where x holds would look like. In deciding whether it is possible for basketballs to be inside of atoms we do not simply make a linguistic determination of whether the proposition is grammatically coherent, we actually think about whether a real world would be able to sustain such a state of affairs. Thus we require a brand of modal realism if we are to use modality at all.

26.4 Details and alternatives

In philosophy possible worlds are usually regarded as real but abstract possibilities, or sometimes as a mere metaphor, abbreviation, or *façon de parler* for sets of counterfactual propositions.

Lewis himself not only claimed to take modal realism seriously (although he did regret his choice of the expression *modal realism*), he also insisted that his claims should be taken literally:

> By what right do we call possible worlds and their inhabitants disreputable entities, unfit for philosophical services unless they can beg redemption from philosophy of language? I know of no accusation against possibles that cannot be made with equal justice against sets. Yet few philosophical consciences scruple at set theory. Sets and possibles alike make for a crowded ontology. Sets and possibles alike raise questions we have no way to answer. [...] I propose to be equally undisturbed by these equally mysterious mysteries.[1]

> How many [possible worlds] are there? In what respects do they vary, and what is common to them all? Do they obey a nontrivial law of identity of indiscernibles? Here I am at a disadvantage compared to someone who pretends as a figure of speech to believe in possible worlds, but really does not. If worlds were creatures of my imagination, I could imagine them to be any way I liked, and I could tell you all you wished to hear simply by carrying on my imaginative creation. But as I believe that there really are other worlds, I am entitled to confess that there is much about them that I do not know, and that I do not know how to find out.[2]

26.5 Criticisms

While it may appear to be a simply extravagant account of modality, modal realism has proven to be historically quite resilient. Nonetheless, a number of philosophers, including Lewis himself, have produced criticisms of (what some call) "extreme realism" about possible worlds.

26.5.1 Lewis's own critique

Lewis's own extended presentation of the theory (*On the Plurality of Worlds*, 1986) raises and then counters several lines of argument against it. That work introduces not only the theory, but its reception among philosophers. The many objections that continue to be published are typically variations on one or other of the lines that Lewis has already canvassed.

Here are some of the major categories of objection:

- **Catastrophic counterintuitiveness** The theory does not accord with our deepest intuitions about reality. This is sometimes called "the incredulous stare", since it lacks argumentative content, and is merely an expression of the affront that the theory represents to "common sense" philosophical and pre-philosophical orthodoxy. Lewis is concerned to support the deliverances of common sense in general: "Common sense is a settled body of theory — unsystematic folk theory — which at any rate we do believe; and I presume that we are reasonable to believe it. (Most of it.)" (1986, p. 134). But *most of it* is not *all of it* (otherwise there would be no place for philosophy at all), and Lewis finds that reasonable argument and the weight of such considerations as theoretical efficiency compel us to accept modal realism. The alternatives, he argues at length, can themselves be shown to yield conclusions offensive to our modal intuitions.

- **Inflated ontology** Some[3] object that modal realism postulates vastly too many entities, compared with other theories. It is therefore, they argue, vulnerable to Occam's razor, according to which we should prefer, all things being equal, those theories that postulate the smallest number of entities. Lewis's reply is that all things are *not* equal, and in particular competing accounts of possible worlds themselves postulate more *classes* of entities, since there must be not only one real "concrete" world (the actual world), but many worlds of a different class altogether ("abstract" in some way or other).

- **Too many worlds** This is perhaps a variant of the previous category, but it relies on appeals to mathematical propriety rather than Occamist principles. Some argue that Lewis's principles of "worldmaking" (means by which we might establish the existence of further worlds by recombination of parts of worlds we already think exist) are too permissive. So permissive are they, in fact, that the total number of worlds must exceed what is mathematically coherent. Lewis allows that there are difficulties and subtleties to address on this front (1986, pp. 89–90). Daniel Nolan ("Recombination unbound", *Philosophical Studies*, 1996, vol. 84, pp. 239–262) mounts a sustained argument against certain forms of the objection; but variations on it continue to appear.

- **Island universes** On the version of his theory that Lewis strongly favours, each world is distinct from every other world by being spatially and temporally isolated from it. Some have objected that a world in which spatio-temporally isolated universes ("island universes") coexist is therefore not possible, by Lewis's theory (see for example Bigelow, John, and Pargetter, Robert, "Beyond the blank stare", *Theoria*, 1987, Vol. 53, pp. 97–114). Lewis's awareness of this difficulty discomforted him; but he could have replied that other means of distinguishing worlds may be available, or alternatively that sometimes there will inevitably be further surprising and counterintuitive consequences — beyond what we had thought we would be committed to at the start of our investigation. But this fact in itself is hardly surprising.

- **Mathematical versus physical reality** Another criticism levelled against modal realism, specifically applied to the mathematical expression of it, Max Tegmark's ultimate ensemble, is that it equates mathematical reality with physical reality:

> Physical existence is something that we have some experience of. We probably can't define it but, like many things we have difficulty defining, we know it when we see it. Mathematical existence is a far weaker thing, but much easier to define. Mathematical existence just means logical self-consistency: this is all that is needed for a mathematical statement to be "true". (Barrow, 2002, pp. 279–80)

A pervasive theme in Lewis's replies to the critics of modal realism is the use of tu quoque argument: *your* account would fail in just the same way that you claim mine would. A major heuristic virtue of Lewis's theory is that it is sufficiently definite for objections to gain some foothold; but these objections, once clearly articulated, can then be turned equally against other theories of the ontology and epistemology of possible worlds.

26.5.2 Stalnaker's response

Robert Stalnaker, while he finds some merit in Lewis's account of possible worlds finds the position to be ultimately untenable. He himself advances a more "moderate" realism about possible worlds, which he terms **modal actualism** (since it holds that all that exists is in fact actual, and that there are no "merely possible" entities."[4] In particular, Stalnaker does not accept Lewis's attempt to argue on the basis of a supposed analogy with the epistemological objection to mathematical Platonism that believing in possible worlds as he (Lewis) imagines them is no less reasonable than believing in mathematical entities such as sets or functions.[5]

26.5.3 Kripke's response

Saul Kripke described modal realism as "totally misguided", "wrong", and "objectionable".[6] Kripke argued that possible worlds were not like distant countries out there to be discovered; rather, we stipulate what is true according to them. Kripke also criticized modal realism for its reliance on counterpart theory, which he regarded as untenable.

26.6 See also

- Counterpart theory
- Impossible world
- Linguistic modality
- Mathematical universe hypothesis
- Multiverse
- Brane cosmology
- J. B. Priestley's Time Plays

26.7 References

[1] David Lewis, *Convention*, 1968, p. 208

[2] David Lewis, *Counterfactuals*, 1973, pp. 87–88

[3] W. V. O. Quine, *"Proportional Objects" in Ontological Relativity and Other Essays'*, 1969, pp.140-147

[4] Stalnaker (1976,1996 both reprinted in Stalnaker 2003)

[5] Stalnaker (1996)

[6] Kripke (1972)

26.8 Bibliography

- David Lewis, *Counterfactuals*, (1973 [revised printing 1986]; Blackwell & Harvard U.P.)

- David Lewis, *Convention: A Philosophical Study*, (1969; Harvard University Press)

- David Lewis, *On the Plurality of Worlds* (1986; Blackwell)

- Saul Kripke, "Identity and Necessity". (*Semantics of Natural Language*, D. Davidson and G. Harman [eds.], [Dordrecht: D. Reidel, 1972]

- David Armstrong, *A Combinatorial Theory of Possibility* (1989; Cambridge University Press)

- John D. Barrow, *The Constants of Nature* (2002; published by Vintage in 2003)

- Colin McGinn, "Modal Reality" (*Reduction, Time, and Reality*, R. Healey [ed.]; Cambridge University Press)

- Stalnaker, Robert (2003). *Ways a world might be: metaphysical and anti-metaphysical essays.* Oxford: Clarendon. ISBN 0-19-925149-5.

- Andrea Sauchelli, "Concrete Possible Worlds and Counterfactual Conditionals", *Synthese*, 176, 3 (2010), pp. 345-56.

Chapter 27

Counterpart theory

In philosophy, specifically in the area of modal metaphysics, **counterpart theory** is an alternative to standard (Kripkean) possible-worlds semantics for interpreting quantified modal logic. Counterpart theory still presupposes possible worlds, but differs in certain important respects from the Kripkean view. The form of the theory most commonly cited was developed by David Lewis, first in a paper and later in his book *On the Plurality of Worlds*.

27.1 Differences from the Kripkean View

Counterpart theory (hereafter "CT"), as formulated by Lewis, requires that individuals exist in only one world. The standard account of possible worlds assumes that a modal statement about an individual (e.g., "it is possible that x is y") means that there is a possible world, W, where the individual x has the property y; in this case there is only one individual, x, at issue. On the contrary, counterpart theory supposes that this statement is really saying that there is a possible world, W, wherein exists an individual that is not x itself, but rather a distinct individual 'x' different from but nonetheless similar to x. So, when I state that I might have been a banker (rather than a philosopher) according to counterpart theory I am saying not that I exist in another possible world where I am a banker, but rather my counterpart does. Nevertheless, this statement about my counterpart is still held to ground the truth of the statement that I might have been a banker. The requirement that any individual exist in only one world is to avoid what Lewis termed the "problem of accidental intrinsics" which (he held) would require a single individual to both have and simultaneously not have particular properties.

In its formalization, counterpart theoretic formalization of modal discourse also departs from the standard formulation by eschewing use of modality operators (Necessarily, Possibly) in favor of quantifiers that range over worlds and 'counterparts' of individuals in those worlds. Lewis put forth a set of primitive predicates and a number of axioms governing CT and a scheme for translating standard modal claims in the language of quantified modal logic into his CT.

In addition to interpreting modal claims about objects and possible worlds, CT can also be applied to the identity of a single object at different points in time. The view that an object can retain its identity over time is often called endurantism, and it claims that objects are 'wholly present' at different moments (see the counterpart relation, below). An opposing view is that any object in time is made up of temporal parts or is perduring.

David Lewis' view on possible worlds is sometimes called modal realism.

27.1.1 The basics

The possibilities that CT is supposed to describe are "ways a world might be" (Lewis 1986:86) or more exactly:

(1) absolutely every way that a world could possibly be is a way that some world is, and

(2) absolutely every way that a part of a world could possibly be is a way that some part of some world is. (Lewis 1986:86.)

Add also the following "principle of recombination," which Lewis describes this way: "patching together parts of different possible worlds yields another possible world [...]. [A]nything can coexist with anything else, [...] provided they occupy distinct spatiotemporal positions." (Lewis 1986:87-88). But these possibilities should be restricted by CT.

27.2 The counterpart relation

The counterpart relation (hereafter C-relation) differs from the notion of identity. Identity is a reflexive, symmetric, and transitive relation. The counterpart relation is only a similarity relation; it needn't be transitive or symmetric. The C-relation is also known as genidentity (Carnap 1967), I-relation (Lewis 1983), and the unity relation (Perry 1975).

If identity is shared between objects in different possible worlds then the same object can be said to exist in different possible worlds (a *trans-world* object, that is, a series of objects sharing a single identity).

27.2.1 Parthood relation

An important part of the way Lewis's worlds deliver possibilities is the use of the parthood relation. This gives some neat formal machinery, mereology. This is an axiomatic system that uses formal logic to describe the relationship between parts and wholes, and between parts within a whole. Especially important, and most reasonable, according to Lewis, is the strongest form that accepts the existence of mereological sums or the thesis of unrestricted mereological composition (Lewis 1986:211-213).

27.3 The formal theory

As a formal theory, counterpart theory can be used to translate sentences into modal quantificational logic. Sentences that seem to be quantifying over possible individuals should be translated into CT. (Explicit primitives and axioms have not yet been stated for the temporal or spatial use of CT.) Let CT be stated in quantificational logic and contain the following primitives:

Wx (x is a possible world)

Ixy (x is in possible world y)

Ax (x is actual)

Cxy (x is a counterpart of y)

We have the following axioms (taken from Lewis 1968):

A1. $Ixy \rightarrow Wy$
(Nothing is in anything except a world)

A2. $Ixy \wedge Ixz \rightarrow y=z$
(Nothing is in two worlds)

A3. $Cxy \rightarrow \exists z Ixz$
(Whatever is a counterpart is in a world)

A4. $Cxy \rightarrow \exists z Iyz$
(Whatever has a counterpart is in a world)

A5. $Ixy \wedge Izy \wedge Cxz \rightarrow x=z$
(Nothing is a counterpart of anything else in its world)

A6. $Ixy \rightarrow Cxx$
(Anything in a world is a counterpart of itself)

A7. $\exists x (Wx \wedge \forall y (Iyx \leftrightarrow Ay))$
(Some world contains all and only actual things)

A8. $\exists x Ax$
(Something is actual)

It is an uncontroversial assumption to assume that the primitives and the axioms A1 through A8 make the standard counterpart system.

27.3.1 Comments on the axioms

- A1 excludes individuals that exist in no world at all. The way an individual is in a world is by being a part of that world, so the basic relation is mereological.

- A2 excludes individuals that exist in more than one possible world. But because David Lewis accepts the existence of arbitrary mereological sums there are individuals that exist in several possible worlds, but they are not possible individuals because none of them have the property of being actual. And that is because it is not possible for such a whole to be actual.

- A3 and A4 make counterparts worldbound, excluding an individual that has a non-worldbound counterpart.

- A5 and A6 restrict the use of the CT-relation so that it is used within a possible world when and only when it is stood in by an entity to itself.

- A7 and A8 make one possible world the unique actual world.

27.3.2 Principles that are not accepted in normal CT

R1 $Cxy \rightarrow Cyx$
(Symmetry of the counterpart relation)

R2 $Cxy \wedge Cyz \rightarrow Cxz$
(Transitivity of the counterpart relation)

R3 $Cy_1x \land Cy_2x \land Iy_1w_1 \land Iy_2w_2 \land y_1 \neq y_2 \rightarrow w_1 \neq w_2$

(Nothing in any world has more than one counterpart in any other world)

R4 $Cyx_1 \land Cyx_2 \land Ix_1w_1 \land Ix_2w_2 \land x_1 \neq x_2 \rightarrow w_1 \neq w_2$

(No two things in any world have a common counterpart in any other world)

R5 $Ww_1 \land Ww_2 \land Ixw_1 \rightarrow \exists y (Iyw_2 \land Cxy)$

(For any two worlds, anything in one is a counterpart of something in the other)

R6 $Ww_1 \land Ww_2 \land Ixw_1 \rightarrow \exists y (Iyw_2 \land Cyx)$

(For any two worlds, anything in one has some counterpart in the other)

27.4 Motivations for Counterpart theory

CT can be applied to the relationship between identical objects in different worlds or at different times. Depending on the subject, there are different reasons for accepting CT as a description of the relation between different entities.

27.4.1 In possible worlds

David Lewis defended Modal realism. This is the view that a possible world is a concrete, maximal connected spatio-temporal region. The actual world is one of the possible worlds; it is also concrete. Because a single concrete object demands spatio-temporal connectedness, a possible concrete object can only exist in one possible world. Still, we say true things like: It is possible that Hubert Humphrey won the 1968 US presidential election. How is it true? Humphrey has a counterpart in another possible world that wins the 1968 election in that world.

Lewis also argues against three other alternatives that might be compatible with possibilism: overlapping individuals, trans-world individuals, and haecceity.

Some philosophers, such as Peter van Inwagen (1985), see no problem with identity within a world. Lewis seems to share this attitude. He says:

> "... like the Holy Roman Empire, it is badly named. [...] In the first place we should bear in mind that Trans-World Airlines is an intercontinental, but not as yet an interplanetary carrier. More important, we should not suppose that we have here any problem with *identity*.
>
> We never have. Identity is utterly simple and unproblematic. Everything is identical to itself; nothing is ever identical to anything else except itself. There is never any problem about what makes something identical to itself; nothing can ever fail to be. And there is never any problem about what makes two things identical; two things never can be identical.
>
> There might be a problem about how to define identity to someone sufficiently lacking in conceptual resources — we note that it won't suffice to teach him certain rules of inference — but since such unfortunates are rare, even among philosophers, we needn't worry much if their condition is incurable.
>
> We *do* state plenty of genuine problems in terms of identity. But we *needn't* state them so." (Lewis 1986:192-193)

Overlapping individuals

An overlapping individual has a part in the actual world and a part in another world. Because identity is not problematic, we get overlapping individuals by having overlapping worlds. Two worlds overlap if they share a common part. But some properties of overlapping objects are, for Lewis, troublesome (Lewis 1986:199-210).

The problem is with an object's accidental intrinsic properties, such as shape and weight, which supervene on its parts. Humphrey could have the property of having six fingers on his left hand. How does he do that? It can't be true that Humphrey has both the property of having six fingers and five fingers on his left hand. What we might say is that he has five fingers *at this* world and six fingers *at that* world. But how should these modifiers be understood?

According to McDaniel (2004), if Lewis is right, the defender of overlapping individuals has to accept genuine contradictions or defend the view that every object has all its properties essentially.

How can you be one year older than you are? One way is to say that there is a possible world where you exist. Another way is for you to have a counterpart in that possible world, who has the property of being one year older than you.

Trans-world individuals

Take Humphrey: if he is a trans-world individual he is the mereological sum of all of the possible Humphreys in the different worlds. He is like a road that goes through different regions. There are parts that overlap, but we can also say that there is a northern part that is connected to the southern part and that the road is the mereological sum of these parts. The same thing with Humphrey. One part of him is in one world, another part in another world.

> "It is possible for something to exist iff it is possible for the whole to exist. That is, iff there is a world at which the whole of it exists. That is, iff there is a world such that quantifying only over parts of that world, the whole of it exists. That is, iff the whole of it is among the parts of some world. That is, iff it is part of some world – and hence not a trans-world individual. Parts of worlds are *possible* individuals; trans-world individuals are therefore *impossible* individuals."

Haecceity

A haecceity or individual essence is a property that only a single object instantiates. Ordinary properties, if one accepts the existence of universals, can be exemplified by more than one object at a time. Another way to explain a haecceity is to distinguish between *suchness* and *thisness*, where thisness has a more demonstrative character.

David Lewis gives the following definition of a haecceitistic difference: "two worlds differ in what they represent *de re* concerning some individual, but do not differ qualitatively in any way." (Lewis 1986:221.)

CT does not require distinct worlds for distinct possibilities – "a single world may provide many possibilities, since many possible individuals inhabit it" (Lewis 1986:230). CT can satisfy multiple counterparts in one possible world.

27.4.2 Temporal parts

Main article: Temporal parts

Perdurantism is the view that material objects are not wholly present at any single instant of time; instead, some temporal parts is said to be present. Sometimes, especially in the theory of relativity as it is expressed by Minkowski, the path traced by an object through spacetime. According to Ted Sider, "Temporal parts theory is the claim that time is like space in one particular respect, namely, with respect to parts."[1] Sider associates endurantism with a C-relation between temporal parts. (*See also:* The argument from temporary intrinsics).

Sider defends a revised way of counting. Instead of counting individual objects, timeline slices or the temporal parts of an object are used. Sider discusses an example of counting road segments instead of roads simpliciter. (Sider 2001:188-192). (Compare with Lewis 1993.) Sider argues that, even if we knew that some material object would go through some fission and split into two, "we would not *say*" that there are two objects located at the same spacetime region. (Sider 2001:189)

How can one predicate temporal properties of these momentary temporal parts? It is here that the C-relation comes in play. Sider proposed the sentence: "Ted was once a boy." The truth condition of this sentence is that "there exists some person stage x prior to the time of utterance, such that x is a boy, and x bears the temporal counterpart relation to Ted." (Sider 2001:193)

27.5 Counterpart theory and the necessity of identity

Kripke's three lectures on proper names and identity, (1980), raised the issues of how we should interpret statements about identity. Take the statement that the Evening Star is identical to the Morning Star. Both are the planet Venus. This seems to be an *a posteriori* identity statement. We discover that the names designate the same thing. The traditional view, since Kant, has been that statements or propositions that are necessarily true are a priori. But in the end of the sixties Saul Kripke and Ruth Barcan Marcus offered proof for the necessary truth of identity statements. Here is the Kripkes version (Kripke 1971):

(1) $\forall x \ \Box \ (x = x)$ [Necessity of self-identity]

(2) $\forall x \forall y \ [x = y \rightarrow \forall P(Px \rightarrow Py)]$ [Leibniz law]

(3) $\forall x \forall y \ [x = y \rightarrow (\ \Box \ (x = x) \rightarrow \ \Box \ (x = y))]$ [From (1) and (2)]

(4) $\forall x \forall y \ [x = y \rightarrow \ \Box \ (x = y)]$ [From the following principle $A \rightarrow B \rightarrow C \Rightarrow A \rightarrow C$ and (3)]

If the proof is correct the distinction between the a priori/a posteriori and necessary/contingent becomes less clear. The same applies if identity statements are necessarily true anyway. (For some interesting comments on the proof, see Lowe 2002.) The statement that for instance "Water is identical to H_2O" is (then) a statement that is necessarily true but a posteriori. If CT is the correct account of modal properties we still can keep the intuition that identity statements are contingent and a priori because counterpart theory understands the modal operator in a different way than standard modal logic.

The relationship between CT and essentialism is of interest. (Essentialism, the necessity of identity, and rigid designators form an important troika of mutual interdependence.) According to David Lewis, claims about an object's essential properties can be true or false depending on context (in Chapter 4.5 in 1986 he calls against constancy, because an absolute conception of essences is constant over the logical space of possibilities). He writes:

> But if I ask how things would be if Saul Kripke had come from no sperm and egg but had been brought by a stork, that makes equally good sense. I create a context that makes my question make sense, and to do so it has to be a context that makes origins not be essential. (Lewis 1986:252.)

27.6 Counterpart theory and rigid designators

Kripke interpreted proper names as rigid designators where a rigid designator picks out the same object in every possible world (Kripke 1980). For someone who accepts contingent identity statements the following semantic problem occurs (semantic because we deal with de dicto necessity) (Rea 1997:xxxvii).

Take a scenario that is mentioned in the paradox of coincidence. A statue (call it "Statue") is made by melding two pieces of clay together. Those two pieces are called "Clay". Statue and Clay seem to be identical, they exist at the same time, and we could incinerate them at the same time. The following seems true:

(7) Necessarily, if Statue exists then Statue is identical to Statue.

But,

(8) Necessarily, if Statue exists then Statue is identical to Clay

is false, because it seems possible that Statue could have been made out of two different pieces of clay, and thus its identity to Clay is not necessary.

Counterpart theory, qua-identity, and individual concepts can offer solutions to this problem.

27.6.1 Arguments for inconstancy

Ted Sider gives roughly the following argument (Sider 2001:223). There is inconstancy if a proposition about the essence of an object is true in one context and false in another. C-relation is a similarity relation. What is similar in one dimension is not similar in another dimension. Therefore, the C-relation can have the same difference and express inconstant judgements about essences.

David Lewis offers another argument. The paradox of coincidence can be solved if we accept inconstancy. We can then say that it is possible for a dishpan and a piece of plastic to coincide, in some context. That context can then be described using CT.

Sider makes the point that David Lewis feels he was forced to defend CT, due to modal realism. Sider uses CT as a solution to the paradox of material coincidence.

27.6.2 Counterpart theory compared to qua-theory and individual concepts

We assume that contingent identity is real. Then it is informative to compare CT with other theories about how to handle *de re* representations.

Qua-theory

Kit Fine (1982) and Alan Gibbard (1975) (according to Rea 1997) are defences of qua-theory. According to qua-theory we can talk about some of an object's modal properties. The theory is handy if we don't think it is possible for Socrates to be identical with a piece of bread or a stone. Socrates qua person is essentially a person.

Individual concepts

According to Rudolf Carnap, in modal contexts variables refer to individual concepts instead of individuals. An individual concept is then defined as a function of individuals in different possible worlds. Basically, individual concepts deliver semantic objects or abstract functions instead of real concrete entities as in CT.

27.6.3 Counterpart theory and epistemic possibility

Kripke accepts the necessity of identity but agrees with the feeling that it still seems that it is possible that Phospherus (the Morning Star) is not identical to Hespherus (the Evening Star). For all we know, it could be that they are different. He says:

> What, then, does the intuition that the table might have turned out to have been made of ice or of

anything else, that it might even have turned out not to be made of molecules, amount to? I think that it means simply that there might have been a table looking and feeling just like this one and placed in this very position in the room, which was in fact made of ice, In other words, I (or some conscious being) could have been qualitatively in the same epistemic situation that in fact obtains, I could have the same sensory evidence that I in fact have, about a table which was made of ice. The situation is thus akin to the one which inspired the counterpart theorists; when I speak of the possibility of the table turning out to be made of various things, I am speaking loosely. This table itself could not have had an origin different form the one it in fact had, but in a situation qualitatively identical to this one with respect to all evidence I had in advance, the room could have contained a table made of ice in place of this one. Something like counterpart theory is thus applicable to the situation, but it applies only because we are not interested in what might not be true of a table given certain evidence. It is precisely because it is not true that this table might have been made of ice from the Thames that we must turn here to qualitative descriptions and counterparts. To apply these notions to genuine de re modalities, is from the present standpoint, perverse. (Kripke 1980:142.)

So to explain how the illusion of necessity is possible, according to Kripke, CT is an alternative. Therefore, CT forms an important part of our theory about the knowledge of modal intuitions. (For doubt about this strategy, see Della Roca, 2002. And for more about the knowledge of modal statements, see Gendler and Hawthorne, 2002.)

27.7 Arguments against Counterpart theory

The most famous is Kripke's "Humphrey Objection". Because a counterpart is never identical to something in another possible world Kripke raised the following objection against CT:

"Thus if we say "Humphrey might have won the election (if only he had done such-and-such), we are not talking about something that might have happened to *Humphrey* but to someone else, a "counterpart"." Probably, however, Humphrey could not care less whether someone *else*, no matter how much resembling him, would have been victorious in another possible world. Thus, Lewis's view seems to me even more bizarre than the usual notions of transworld identification that it replaces. (Kripke 1980:45 note 13.)

One way to spell out the meaning of Kripke's claim is by the following imaginary dialogue: (Based on Sider MS)

Against: Kripke means that Humphrey himself doesn't have the property of possibly winning the election, because it is only the counterpart that wins.

For: The property of possibly winning the election is the property of the counterpart.

Against: But they can't be the same property because Humphrey has different attitudes to them: he cares about he himself having the property of possibly winning the election. He doesn't care about the counterpart having the property of possibly winning the election.

For: But properties don't work the same way as objects, our attitudes towards them can be different, because we have different descriptions – they are still the same properties. That lesson is taught by the paradox of analysis.

CT is inadequate if it can't translate all modal sentences or intuitions. Fred Feldman mentioned two sentences (Feldman 1971):

(1) I could have been quite unlike what I in fact am.

(2) I could have been more like what you in fact are than like what I in fact am. At the same time, you could have been more like what I in fact am than what you in fact are.

27.8 See also

- Identity (philosophy)
- Modal logic
- Modal realism
- Many-worlds interpretation

27.9 References

- Balashov, Yuri, 2007, "Defining endurance", Philosophical studies, 133:143-149.

- Carnap, Rudolf, 1967, The Logical Structure of the World, trans. Rolf A. George, Berkeley: University of California Press.

- Della Rocca, Michael, 2002, "Essentialism versus Essentialism", in Gendler and Hawthorne 2002.

- Feldman, Fred, 1971 "Counterparts", Journal of Philosophy 68 (1971), pp. 406–409.

- Fine, Kit, 1982, "Acts, Events and Things.", in W. Leinfellner, E. Kraemer, and J. Schank (eds.) Proceedings of the 6th International Wittgenstein Symposium, pp. 97–105, Wien: Hälder Pichler-Tempsky.

- Gendler, Tamar Szabó and Hawthorne, John, 2002, Conceivability and Possibility, Oxford: Oxford University Press.

- Gibbard, Alan, 1975, "Contingent Identity", Journal of Philosophical Logic 4, pp. 197–221 or in Rea 1997.

- Hawley, Kathrine, 2001, *How Things Persist*, Oxford: Clarendon Press.

- Kripke, Saul, 1971, "Identity and Necessity", in Milton K. Munitz, Identity and Individuation, pp. 135–64, New York: New York University Press.

- Kripke, Saul, 1980, Naming and Necessity, Cambridge: Harvard University Press.

- Lewis, David, 1968, "Counterpart Theory and Quantified Modal Logic", Journal of Philosophy 65 (1968), pp 113–26.

- Lewis, David, 1971, "Counterparts of Persons and Their Bodies", Journal of Philosophy 68 (1971), pp 203–11 and in Philosophical Papers I.

- Lewis, David, 1983, "Survival and Identity", in Amelie O. Rorty [ed.] The Identities of Persons (1976; University of California Press.) and in Philosophical Papers I, Oxford: Oxford University Press.

- Lewis, David, 1986, On the Plurality of Worlds, Blackwell.

- Lewis, David, 1993, "Many, But Almost One", in Keith Campbell, John Bacon and Lloyd Reinhart eds., Ontology, Causality and Mind:Essays in Honour of D.M. Armstrong, Cambridge:Cambridge University Press.

- Lowe, E. J., 2002, A survey of Metaphysics, Oxford: Oxford University Press.

- Mackie, Penelope, 2006, How Things Might Have Been – Individuals, Kinds and essential Properties, Oxford: Clarendon Press.

- McDaniel, Kris, 2004, "Modal Realism with Overlap", The Australasian Journal of Philosophy vol. 82, No. 1, pp. 137–152.

- Merricks, Trenton, 2003, "The End of Counterpart Theory," Journal of Philosophy 100: 521-549.

- Rea, Michael, ed., 1997, Material Constitution – A reader, Rowman & Littlefield Publishers.

- Sider, Ted, 2001, Four-dimensionalism. Oxford: Oxford University Press.

- Sider, Ted, XXXX, Beyond the Humphrey objection.

- Perry, John, ed., 1975, Personal Identity, Berkeley: University of California Press

- van Inwagen, Peter, 1985, "Plantinga on Trans-World Identity", in Alvin Plantina: A Profile, ed. James Tomberlin & Peter van Inwagen, Reidel.

[1] Sider, et al. (2008) *Contemporary Debates in Metaphysics*, "Temporal Parts".

27.10 External links

- Counterpart theory at PhilPapers

- Possible objects entry in the *Stanford Encyclopedia of Philosophy*

Chapter 28

The Fabric of Reality

The Fabric of Reality is a 1997 book by physicist David Deutsch. The text was initially published on August 1, 1997 by Viking Adult and Deutsch wrote a followup book entitled *The Beginning of Infinity*, which was published in 2011.

28.1 Overview

The book expands upon his views of quantum mechanics and its implications for understanding reality. This interpretation, which he calls the *multiverse* hypothesis, is one of a four-strand Theory of Everything (TOE).[1]

28.1.1 The four strands

1. Hugh Everett's many-worlds interpretation of quantum physics, "The first and most important of the four strands".

2. Karl Popper's epistemology, especially its anti-inductivism and its requiring a realist (non-instrumental) interpretation of scientific theories, and its emphasis on taking seriously those bold conjectures that resist falsification.

3. Alan Turing's theory of computation especially as developed in Deutsch's "Turing principle", Turing's Universal Turing machine being replaced by Deutsch's universal quantum computer. ("*The* theory of computation is now the quantum theory of computation.")

4. Richard Dawkins's refinement of Darwinian evolutionary theory and the modern evolutionary synthesis, especially the ideas of replicator and meme as they integrate with Popperian problem-solving (the epistemological strand).

28.1.2 Deutsch's TOE

His theory of everything is (weakly) emergentist rather than reductive. It aims not at the reduction of everything to particle physics, but rather at mutual support among multiverse, computational, epistemological, and evolutionary principles.

28.2 Reception

Critical reception has been positive.[2][3][1][4] The *New York Times* wrote a mixed review for *The Fabric of Reality*, writing that it "is full of refreshingly oblique, provocative insights. But I came away from it with only the mushiest sense of how the strands in Deutsch's tapestry hang together."[5] *The Guardian* was more favorable in their review, stating "This is a deep and ambitious book and there were plenty of moments when I was out of my depth (the Platonic dialogue between Deutsch and a Crypto-inductivist left me with a pronounced sinking feeling). But the sheer adventure of thinking not just out of the envelope but right out of the Newtonian universe is exhilarating."[6]

28.3 See also

- *The Beginning of Infinity*
- *The 4 Percent Universe*
- Simulated reality
- Solipsism

28.4 References

[1] Shankel, Jason. "David Deutsch's The Fabric of Reality connects the spookier elements of quantum mechanics". io9. Retrieved 2 September 2015.

[2] Macfie, Alexander Lyon (20 March 2015). "The fabric of reality (review)". *Rethinking History: The Journal of Theory and Practice*: 1–9. doi:10.1080/13642529.2015.1022997. Retrieved 2 September 2015.

[3] Whitaker, Andrew (2001). "The Fabric of Reality (review)". *Studies in History and Philosophy of Modern Physics* **32** (1): 137–141. doi:10.1016/S1355-2198(00)00032-0. Retrieved 2 September 2015.

[4] Price, Huw (June 1999). "Reviewed Work: The Fabric of Reality by David Deutsch". *The British Journal for the Philosophy of Science* **50** (2): 309–312. Retrieved 2 September 2015.

[5] Johnson, George. "Shadow Worlds". New York Times. Retrieved 2 September 2015.

[6] Radford, Tim. "David Deutsch's multiverse carries us beyond the realms of imagination". The Guardian. Retrieved 2 September 2015.

28.5 Text and image sources, contributors, and licenses

28.5.1 Text

- **Multiverse** *Source:* https://en.wikipedia.org/wiki/Multiverse?oldid=689060672 *Contributors:* Derek Ross, Dan~enwiki, Bryan Derksen, The Anome, Css, BenBaker, Ant, B4hand, Modemac, Hephaestos, Olivier, Patrick, Boud, Michael Hardy, Tim Starling, Modster, Delirium, William M. Connolley, Angela, Ciphergoth, Poor Yorick, [212], Emperorbma, Peter Damian (original account), Dcoetzee, Morwen, Fairandbalanced, BenRG, Sjorford, Tlogmer, Nurg, Rursus, Hadal, Wereon, Johnstone, Giftlite, Smjg, Graeme Bartlett, Mshonle~enwiki, Gtrmp, Mporter, Barbara Shack, ShaunMacPherson, Bfinn, Lefty, Anville, Beardo, Gracefool, Pascal666, Siroxo, Eequor, Pne, Hannes Karnoefel, Fishal, Gdm, Pcarbonn, MadIce, Khaosworks, CSTAR, Slivester, Sam Hocevar, JeffreyN, Soilguy5, Sam, Fermion, Waza, MakeRocketGoNow, Grunt, Mike Rosoft, Freakofnurture, Dbachmann, Paul August, Petersam, BlueNight, Zippedmartin, Livajo, Tverbeek, Haxwell, RoyBoy, Bobo192, Androo, K0hlrabi, Rbj, Reuben, Dlarmore, Thanos6, I9Q79oL78KiL0QTFHgyc, Martg76, Apostrophe, Zetawoof, Dillee1, Alansohn, SnowFire, Walter Görlitz, Arthena, Hu, Titanium Dragon, Knowledge Seeker, Pfahlstrom, DV8 2XL, Pwqn, Netkinetic, Kitch, Falcorian, Ott, Zntrip, Dmitrybrant, Stemonitis, Firsfron, Roboshed, Mel Etitis, FeanorStar7, ApLundell, Netdragon, Jeff3000, Meneth, Rchamberlain, CharlesC, Noetica, Joke137, Ictlogist, Teflon Don, Drbogdan, Rjwilmsi, Electricnet, Salix alba, Mike Peel, Bubba73, Dianelos, Brighterorange, Cfortunato, Cassowary, JeffStickney, FlaBot, Pitamakan, Old Moonraker, Latka, Saswann, Phoenix2~enwiki, Chobot, DVdm, Bgwhite, Whosasking, PointedEars, YurikBot, ThunderPeel2001, Fabartus, Bhny, Admiral Roo, Hellbus, Mpfrank, Eleassar, NawlinWiki, Joncolvin, Mike18xx, Spikehay, Inhighspeed, Taigei, Obey, RonCram, Shadowblade, Tomisti, WAS 4.250, Deville, Joecab, Moogsi, KGasso, Xaxafrad, Gating, JoanneB, Johnpseudo, Ilmari Karonen, Markbenecke, Hathaldir~enwiki, Johnbmckenna, NeilN, Philip Stevens, Eog1916, SmackBot, Reedy, K1234567890y, MJMyers2~enwiki, Vald, Jagged 85, Eskimbot, KelleyCook, Kintetsubuffalo, Alex earlier account, Gilliam, Portillo, Ohnoitsjamie, Kmarinas86, Armeria, Chris the speller, Glastohead, Apeloverage, Foosher, Jerome Charles Potts, Octahedron80, Scwlong, Vanished User 0001, Zaian, Xiner, Rrburke, Wen D House, Nakon, Geoffr, Adrigon, Metamagician3000, Kurrupt3d, Vina-iwbot~enwiki, Ollj, CHBMch05, Ohconfucius, Lambiam, Rory096, Amnuel~enwiki, Police officer, Vampus, JoshuaZ, IronGargoyle, JHunterJ, Makyen, Noah Salzman, BenRayfield, Ehheh, AxG, Z E U S, DCNanney, Clq, SandyGeorgia, Eridani, Doczilla, Dl2000, ShakingSpirit, Iridescent, Joseph Solis in Australia, Newone, Imad marie, Blehfu, Gco, Valoem, Xammer, CRGreathouse, Vyznev Xnebara, Rwflammang, Editorius, K00bine, Sanspeur, Hemlock Martinis, Mchmike, Peterdjones, Gogo Dodo, DarthSidious, Michael C Price, Kozuch, Danogo, Silicianer, UberScienceNerd, JamesAM, Thijs!bot, Daniel Newman, Wylfing, GentlemanGhost, Headbomb, Second Quantization, Parsiferon, Chunminghan, Riction, AntiVandalBot, AbstractClass, Gioto, Caledones, Christinedoby, EdgarCarpenter, QuiteUnusual, Handsaw, Darklilac, Tim Shuba, Ran4, LinkinPark, Jenattiyeh, Geshel, .anacondabot, Magioladitis, VoABot II, Hawkaris, Craze3, Hiplibrarianship, Winterus, Urco, Wikianon, Rickard Vogelberg, Greenguy1090, MartinBot, STBot, Anarchia, The Anonymous One, EdBever, Tgeairn, Trusilver, Multisport3, UBeR, MattB2, Arion 3x3, McSly, Tarotcards, RickardV, Richard D. LeCour, NewEnglandYankee, Mirage GSM, Dorftrottel, VolkovBot, Camrn86, Riraito, Sunwukongmonkeygod, TXiKiBoT, Mercurywoodrose, Kilo39, Kriak, AstroWiki, Albertmin, Anonymous Dissident, Riemann Zeta, Shiritai, Aaron Bowen, Masqline, BotKung, Jamelan, Mjkrol, Parsifal, Roy7, Sapphic, BrianY, Zx-man, Logan, Neparis, Hrafn, Peter.thelander, Silver Spoon, Spartan, Scarian, Malcolmxl5, Richardvanas1, Krawi, Curst Saden, ThisIsRealPuma, Alexbook, Yintan, Bernard Marx, Araignee, Hac13, Flyer22 Reborn, Radon210, Dominik92, Wmpearl, Kriss12, Sedecrem, OKBot, Kurtilein, BigBang616, Escape Orbit, GeorgiaSLee, Vinay Jha, IomPsinso, Athenean, XDanielx, WikipedianMarlith, Martarius, ClueBot, Troublemane, Boing! said Zebedee, Nfenn, K a r n a, Undisputedloser, Chimesmonster, Masterpiece2000, Stupidjassi, SolomonFreer, BobKawanaka, Hans Adler, Rui Gabriel Correia, Aitias, Yboc2000, Stoned philosophaster, Editor2020, Miami33139, DumZiBoT, Bridies, Wednesday Next, XLinkBot, Ajcheema, ErkinBatu, Thatguyflint, Cloudblazer, Santasa99, Mbroderick271, Addbot, Ito123456789, Moosester, Uruk2008, DOI bot, Tcncv, Discrepancy, Morriswa, Longhandle1137, ExternalGazer, Download, Bwikiroa, Richard-r-v, Smurfsider, Tide rolls, MuZemike, Bartledan, SeniorInt, LuK3, Athanatsius, Luckas-bot, Yobot, Ht686rg90, HieronymousCrowley, GateKeeper, Swister-Twister, Naroood, MrH3MinuteMile, Rickybaksh, Kookyunii, AnomieBOT, 1exec1, Kingpin13, Projectyugo, Materialscientist, Citation bot, Allen234, Xqbot, Addihockey10, Smk65536, J JMesserly, Coretheapple, J04n, Foreignshore, Artemka373, Sophus Bie, Keeganp, Another disinterested reader, Steven Avraham Rosten, KronicTOOL, Alxeedo, Steve Quinn, Machine Elf 1735, Citation bot 1, Anthony on Stilts, Osprey9713, Gil987, Pinethicket, Three887, Tom.Reding, Dazedbythebell, Chrisdann, Casimir9999, BlackHades, Tlhslobus, Zhonghuo~enwiki, SkyMachine, Trappist the monk, Lotje, Nickyus, Bacabed, Bluefist, TheGrimReaper NS, Jeffrd10, Skakkle, Jesse V., Twoshotted, Beyond My Ken, Djf2564, DASHBot, EmausBot, WikitanvirBot, RA0808, Annaalexandrae, Wesley J M, Slightsmile, Wikipelli, Slawekb, Solomonfromfinland, Rafi5749, Italia2006, ZéroBot, Naiveandsilly, Daonguyen95, Chasrob, JamesGeddes, Sgerbic, RUBEN TESOLIN, Hyblackeagle22, Red Echidna, AnnSec, AManWithNoPlan, PoisonGM, 09dhummel, Tiiliskivi, Hanjunjade, Crux007, ChuispastonBot, Terraflorin, Spydra, 28bot, Isocliff, Petrb, ClueBot NG, Thanose, Satellizer, Kikichugirl, Movses-bot, Tabletrack, Primergrey, O.Koslowski, Dream of Nyx, Parthdu, North Atlanticist Usonian, Helpful Pixie Bot, Bibcode Bot, Jeraphine Gryphon, BG19bot, Akash pagla, MusikAnimal, Rnm102696, Future kosmos, In another multiverse, Meabandit, Achowat, GreenUniverse, OCCullens, BattyBot, Ha33 blondechick, Mdann52, Nerudography, ChrisGualtieri, Mediran, EuroCarGT, BorealisXIV, Quafman, Dexbot, MarioCGR, Lscagle, Cwobeel, Sminthopsis84, Ddddurga, CuriousMind01, Lugia2453, Graphium, Vibhor.sen, Marvelous8, Asdf62872, Reatlas, Maniesansdelire, Martinsolis1970, Epicgenius, Debouch, Draviste, Sneazy, Jakec, Dustin V. S., Rolf h nelson, CensoredScribe, Tlönorbis, Matthewunthird, Christopherryan100, Jwratner1, Ginsuloft, Kogge, DrRNC, Aubreybardo, Cr.marthi, Charizard5904, MagicatthemovieS, Qwertyuiopasdfghjkl111, Romponu, RLight1, Tim001Jack, Bluesnickers, Samstan2100, Monkbot, DR177, Jamekam, Ffioninwonderland, Johnevella, Thundergodz, MarquisMartinez, DanBalance, Anunaki truth, Tetra quark, Smalegander5, Isambard Kingdom, Sleepy Geek, Anand2202, Kripmo, Thomasberggren, Multiverse Guy, SocraticOath, Nøkkenbuer, Sweggysweg, Ksb1231, Daveptt, Phseek, Youknowwhatimsayin, Algaks236, Sethrhales and Anonymous: 641

- **Universe** *Source:* https://en.wikipedia.org/wiki/Universe?oldid=690140309 *Contributors:* AxelBoldt, Lee Daniel Crocker, CYD, Bryan Derksen, Ed Poor, Wayne Hardman, XJaM, Aldie, William Avery, Montrealais, Stevertigo, Lir, Nealmcb, Patrick, Boud, Michael Hardy, Nixdorf, Liftarn, MartinHarper, Ixfd64, Minesweeper, Looxix~enwiki, Ahoerstemeier, Mac, CatherineMunro, Suisui, Angela, JWSchmidt, Setu, Glenn, AugPi, Andres, Evercat, Hectorthebat, Samuel~enwiki, Pizza Puzzle, Timwi, Janko, Hydnjo, DJ Clayworth, Tpbradbury, DW40, Maximus Rex, Furrykef, Saltine, Xevi~enwiki, Joseaperez, Topbanana, Fvw, Gakrivas, Pakaran, BenRG, Frazzydee, Jni, Phil Boswell, Rossnixon, Paranoid, Astronautics~enwiki, Fredrik, Altenmann, Peak, Yelyos, Nurg, Romanm, Gandalf61, Mirv, Academic Challenger, DHN, Borislav, Johnstone, Aetheling, Guy Peters, Mattflaschen, Cordell, SpellBott, Giftlite, Dbenbenn, Graeme Bartlett, Awolf002, Barbara Shack, Herbee, Mark.murphy, Peruvianllama, Everyking, No Guru, Michael Devore, Bensaccount, Jcobb, Pascal666, Eequor, Matt Crypto, Python eggs, Jackol, Bobblewik, Golbez, Chowbok, Gadfium, Utcursch, Andycjp, R. fiend, Gdm, Zeimusu, Quadell, Antandrus, Beland, OverlordQ, Lesgles, Kaldari, Yafujifide, ShakataGaNai, Karol Langner, JimWae, Latitude0116, Mike Storm, Kevin B12, Jawed, Icairns, CesarFelipe, Zfr, Karl-Henner, Neutrality, TJSwoboda, Trevor MacInnis, Randwicked, Canterbury Tail, RevRagnarok, Mike Rosoft, Shahab, Freakofnurture, A-giau, Naryathegreat, Discospinster, Eb.hoop, Rich Farmbrough, Vsmith, Jpk, Florian Blaschke, StephanKetz, YUL89YYZ, Jordancpeterson, Zamfi, Dbachmann, Nchaimov, Martpol, Aardark, SpookyMulder, ESkog, Brian0918, RJHall, El C, Rgdboer, Art LaPella, Bjoern~enwiki, Orlady, Sajt, Adambro, Guettarda, Bobo192, Army1987, Androo, Smalljim, K0hlrabi, Maureen, ParticleMan, I9Q79oL78KiL0QTFHgyc, Juzeris, Sawadeekrap, Joe Jarvis, Acjelen, Nk, VBGFscJUn3, Ardric47, MPerel, Sam Korn, Thialfi, Krellis, Storm Rider, Stephen G. Brown, Honeycake, Alansohn, Atlant, Mr Adequate, Jeltz, WTGDMan1986, Andrewpmk, Plumbago, Bblackmoor, AzaToth, Lightdarkness, Fritzpoll, Kel-nage, Malo, Snowolf, Magnoliasucks, Wtmitchell, Velella, ProhibitOnions, Knowledge Seeker, RaiderRobert, Vcelloho, Eddie Dealtry, Bsadowski1, Reaverdrop, SteinbDJ, MIT Trekkie, Johntex, HenryLi, Kazvorpal, Oleg Alexandrov, Quirkie, WilliamKF, Jeffrey O. Gustafson, Woohookitty, Mindmatrix, FeanorStar7, Shreevatsa, Daniel Case, Uncle G, Savantnavas, Ruud Koot, WadeSimMiser, JeremyA, Chochopk, MONGO, Jok2000, Uris, Trevor Andersen, Jleon, Bbatsell, Sengkang, GregorB, Andromeda321, SDC, CharlesC, TheAlphaWolf, Joke137, Christopher Thomas, Palica, Dysepsion, GSlicer, Wulfila, Rnt20, Graham87, Deltabeignet, Magister Mathematicae, BD2412, Zeroparallax, FreplySpang, Zoz, Canderson7, Drbogdan, Jorunn, Rjwilmsi, Mayumashu, Joe Decker, P3Pp3r, Nightscream, Koavf, Zbxgscqf, Jake Wartenberg, Rillian, SMC, Mike Peel, The wub, DoubleBlue, MarnetteD, Yamamoto Ichiro, Dionyseus, FayssalF, Old Moonraker, Chanting Fox, RexNL, Gurch, TheDJ, Gakon5, Thewolrab, TeaDrinker, Dsewell, Butros, King of Hearts, Chobot, Sharkface217, DVdm, Citizen Premier, Scoo, Napate, Gwernol, Wjfox2005, The Rambling Man, Siddhant, Siddharth Prabhu, YurikBot, Wavelength, Spacepotato, Err0neous, Vedranf, Splintercellguy, Sceptre, Blightsoot, Nipponese, Jimp, Retodon8, StuffOfInterest, RussBot, Petiatil, Hyad, Anonymous editor, Bhny, SpuriousQ, Stephenb, Chaos, Bullzeye, NawlinWiki, SEWilcoBot, Neural, Grafen, Erielhonan, Jaxl, RazorICE, Nsmith 84, Irishguy, Randolf Richardson, Chrisbrl88, Matticus78, Rmky87, Haoie, Saggipie, Iamnotanorange~enwiki, Epipelagic, SFC9394, Roy Brumback, DeadEyeArrow, Psy guy, Martinwilke1980, Nlu, Dna-webmaster, Dv82matt, Jpmccord, 2over0, Lt-wiki-bot, Ageekgal, Breakfastchief, Theda, Closedmouth, Arthur Rubin, KGasso, Nemu, Th1rt3en, Reyk, Exodio, GraemeL, JoanneB, LeonardoRob0t, Leeannedy, ArielGold, Caco de vidro, Aeosynth, RG2, JuniorMuruin, Serendipodous, DVD R W, Eog1916, Dupz, Kicking222, Sardanaphalus, SmackBot, Roger Davies, Mehranwahid, Ashill, Kurochka, Zazaban, KnowledgeOfSelf, Olorin28, McGeddon, Unyoyega, Jacek Kendysz, Jagged 85, Davewild, WookieInHeat, Delldot, Hardyplants, ZerodEgo, Shai-kun, DreamOfMirrors, Gaff, Onsly, JFHJr, Gilliam, Ohnoitsjamie, Wlmg, Skizzik, BirdValiant, Saros136, Amatulic, Rrscott, Persian Poet Gal, Telempe, Exploreuniverse, Miquonranger03, MalafayaBot, Silly rabbit, SchfiftyThree, Hibernian, Hurdygurdyman1234, Octahedron80, EdgeOfEpsilon, Patriarch, DHN-bot~enwiki, Sbharris, Sahsan~enwiki, Darth Panda, Firetrap9254, Bangarangmanchester, Diyako, Scwlong, Tsca.bot, Shalom Yechiel, Hve, Vanished User 0001, Vere scatman, Yidisheryid, Xiner, Rrburke, Addshore, Lobner, SundarBot, UU, Madman2001, Aldaron, Krich, Tvaughn05, Cybercobra, Kntrabssi, John D. Croft, Craner Murdock, Dreadstar, RandomP, Lcarscad, Alasdair Routh, BullRangifer, Drooling Sheep, Orczar, Kotjze, Iamorlando, Evlekis, Bejnar, Kukini, Ollj, Ged UK, DorJ, Weatherman1126, SashatoBot, Lambiam, Danielrcote, Dr. Sunglasses, 007david, Abob6, Kuru, John, T g7, MagnaMopus, N3bulous, Buchanan-Hermit, Kipala, SilkTork, Erdelyiek, Sir Nicholas de Mimsy-Porpington, JorisvS, Mgiganteus1, IronGargoyle, Ekrub-ntyh, Ckatz, Ian Dalziel, The Bread, Smith609, Beetstra, Hypnosifl, Waggers, Doczilla, Dr.K., EEPROM Eagle, Jose77, Yresh, MarkThomas, Autonova, Hu12, ThuranX, Nehrams2020, Iridescent, K, Hurricanefloyd, Shoeofdeath, Newone, NativeForeigner, J Di, Aeternus, AGK, Courcelles, Tawkerbot2, JRSpriggs, Firewall62, Chetvorno, Xammer, Uq, MarkTB, JForget, Friendly Neighbour, Ale jrb, Insanephantom, Dycedarg, Cytocon, Scohoust, Albert.white, Woudloper, JohnCD, Dub8lad1, Mr plant420, Runningonbrains, Lawnchair On Jupiter, CuriousEric, MarsRover, Geniustwin, Joelholdsworth, WeggeBot, Awesome streak, Lokal Profil, Karenjc, Myasuda, Phase Theory, Gregbard, Icarus of old, Cydebot, Ryan, Treybien, WillowW, Grahamec, Perfect Proposal, Steel, Peterdjones, UncleBubba, Gogo Dodo, Travelbird, FellowWikipedian, Frosty0814snowman, Llort, ST47, Scroggie, Eu.stefan, Wildnox, Tawkerbot4, Doug Weller, Moingv, Dchristle, DumbBOT, Hontogaichiban, Kozuch, Omicronpersei8, RotaryAce, Satori Son, Mattisse, Malleus Fatuorum, Joernderschlaue, Thijs!bot, Epbr123, Barticus88, Mbell, 271828182, Ramananrv123, Hazmat2, Keraunos, Timo3, Mojo Hand, The Dark Side, Headbomb, Marek69, West Brom 4ever, Tapir Terrific, Kathovo, Peter Gulutzan, Picus viridis, Tellyaddict, Cool Blue, Dfrg.msc, AgentPeppermint, Pure maple sugar, Elert, Futurebird, Escarbot, Stannered, Mentifisto, WikiSlasher, AntiVandalBot, Majorly, Yonatan, Kba, Seaphoto, Quintote, Voortle, Nseidm1, Mal4mac, Jj137, Scepia, Geogeogeo, Dylan Lake, Danger, Spencer, Larry Lawrence, Legare, Myanw, PresN, Canadian-Bacon, JAnDbot, Jimothytrotter, Vorpal blade, Davewho2, Barek, MER-C, Epeefleche, The Transhumanist, DarkLouis, Fetchcomms, Andonic, Hut 8.5, Rdht, Snibbe, Badacmw90, Schmackity, ILSS, Murgh, Bongwarrior, VoABot II, MartinDK, Sushant gupta, AuburnPilot, JNW, Mclay1, Jéské Couriano, Think outside the box, Rivertorch, Depolarizer, Nyttend, Sruk77, Aka002, SparrowsWing, Bubba hotep, BrianGV, Fabrictramp, Catgut, Animum, Spacegoat, Bloodredrover, JJ Harrison, Mlhooten, Just James, DerHexer, Floria L, Dirtyharry2, Patstuart, Jdorwin, Sjtarr, NatureA16, B9 hummingbird hovering, Blacksqr, Sonikkua, Jackson Peebles, Hdt83, MartinBot, NAHID, Meduban, Jim.henderson, GomeonaFinnigan, Rettetast, Keith D, CommonsDelinker, AlexiusHoratius, PrestonH, WelshMatt, Ssolbergj, AlphaEta, Watch37264, J.delanoy, Pharaoh of the Wizards, JEREMYBB, Tom Kitt, Ali, MikeBaharmast, Uncle Dick, Ciotog, Maurice Carbonaro, Brest, All Is One, G. Campbell, Q2op, Barts1a, Katalaveno, Ncmvocalist, McSly, Mikael Häggström, Gurchzilla, Bilbobee, Pyrospirit, Qazwsx197966, Spig a digs, Vanished User 4517, TomasBat, NewEnglandYankee, Djambalawa, Rebel700, Trilobitealive, SJP, Bobianite, LeighvsOptimvsMaximvs, Jorfer, Zojj, Mufka, Student7, Rickmeister~enwiki, Terik brunson, MetsFan76, KylieTastic, Mattu00, Remember the dot, Gwen Gale, Sinep2, Vanished user 39948282, DavyJonesGSB, Robbiemasters89, HiEv, Cuckooman4, Bonadea, Rickmeister6, Dude00311, JavierMC, Izno, Martial75, Xiahou, Squids and Chips, Steel1943, CardinalDan, Idioma-bot, Funandtrvl, Thedjatclubrock, ABF, Jeff G., Fences and windows, Soriano9, Abhiag, Philip Trueman, Jhon montes24, TXiKiBoT, Dang3210, Cosmic Latte, Kip the Dip, Vipinhari, Canuckle, Hqb, GDonato, DarrynJ, Ridernyc, Anonymous Dissident, GcSwRhIc, Vishal144, Qxz, Someguy1221, Trahern1994, Anna Lincoln, Lradrama, Clarince63, John haley, Patssle, JhsBot, Bob Andolusorn, Abdullais4u, Fbs. 13, LeaveSleaves, Manchurian candidate, UnitedStatesian, Dantheman2008, Geometry guy, Saturn star, Nighthawk380, Knightshield, SheffieldSteel, Billinghurst, Lamro, Maethordaer, Perníček~enwiki, Falcon8765, Enviroboy, FKmailliW, Seresin, Someguyonthestreet, Agüeybaná, Brianga, Flyingostrich, Thealltruth, Wavehunter, AlleborgoBot, Baaleos, Logan, Domi33, Neilk9393, Scottywong, NHRS2010, Hcagri, EmxBot, Deconstructhis, Futuristcorporation, Blah987654, TimProof, Theoneintraining, HarryMcDeath, Shroitman, Brooktree, SieBot, Dusti, Ttony21, K. Annoyomous, Renil~enwiki, Gerakibot, Claus Ableiter, Caltas, Muraabi6, ConfuciusOrnis, Yintan, Kiefer100, Srushe, Mookiefurr, Keilana, Who3, Flyer22 Reborn, TitanOne, The Evil Spartan, MinorContributor, Xingzeng, Oda Mari, ShinobiX200, Terper, Jojalozzo, Pontoots9, Jamiepgs, JetLover, Myotis, Teles.ME, Aperseghin, Wmpearl, Oxymoron83, RobertMel, Faradayplank, Nuttycoconut, Zharradan.angelfire, Vijinjain, KibaKibbles, Lightmouse, Beej175560, Jhacob, Ks0stm, RyanParis, Fratrep, Sunrise, Pediainsight, Silvergoat, LonelyMarble, Reneeholle, StaticGull, Mygerardromance, X31forest, Ascidian, Nathan1991nathan1991, LAS1180, Neo., Brave warrior, Moomoomoostwos, Pinkadelica, Pyrophotographer, Supraboy001, Denisarona, Escape Orbit, Benjamin Nicholas Johnston, Marmenta, Athenean, WikipedianMarlith, Jimmyjkjr, Loren.wilton, Martarius, Sfan00 IMG, FlamingSilmaril, Elassint, ClueBot, Compdude47, Rumping, GorillaWarfare, The Thing That Should Not Be, Topher208, Eusticeconway, IceUnshattered, Plastikspork, EMC125, Wysprgr2005, Synthiac, Johnny4netglimse, Gopher65, Alecsdaniel, Aj767, Drmies, VQuakr, Ceris2, Uncle Milty, J8079s, Shadowdemon936, CounterVandalismBot, Wirelesspp, 03ctodd, Blanchardb, Agge1000, Kbev, Neverquick, ChandlerMapBot, Deadman3215, Phobiaphobia, Awickert, Excirial, Pumpmeup, Alan268, -Midorihana-, Alexbot, Jusdafax, M4gnum0n, Sinteractive, Robbie098, Erebus Morgaine, McLovin123459, Alejandrocaro35, SeeYouNextTuesday999, Adric Kearney, Parresh, Prancibaldfpants, NuclearWarfare, Cenarium, Promethean, Vendeka, Hans Adler, Kentgen1, Lilbrew369, Razorflame, Noosentaal, Olliegodwin, La Pianista, Panos84, Thingg, Lindberg, Friedpotatoeparty, Darren23, Porthos0, Aitias, Galor612, M karthikkannan, Versus22, Mancini141, SoxBot III, Apparition11, Goodvac, B15nes7, Vanished user uih38riiw4hjlsd, Vanished User 1004, DumZiBoT, Jolekweatin, TimothyRias, InternetMeme, Williamrlinden, Anubad95, Skunkboy74, BarretB, Paidoantonio, Arianewiki1, XLinkBot, DrOxacropheles, Popol0707, Jovianeye, Rror, Gwark, Grubtatorship, Purnajitphukon, Kasyap.d, Coldplayrock08, Ilikepie2221, Denton22, Coreylook, NellieBly, Mifter, Maruthi Achsara, Noctibus, Aunt Entropy, Whizmd, ZooFari, Abomasnow, MaizeAndBlue86, Truthnlove, Dbannie07, Lemmey, HexaChord, Bigfatmonkeyturd, Houdabouda123, King Pickle, Addbot, Willking1979, Tonezilla88, Moosester, Some jerk on the Internet, DOI bot, Tcncv, Shitonmydick1234, Fyrael, Landon1980, Vedjain, Captain-tucker, Rashaani, PaterMcFly, GSMR, Amirazemi, SpellingBot, Reidlophile, Ronhjones, TutterMouse, NiallJones, Njaelkies Lea, Ajp4, Xlec, Vishnava, CanadianLinuxUser, Leszek Jańczuk, Fluffernutter, NjardarBot, Ka Faraq Gatri, Looie496, Skyezx, Xxuberzang, Download, SoSaysChappy, Jeffo223344, CarsracBot, Gifðas, Mjr162006, PFSLAKES1, Muffin123456789, Punkrockpiper, Debresser, Scopesmonkeys, Favonian, Azazeel, Sappho'd, CuteHappyBrute, Xoxoxoxoxxx3, Cranberry5553, Brainmachine, Dath Dath Binks, Numbo3-bot, Debashish Mahapatra., Ljay9206, Tide rolls, Lightbot, OlEnglish, Libertype, Sky83, Teles, Lrrasd, Smallman21, Gail, SeniorInt, LuK3, Alfie66, Joshua098, Ben Ben, Legobot, Luckas-bot, Yobot, 2D, Felixmonk1, Ptbotgourou, Oscardove, Legobot II, Justintan88, SashaTheAwesome, Deadly Matty, II MusLiM HyBRiD II, Adi, Aldebaran66, WikiworldJ, THEN WHO WAS PHONE?, Zkczkc, Godzilla 2002, PoizonMyst, AnakngAraw, Thulasidharan2k8, Donthegon101, Echtner, Azcolvin429, Darkowlf8592, Madan.sreddy, Tempodivalse, Kookyunii, Synchronism, AnomieBOT, AndrooUK, Jaimeescobar, Hermanschrader, Killiondude, Jim1138, AdjustShift, Bobsexual, Cheese12345cheese, P123567890p, Georgiocj, Kingpin13, Sz-iwbot, Dr. Günter Bechly, Ulric1313, GWPFBE, Mann jess, Materialscientist, Hellopunish3r, RobertEves92, The High Fin Sperm Whale, Citation bot, Srinivas, OllieFury, Mr.Kassner, Stevemanjones, Xeonxeon12, Berkeley626, Neurolysis, ArthurBot, Phlembowper99, Obersachsebot, Xqbot, Samoboow, Jstana, Typuifre, Animonster, S h i v a (Visnu), Intelati, JimVC3, Capricorn42, Oraculo miraculoso, Wperdue, Renaissancee, Jeffrey Mall, Nothingisayisreal, Smk65536, Nickkid5, Pwnage09, Jsharpminor, Imstillhereyoh, Light-assasin, Grim23, Jzhuo, Superbrainr, Tomwsulcer, Br77rino, Rootswailer, Mlpearc, Gap9551, Logan6362, J04n, GrouchoBot, Nayvik, Solphusion~enwiki, Call me Bubba, Mario777Zelda, False vacuum, Omnipaedista, Shirik, Wooitscaroline, Akshat2, Prunesqualer, RibotBOT, SassoBot, Amaury, Permafry42, The Wiki ghost, GhalyBot, Someone0707, Jrossr, DarkElrad, LORDGOD7777, Shadowjams, AlimanRuna, Peter470, Babebait227, BoomerAB, Captain-n00dle, Joel grover, Nagualdesign, George585, FrescoBot, AtomsOrSystems, Jbvjkhgvkjukjhvkj, Tobby72, Bluto7, Dantesparda271191, Dannyat43, Endofskull, Tylergriswold, Machine Elf 1735, Ruben3186, Airborne84, Amicaveritas, Wireless Keyboard, Dalekian, HamburgerRadio, Citation bot 1, Familyguylover64, Rommopaula, Redrose64, Cloroplast horse, Breadhead23, Pekayer11, Pinethicket, Jinsubpretzels, Tom.Reding, قربان‌علی بی‌ک, Calmer Waters, StrawberryPink, Dazedbythebell, Procatcher31, Jschnur, RedBot, KnowYourCosmos, SpaceFlight89, VinnyXY, Vitaliy skrynnik, Trec'hlid mitonet, Tamsier, Newgrounder, Infinitesolid, December21st2012Freak, Utility Monster, IVAN3MAN, Meier99, TobeBot, Trappist the monk, Bibleboy14, Jkveton08, Jules93, Hahaho~enwiki, AHeneen, Gafferuk, Jonkerz, Nickyus, GregKaye, Vrenator, علی ویکی, Begoon, Sidoburn, Canuck100, Dcs002, Zachareth, Diannaa, Michael.goulais, Earthandmoon, Tbhotch, Reach Out to the Truth, Hmmwhatsthisdo, Minimac, Brambleclawx, Jpabian01, DARTH SIDIOUS 2, JedediahBaugh, Jeglikerdeg, Mean as custard, Nate5713, Ripchip Bot, Citationeeded, Bhawani Gautam, Beyond My Ken, LucChickenfingers1873, Wintonian, Crazyj922, Slon02, Dumbdarwin, EmausBot, John of Reading, Mzilikazi1939, WikitanvirBot, Eekerz, Akhil.aggarwal2, Katherine, Ihatejustinbeiber, GoingBatty, RA0808, JacobParker100, Strane1991, Kamakazii101, Bt8257, Miladragon3, Pcorty, Slightsmile, Phil2324, Tommy2010, Leemadd92, Alicam8, Wikipelli, Aynan678, TeleComNasSprVen, Ryuzakihateskira, Italia2006, Hhhippo, Yddam, Susfele, Tnvkumar, WelcomeBackWinter, Mattsarahcat, Xunknownxx, HyperSonic X, Jhum~enwiki, Aeonx, Bilbo571, H3llBot, Git2010, Confession0791, Kirstylovesatl, Kjgiant, David J Johnson, Weirdo1990, Tolly4bolly, Openstrings, Butterfly 25007, Tiiliskivi, Capricorn4049, IGeMiNix, Coasterlover1994, Purduecit, L Kensington, Quantumor, בני35, Inswoon, Farheen1973, Wikiloop, Bulwersator, Harmi.banik, Vanfug, Orange Suede Sofa, ChuispastonBot, Matthewrbowker, Bubble queen, Iketsi, LikeLakers2, 28bot, H1tchh1ker4, Mjbmrbot, Jesuschristlover, Petrzak, ClueBot NG, Snoopdoug22, Dr A Thompson, VLDG123, Thematrix007, JustKiddingPro2, Ulflund, 8732Spacefish, ALovelyOne1, Gilderien, EricWaltonBall, Sdht, Deepvalley, Joefromrandb, Three geeks, Hermes the Wise, Zenithfel, Braincricket, Rezabot, Sudhir.routray, Widr, Karl 334, Squirtle1994, JackizCool69, Gunfists, Ale Ronzani, Pluma, North Atlanticist Usonian, Mightymights, Helpful Pixie Bot, Art and Muscle, Rryswny, Faus, Jumpingjacksparrow, Nikhil Sasi, Winegum74, I LIVE 2 TROLL, UT200100, Bibcode Bot, Olso12345, Lowercase sigmabot, BG19bot, Island Monkey, Sahuanimesh, Gaithdroby500, Furkhaocean, Apeconmyth, Quarkgluonsoup, MongolWiki, Xzuiko, AvocatoBot, Inkpot80, Davidiad, J991, Abhinandan27, Dan653, Tracy49, Mariraja2007, Cadiomals, Joydeep, Omeed Was Here, Schmooble, Chretienorthodoxe, Maharding, Hotturbos, Stimulieconomy, Hamish59, Eguinto, Theone70, Glacialfox, Andi2011, Mozerella8, Oleg-ch, Ghostillnses, Boogawooga02, Anbu121, Amphibio, Adamlewis157, BattyBot, Tutelary, Hihihihibye, Priyamd, SFaddict42, Stigmatella aurantiaca, Th4n3r, Lawlessballer23, Cyberbot II, Atularunpandey, Craigc29, Khazar2, Dexbot, Webclient101, CuriousMind01, Lugia2453, Frosty, Hair, Graphium, Philipandrew, Rmoole, Paryinmahpantz, LemonsWillWin, Keefyj, Alexfrench70, Guruparanjothijason, Bccmac14, Reatlas, 123yokomo, Heskey eats soup with a fork, Ihatefindingusernamestheyneverwork, USEFUL MENACE, Faizan, Supercool900, Godot13, Ianreisterariola, Vcfahrenbruck, Redshiftimprove, A Wolfgang, Christofkopera, Extremind, Alihosseinisg, Lingzhi, MarioVSYoshi, KermitTheFrogOwns, ???, Tentinator, Sakhail, Boofle1, Infamous Castle, Jayminsiple, Leave61, Seps123, Necron681, Dorrphilip, Ventripotent13, Alaudduin, CuirassierX, Pmacclain, Ugog Nizdast, FrogySK, Prokaryotes, Eagle3399, Jwratner1, Kogge, Damián A. Fernández Beanato, OccultZone, Jackmerius, Anrnusna, Asterixasterix, Mikey Camarista, Beenybobby, Aleksa is cool, LVL.6, Maderthaner, Goodkushandalcohol, BubbaLAquatics10, Concord hioz, Monkbot, Apipia, Sofia Koutsouveli, Pinkpills, Thatoneguy08, Astronomnom, PrivateNoPrivates, SkyFlubbler, Mannerheimcross, SarahTehCat, Analfistingjanick, SeaMosse, Corneel1, BenVinnie Tennieson, Squidzy Mcgavin, Jojoisawsome, EoRdE6, Julietdeltalima, Tonathan100, Satki, Atomic bacon, Snabbkaffe, I'm your Grandma., Tetra quark, Isambard Kingdom, Anand2202, Mahad Asif khan, Zxzxzxzx29, Warrerrrrrrfjgfufgvgjfjfjjfjfggfu, Catman3659, KasparBot, Youknowwhatimsayin, Outedexits and Anonymous: 1854

- **Many-worlds interpretation** *Source:* https://en.wikipedia.org/wiki/Many-worlds_interpretation?oldid=689396558 *Contributors:* AxelBoldt, Derek Ross, LC~enwiki, CYD, Zundark, Timo Honkasalo, The Anome, WillWare, Eclecticology, Josh Grosse, Darius Bacon, Nate Silva, Roadrunner, Maury Markowitz, FvdP, Stevertigo, Michael Hardy, JakeVortex, Oliver Pereira, Gabbe, Ixfd64, Alfio, William M. Connolley, Aarchiba, Evercat, HPA, Richj, Charles Matthews, Timwi, Terse, Grendelkhan, Phys, Fairandbalanced, Bevo, Andy Fugard, Nnh, BenRG, Fredrik, Vespristiano, Altenmann, Arkuat, Rholton, Bkell, ElBenevolente, Carnildo, Pablo-flores, Giftlite, Gtrmp, Barbara Shack, Wolfkeeper, Lethe,

Bfinn, Wwoods, Ebonmuse, Ryanaxp, Wmahan, Mmm~enwiki, ChicXulub, Quadell, Piotrus, MadIce, Khaosworks, CSTAR, Togo~enwiki, Latitude0116, Sam Hocevar, Lumidek, Jcorgan, Eyv, Robin klein, Acsenray, Chris Howard, D6, Freakofnurture, Rich Farmbrough, Guanabot, Leibniz, Florian Blaschke, Dbachmann, Pavel Vozenilek, Nchaimov, Ben Standeven, El C, Haxwell, Kotuku33, SamRushing, Ehaque, Danski14, Plumbago, SlimVirgin, Batmanand, Hdeasy, Schaefer, Fourthords, Count Iblis, Pauli133, DV8 2XL, Kromozone, Dan100, Falcorian, Firsfron, Woohookitty, Linas, Dandv, JonBirge, Thruston, SDC, Joke137, Pfalstad, Marudubshinki, Mandarax, QuaestorXVII, Graham87, Grammarbot, Rjwilmsi, Ckoenigsberg, Eyu100, Aero66, HappyCamper, Theodork, Wragge, Arnero, ZoneSeek, Diza, Saswann, Chobot, DVdm, Korg, McGinnis, YurikBot, NTBot~enwiki, RussBot, Fabartus, Bhny, Gaius Cornelius, Vincej, Bboyneko, Thane, Draeco, David R. Ingham, Joncolvin, Nti2005, Schlafly, Brian Olsen, Kxjan, Inhighspeed, Pnrj, Emersoni, Zwobot, Kkmurray, Light current, Geoffrey.landis, Radioflux, Nixer, Ilmari Karonen, NeilN, Teo64x, SmackBot, Rex the first, JohnSankey, C.Fred, Jrockley, Dave Kielpinski, MalafayaBot, Silly rabbit, Imaginaryoctopus, Salmar, Shantrika, Calbaer, DenisDiderot, Savidan, DavidBoden, Ryan Roos, Hunter2005, DJIndica, Lambiam, Ser Amantio di Nicolao, Dr. Sunglasses, Mgiganteus1, Ckatz, Santa Sangre, Dicklyon, Hypnosifl, Eridani, Stephen B Streater, HisSpaceResearch, Dreftymac, Joseph Solis in Australia, Brian Wowk, Courcelles, Laplace's Demon, Bstepp99, Mustbe, 8754865, JForget, Will314159, CmdrObot, Vyznev Xnebara, AshLin, Denis MacEoin, Rgonsalv, Myasuda, Gregbard, Dragon's Blood, Cydebot, Peterbyrne, Peterdjones, Michael C Price, Alexnye, Geewee, Michael D. Wolok, Mbell, Headbomb, Parsiferon, Cj67, Stannered, Cyclonenim, AntiVandalBot, Mukake, Czj, MER-C, CosineKitty, Txomin, Magioladitis, SHCarter, Skew-t, Brusegadi, Torchiest, Duendeverde, Plexos, Sm8900, R'n'B, J.delanoy, Maurice Carbonaro, Kevin aylward, 5Q5, Coastal593, Shawn in Montreal, Gill110951, Tarotcards, Eli the Barrow-boy~enwiki, Sarge009, Phatius McBluff, Ajfweb, Inwind, Sheliak, Neuromath, VolkovBot, Jdcaust, Eve Hall, Xnquist, Aymatth2, Don4of4, TBond, Ar-wiki, Sapphic, Spinningspark, Hrimpurstala, AlleborgoBot, Quantum Person, Yintan, Wing gundam, Likebox, Flyer22 Reborn, JohnSawyer, JohnnyMrNinja, Longi 93, AussieScribe, VanishedUser sdu9aya9fs787sads, Martarius, ClueBot, EoGuy, General Epitaph, EMC125, Djr32, Tianasez, Estirabot, Vanhoabui, Sun Creator, Resuna, SchreiberBike, Carriearchdale, Thingg, Ctkohl, DumZiBoT, BarretB, XLinkBot, Sanchoquixote, Kbdankbot, Addbot, Miskaton, Gregz08, Zahd, Tanhabot, Thirteenangrymen, Mac Dreamstate, Proxima Centauri, Ld100, Lightbot, Taketa, زرشک, Zorrobot, Whitneyz32, Legobot, MichelCPrice, Yobot, Pink!Teen, AnomieBOT, Unara, Citation bot, DirlBot, LilHelpa, Measles, Omnipaedista, Targeran, WaysToEscape, FrescoBot, ایلی2010, Machine Elf 1735, MorphismOfDoom, Citation bot 1, I dream of horses, Jonesey95, Chatfecter, Grok42, Jusses2, Fredkinfollower, SkyMachine, Dc987, Duke159, Lhollo, GregWooledge, Jamesabloom, George Richard Leeming, Rekcana, Feelingsman22, Slightsmile, Italia2006, ZéroBot, Cogiati, Gahr gardner, Quantumavik, RUBEN TESOLIN, H3llBot, Quondum, Quantholic, L Kensington, Senjuto, Surajt88, Ihardlythinkso, Kartasto, Teapeat, Rememberway, ClueBot NG, Nobody60, Incompetence, Kazzie1995, Kasirbot, Helpful Pixie Bot, Bibcode Bot, BG19bot, Boriaj, Stelpa, FiveColourMap, AIMW32, Huntingg, Eqb1987, Willempramschot, Brendan.Oz, Harizotoh9, Dule1101, Modalizer, Khazar2, JYBot, Mogism, Corn cheese, Reatlas, Epicgenius, Jamesmcmahon0, A Certain Lack of Grandeur, Thevideodrome, Arfæst Ealdwrítere, Bardoligneo, Stamptrader, Fixuture, Ismael755, 2PeterElls, Monkbot, Kalipsos, MazeHatter, Jonkirstenhof, Azealia911, Fixing.your.problems. and Anonymous: 271

- **Eternal inflation** Source: https://en.wikipedia.org/wiki/Eternal_inflation?oldid=686406622 Contributors: The Anome, Jagged, Dcljr, Everyking, Eteq, Rjwilmsi, RussBot, Romanc19s, InverseHypercube, Robin Williams, Lambiam, Myasuda, Cydebot, Michael C Price, Headbomb, Peter Gulutzan, Geostar1024, Mufka, Weburbia, Speaker to wolves, Avoided, Addbot, Proxima Centauri, Luckas-bot, AnomieBOT, J JMesserly, Omnipaedista, Waleswatcher, JLincoln, Bj norge, Slightsmile, Italia2006, Hhhippo, ClueBot NG, KLBot2, Bibcode Bot, Prokaryotes, Anrnusna, Tetra quark, Sleepy Geek, Stewi101015 and Anonymous: 18

- **Cosmological principle** Source: https://en.wikipedia.org/wiki/Cosmological_principle?oldid=685144706 Contributors: General Wesc, XJaM, Roadrunner, Pandora, Looxix~enwiki, Schneelocke, Timwi, Jni, Robbot, Rursus, Wlievens, Gzornenplatz, SoWhy, Dreamtheater, Karol Langner, Chmod007, Vsmith, Mdhowe, I9Q79oL78KiL0QTFHgyc, Giraffedata, Enirac Sum, Hackwrench, Wricardoh~enwiki, Gpvos, Woohookitty, Linas, Jok2000, Btyner, Rjwilmsi, Everton137, Margosbot~enwiki, McGinnis, Russell C. Sibley, Salsb, Noosfractal, Light current, Enormousdude, Jules.LT, Jasonuhl, Rtc, Kmarinas86, MalafayaBot, Colonies Chris, Infovoria, Titus III, Euchiasmus, Mgiganteus1, Adodge, Paul venter, PetaRZ, Stanlekub, Memetics, Friendlystar, Michael C Price, BishopOcelot, Thijs!bot, Oerjan, Headbomb, Peter Gulutzan, JAnDbot, .anacondabot, Avjoska, Mowafag, Drollere, Nyttend, Geboy, Maxeng, Trilobitealive, TXiKiBoT, BotKung, Wmpearl, Sunrise, Cosmo0, Wyattmj, ArepoEn, ClueBot, Plastikspork, Niceguyedc, ChandlerMapBot, DragonBot, Brews ohare, Truth is relative, understanding is limited, Addbot, Eivindbot, Luckas-bot, Yobot, Amirobot, Citation bot, Spring Back, Lithopsian, RibotBOT, C3lticmatt, Mnmngb, D'ohBot, Machine Elf 1735, Citation bot 1, Ahnoneemoos, Tom.Reding, Ole.Holm, Full-date unlinking bot, RockSolidCosmo, Trappist the monk, Jesse V., Nistra, Tuxedo junction, Italia2006, Helpful Pixie Bot, Bibcode Bot, Levelswung, TheTahoeNatrLuvnYaho, Johndric Valdez, Anrnusna, Monkbot, Tetra quark, StraboVarenius and Anonymous: 46

- **Inflation (cosmology)** Source: https://en.wikipedia.org/wiki/Inflation_(cosmology)?oldid=690243392 Contributors: Bryan Derksen, The Anome, Diatarn iv~enwiki, Roadrunner, David spector, Hephaestos, Stevertigo, Edward, Nealmcb, Boud, Michael Hardy, Tim Starling, Dcljr, Cyde, Ellywa, William M. Connolley, Theresa knott, Jeff Relf, Mxn, Timwi, Rednblu, Bartosz, Pierre Boreal, Raul654, Chuunen Baka, Robbot, Gandalf61, Rursus, Ancheta Wis, Giftlite, Barbara Shack, Mikez, Lethe, Dratman, Curps, Jcobb, Just Another Dan, Andycjp, HorsePunchKid, Beland, Elroch, JDoolin, Burschik, Shadypalm88, Eep[2], Mike Rosoft, DanielCD, Noisy, Rich Farmbrough, FT2, Pjacobi, Luxdormiens, Dbachmann, Bender235, AdamSolomon, Pt, Worldtraveller, Art LaPella, Orlady, Drhex, Guettarda, I9Q79oL78KiL0QTFHgyc, Jeodesic, Rsholmes, Anthony Appleyard, Plumbago, JHG, Schaefer, EmmetCaulfield, Cgmusselman, Dirac1933, Oleg Alexandrov, Matevzk, Yeastbeast, StradivariusTV, BillC, Bluemoose, Wdanwatts, Joke137, Rnt20, Malangthon, Ketiltrout, Drbogdan, Rjwilmsi, Zbxgscqf, Mattmartin, Strait, Eyu100, Jehochman, Ems57fcfva, Bubba73, FlaBot, Nihiltres, Itinerant1, Phoenix2~enwiki, Chobot, Hermitage, Bgwhite, YurikBot, Wavelength, Supasheep, Ytrottier, Gaius Cornelius, Anomalocaris, NawlinWiki, LiamE, Davemck, JonathanD, Enormousdude, 2over0, Arthur Rubin, Argo Navis, Physicsdavid, Profero, Luk, SmackBot, Haza-w, KnowledgeOfSelf, Lawrencekhoo, Onsly, Jdthood, Salmar, Jefffire, Hve, QFT, Vanished User 0001, Stevenmitchell, BIL, Lostart, Ligulembot, Yevgeny Kats, Byelf2007, Lambiam, Rcapone, JorisvS, Heliogabulus, Dan Gluck, Spebudmak, JoeBot, UncleDouggie, Fsotrain09, Oshah, JRSpriggs, Chetvorno, Friendly Neighbour, Drinibot, Vanished user 2345, Brownlee, SuperMidget, Cydebot, BobQQ, Mortus Est, Cyhawk, Ttiotsw, Julian Mendez, Dr.enh, Michael C Price, Kozuch, LilDice, Thijs!bot, Headbomb, Z10x, Jklumker, Alfredr, Dawnseeker2000, Pollira, Rico402, Lfstevens, Gmarsden, JAnDbot, Olaf, LinkinPark, GurchBot, Magioladitis, Jpod2, Vanished user ty12k189jq10, Rickard Vogelberg, Dr. Morbius, Bhenderson, TomS TDotO, Tarotcards, Wesino, Student7, Potatoswatter, Ollie 9045, Ja 62, Useight, Idioma-bot, Sheliak, Tokenhost, VolkovBot, ABF, ColdCase, Philip Trueman, TXiKiBoT, Calwiki, Thrawn562, Gobofro, SwordSmurf, Northfox, PaddyLeahy, SieBot, Wing gundam, OpenLoop, Likebox, Flyer22 Reborn, Mimihitam, Hockeyboi34, Lightmouse, Sunrise, Southtown, Hamiltondaniel, Epistemion, ClueBot, Niceguyedc, ChandlerMapBot, Jusdafax, ResidueOfDesign, Ploft, Scog, SchreiberBike, TimothyRias, Katsushi, MidwestGeek, Addbot, Roentgenium111, DOI bot, Blethering Scot, Ronhjones, Glane23, Deamon138, TStein,

28.5. TEXT AND IMAGE SOURCES, CONTRIBUTORS, AND LICENSES

Barak Sh, Tassedethe, Zorrobot, Ben Ben, Legobot, Yinweichen, Luckas-bot, Amirobot, Aldebaran66, Amble, Isotelesis, Magog the Ogre, AnomieBOT, Pyrrhon8, Rubinbot, Piano non troppo, Collieuk, Ulric1313, Citation bot, Xqbot, Plastadity, Capricorn42, P14nic997, False vacuum, Waleswatcher, Ignoranteconomist, Bigger digger, Chatul, ??, CES1596, FrescoBot, Mesterhd, Paine Ellsworth, Schnufflus, Charles Edwin Shipp, Bbhustles, Ahnoneemoos, Pinethicket, Tom.Reding, Σ, Aknochel, Mercy11, Trappist the monk, Jordgette, Wdanbae, Aabaakawad, Michael9422, CobraBot, Deathflyer, Mathewsyriac, EmausBot, Thucyd, GoingBatty, Wikipelli, Kiatdd, Italia2006, Werieth, ZéroBot, Chasrob, Wackywace, Bamyers99, Suslindisambiguator, AManWithNoPlan, RaptureBot, Maschen, HCPotter, Crux007, RockMagnetist, Whoop whoop pull up, ClueBot NG, J kay831, Law of Entropy, Supermint, Helpful Pixie Bot, Bibcode Bot, Lowercase sigmabot, BG19bot, Negativecharge, MSgtpotter, Badon, BML0309, Zedshort, Hamish59, Minsbot, BattyBot, SupernovaExplosion, ChrisGualtieri, JYBot, Rfassbind, Ikjyotsingh, Astroali, Lepton01, Pkanella, Chwon, Rolf h nelson, Comp.arch, Kogge, Hilmer B, Anrrusna, Stamptrader, Dodi 8238, Epaminondas of Thebes, Man of Steel 85, Abitslow, Monkbot, Accnln, BradNorton1979, YeOldeGentleman, Tetra quark, Isambard Kingdom, Sleepy Geek, Anand2202, Quasiopinionated, EnigmaLord515, Phseek and Anonymous: 212

- **Spontaneous symmetry breaking** *Source:* https://en.wikipedia.org/wiki/Spontaneous_symmetry_breaking?oldid=678321175 *Contributors:* AxelBoldt, Bryan Derksen, XJaM, Edward, Michael Hardy, Lexor, Charles Matthews, Timwi, Reddi, Phys, Bevo, Dusik, Nagelfar, Giftlite, Lethe, Alison, JeffBobFrank, Jcobb, Gotanda, Gadfium, DragonflySixtyseven, Lumidek, FT2, Hidaspal, Ascánder, Mal~enwiki, Bender235, Clement Cherlin, PhilHibbs, MPS, Shenme, Physicistjedi, Kocio, StuTheSheep, Linas, Jmhodges, Dennis Estenson II, Salix alba, Jehochman, BjKa, Chobot, YurikBot, Ugha, Bambaiah, Archelon, Zzuuzz, Reyk, Roques, RupertMillard, SmackBot, Maksim-e~enwiki, Complexica, Colonies Chris, Jmnbatista, Lagrangian, Akriasas, P199, JarahE, Hetar, Dan Gluck, JMK, Harej bot, Ezrakilty, Thijs!bot, Barticus88, Headbomb, Arcresu, Hillarryous, Dougher, Gökhan, JAnDbot, Yuksing, Attarparn, Jpod2, R'n'B, Natsirtguy, Lseixas, BernardZ, Cuzkatzimhut, Holme053, TXiKiBoT, Red Act, Michael H 34, Pamputt, Moose-32, SieBot, Wing gundam, Renatops, Denisarona, Mastertek, BlueDevil, MelonBot, Truthnlove, Addbot, Yakiv Gluck, Zahd, LaaknorBot, SpBot, OlEnglish, Luckas-bot, Yobot, Yotcmdr, Christopher Pritchard, Zimboz Montizawooba, Obersachsebot, False vacuum, Waleswatcher, Gsard, A. di M., ??, CES1596, Freddy78, Pmokeefe, RobinK, Mary at CERN, Marie Poise, Slightsmile, Quondum, Shovkovy, Maschen, Boris Breuer, Vatsal19, Helpful Pixie Bot, Bibcode Bot, Ahhaha, Kalmiopsiskid, Fraulein451, Dexbot, Lugia2453, CMTdrew, Mparisi90, LudicrousTripe and Anonymous: 85

- **Cyclic model** *Source:* https://en.wikipedia.org/wiki/Cyclic_model?oldid=684561491 *Contributors:* ChangChienFu, Michael Hardy, Dante Alighieri, Ahoerstemeier, J'raxis, Charles Matthews, Reddi, Rursus, Tobias Bergemann, Herbee, DÅ‚ugosz, Elroch, Dmr2, AdamSolomon, El C, RoyBoy, Sole Soul, I9Q79oL78KiL0QTFHgyc, Clotten, Mpatel, Tabletop, Joke137, Drbogdan, Rjwilmsi, CraigDuncan, DVdm, Wavelength, Angus Lepper, JocK, CWenger, SmackBot, Hongooi, Glloq, Wen D House, Radagast83, Ne0Freedom, ShelfSkewed, Robertinventor, Headbomb, Peter Gulutzan, Ozanyarman, Tim Shuba, DagosNavy, Destroyer000, Skylights76, Rickard Vogelberg, Gowish, Melamed katz, Lithfo, Kenneth M Burke, AzureCitizen, Master z0b, HowardFrampton, TXiKiBoT, Kennison, Monty845, Hrafn, SieBot, Jojalozzo, OKBot, Jdkessler, Coinmanj, Editor2020, Agentxyz, XLinkBot, Addbot, Lightbot, Luckas-bot, Yobot, JohnHarold, Wireader, Arjun G. Menon, Materialscientist, Citation bot, Grifterlake, Imushfiq, False vacuum, Omnipaedista, Holversb, Citation bot 1, Tom.Reding, Xaoyin, SkyMachine, LuftWaffle0, Chronulator, WikitanvirBot, TuHan-Bot, Hhhippo, Vramasub, Mattedia, Davidaedwards, ChuispastonBot, EdoBot, Germanviscuso, ClueBot NG, Helpful Pixie Bot, Bibcode Bot, Raymond1922A, TwoTwoHello, Jp4gs, IvanderClarent, Anrrusna, CarnivorousBunny, Tetra quark, ASavantDude and Anonymous: 60

- **Lee Smolin** *Source:* https://en.wikipedia.org/wiki/Lee_Smolin?oldid=683933211 *Contributors:* The Anome, Toby Bartels, Anthere, Hephaestos, Cyde, TakuyaMurata, Mcarling, Arpingstone, Radicalsubversiv, RodC, Goethean, Nurg, Gidonb, Timrollpickering, Giftlite, Christofurio, Lumidek, ELApro, Florian Blaschke, Cagliost, David Schaich, Bender235, Art LaPella, I9Q79oL78KiL0QTFHgyc, Toh, 4v4l0n42, Nurban, Plumbago, SlimVirgin, RainbowOfLight, Jvajda, Ringbang, Woohookitty, Tabletop, Twthmoses, GregorB, Parudox, Ashmoo, Josh Parris, Koavf, Jweiss11, FlaBot, Harmil, Tjligocki, Chobot, DVdm, Bgwhite, RussBot, Bhny, Alvo~enwiki, Jredwards, Evrik, Arthur Rubin, Paul D. Anderson, SmackBot, YellowMonkey, Lestrade, Vald, GaeusOctavius, Bluebot, Rick7425, Frap, Fiziker, Wen D House, Vincenzo.romano, Robofish, Tarcieri, ATren, Dicklyon, Levj, Pjrm, JohnSmart, Joseph Solis in Australia, Courcelles, Valoem, Albertod4, Cydebot, Quajafrie, Pingku, Sweetmoose6, PKT, Thijs!bot, Jm3, Peter Gulutzan, Cj67, Mmortal03, Julia Rossi, Tim Shuba, NBeale, Epeefleche, Hut 8.5, Magioladitis, Andrewthomas10, JCNSmith, Waacstats, LeeSmolin, Connor Behan, Rickard Vogelberg, Gwern, Mark387533, Vanished user 342562, Silas S. Brown, Tarotcards, OriEri, Fran Rogers, Tzahy, Ask123, BotKung, Duncan.Hull, Lamro, Lakunoc, BOTijo, Hrafn, Ufinne, Mr. Stradivarius, Zeyn1, DFRussia, Boing! said Zebedee, Skewyou, Muro Bot, Qwfp, DumZiBoT, Tenner47, Freddy engels, Good Olfactory, Kbdankbot, Addbot, Taktoa, Iguana2, Olga Sala~enwiki, AndersBot, Debresser, SamatBot, Scientryst, Yobot, AnomieBOT, Tripodian, Jockocampbell, Paliwikiuser, J JMesserly, Abce2, Omnipaedista, RCraig09, Thehelpfulbot, FrescoBot, MandelBot, Nageh, Adam9389, Machine Elf 1735, Redrose64, Xaviertan, Foobarnix, Megzor, RjwilmsiBot, Beyond My Ken, Afteread, Verbapple, Golumbo, EleferenBot, Stringtheoryrocks, Suslindisambiguator, Jhcapps, Gum375, Dru of Id, Ben morphett, Helpful Pixie Bot, Brad7777, Ageplus, Hmainsbot1, Mogism, VIAFbot, JulianBrabour, Pinocchio3000, HMHBooks, Gnoldog, Jonarnold1985, Tetra quark, KasparBot and Anonymous: 97

- **Multiple histories** *Source:* https://en.wikipedia.org/wiki/Multiple_histories?oldid=650765730 *Contributors:* The Anome, Charles Matthews, RickK, Icairns, RJHall, Jag123, RJFJR, Linas, DoubleBlue, Conscious, Tocorrode, Marcus Brute, Amalas, Peterdjones, Peter Gulutzan, Magioladitis, Lordvolton, EMC125, Djr32, Addbot, Luckas-bot, Yobot, AnomieBOT, Italia2006, BattyBot, Tetra quark and Anonymous: 6

- **Many-minds interpretation** *Source:* https://en.wikipedia.org/wiki/Many-minds_interpretation?oldid=679625110 *Contributors:* Bryan Derksen, Josh Grosse, Maury Markowitz, Stevertigo, Looxix~enwiki, Timwi, Anakolouthon, Doradus, Fredrik, Goethean, Vacuum, CSTAR, Fenice, Gary, Burn, DV8 2XL, Loxley~enwiki, Woohookitty, Linas, Rjwilmsi, Ewlyahoocom, Mentat37, KSchutte, Neutron, Incnis Mrsi, John, Dreftymac, Michael C Price, Tarotcards, Samlyn.josfyn, Lisatwo, JL-Bot, Resuna, Truth is relative, understanding is limited, Addbot, Gregz08, Zahd, AnomieBOT, Aaron Kauppi, Jordgette, EmausBot, MrBill3 and Anonymous: 23

- **Mathematical universe hypothesis** *Source:* https://en.wikipedia.org/wiki/Mathematical_universe_hypothesis?oldid=687371541 *Contributors:* Zundark, Fubar Obfusco, Michael Hardy, Sverdrup, Dbachmann, Bender235, Jhertel, DanielVallstrom, BRW, Woohookitty, GregorB, Rjwilmsi, MarSch, Dianelos, Jfraatz, Hillman, Salsb, Joncolvin, Banus, Bluebot, Lambiam, Dark Formal, John H, Morgan, Billgunn, George100, Mbell, Headbomb, Parsiferon, Tim333, Magioladitis, NerdyNSK, Maxim, Lamro, Martarius, Silent Key, XLinkBot, MystBot, Addbot, Discrepancy, Favonian, Luckas-bot, Yobot, AnomieBOT, Omnipaedista, IO Device, Machine Elf 1735, Randalliser, Kartasto, Bibcode Bot, BG19bot, Erik.Bjareholt, John Aiello, Sol1, Jwratner1, Monkbot, Frettled Gruntbuggly, HouseOfChange, Quantalogos and Anonymous: 43

- **Brane cosmology** *Source:* https://en.wikipedia.org/wiki/Brane_cosmology?oldid=665167986 *Contributors:* AxelBoldt, Camembert, Bth, Twilsonb, DIG~enwiki, AugPi, Pedant17, Rursus, David Gerard, Barbara Shack, Herbee, LeYaYa, Tomothy, H0riz0n, Pjacobi, El C, Constantine, I9Q79oL78KiL0QTFHgyc, Oliphaunt, Joke137, Strait, Algri, Eric B, YurikBot, Hairy Dude, Gaius Cornelius, Salsb, Bucketsofg, BOT-Superzerocool, Bmju, Closedmouth, Tim314, SmackBot, Kurochka, SSJemmett, TimBentley, Jprg1966, Dugodugo, Colonies Chris, Stevenmitchell, Savidan, Ligulembot, Yevgeny Kats, Quaeler, CapitalR, George100, Verdy p, Michael C Price, Headbomb, Escarbot, VictorAnyakin, Skylights76, Robin S, FlieGerFaUstMe262, HiEv, Afluent Rider, Vendrov, PlanetStar, James Banogon, Fjados, Masterpiece2000, MelonBot, Thinking Stone, Mixen Dixon, The Thin Man Who Never Leaves, Yobot, Almabot, Louperibot, DrilBot, Full-date unlinking bot, Tkachyk, Solomonfromfinland, MerlIwBot, Bibcode Bot, Davidiad, Ownedroad9, PamFromMD2, Harsh 2580, Twhitguy15, E8xE8, Monkbot, BradNorton1979, Tetra quark, Srednuas Lenoroc and Anonymous: 28

- **Brane** *Source:* https://en.wikipedia.org/wiki/Brane?oldid=678165535 *Contributors:* Bth, Michael Hardy, DIG~enwiki, Samuelsen, JWSchmidt, Silverfish, Wetman, Bcorr, Blainster, Fropuff, Just Another Dan, D3, Lumidek, Yuriz, Rhobite, H0riz0n, Ben Standeven, RoyBoy, Mairi, Constantine, GatesPlusPlus, Kocio, Agquarx, Mpatel, Liface, BD2412, Quiddity, R.e.b., Mathbot, BradBeattie, Metropolitan90, YurikBot, Wavelength, NawlinWiki, Wknight94, Closedmouth, SmackBot, Kurochka, Jwestbrook, Autarch, Seanor32, Silly rabbit, Colonies Chris, Nsmith4658, Mesons, Monotonehell, TheVikingRaider, Yevgeny Kats, Spiritia, PaddyM, Czoller, Calmargulis, BeenAroundAWhile, Adailton, Julius M-D, J. W. Love, Julia Rossi, Chrisjj3, MER-C, Steveprutz, Just H, N.Nahber, Urco, Alexrussell101, Cyborg Ninja, Idioma-bot, VolkovBot, Anonymous Dissident, Paucabot, Drschawrz, SieBot, Tresiden, OKBot, ClueBot, The Thing That Should Not Be, SilvonenBot, NonvocalScream, Addbot, Jujutsuka, Royote, LilHelpa, Patmethenyfan, Omnipaedista, Nagualdesign, Kgrad, Tbhotch, Tesseract2, EmausBot, MathMaven, ClueBot NG, Mikeflem, Gilderien, Baseball Watcher, Frietjes, DBigXray, Lowercase sigmabot, Solomon7968, OCCullens, BattyBot, Brirush, E8xE8, Dimension10, Polytope24 and Anonymous: 50

- **D-brane** *Source:* https://en.wikipedia.org/wiki/D-brane?oldid=676567255 *Contributors:* Zundark, TakuyaMurata, Karada, JWSchmidt, AugPi, Smack, Schneelocke, Gandalf61, Michael Snow, Fropuff, Anville, Just Another Dan, Phe, Lumidek, Rgrg, H0riz0n, El C, Constantine, I9Q79oL78KiL0QTFHgyc, FlaBot, Bhny, Nick, Zwobot, Sardanaphalus, KnightRider~enwiki, Teemu Ruskeepää, Colonies Chris, Scwlong, QFT, Eric Olson, Fughghettaboutit, Vampus, JarahE, Twyder, Eewild, 345Kai, Cydebot, Headbomb, J. W. Love, Nick Number, Magioladitis, Jpod2, STBot, HEL, VolkovBot, TXiKiBoT, PhysPhD, Jonathanrcoxhead, Excirial, Alexbot, ResidueOfDesign, Addbot, LaaknorBot, Tassedethe, Lightbot, Luckas-bot, Yobot, Amirobot, Azcolvin429, Royote, Citation bot, Twri, Omnipaedista, Galaktiker, Mentibot, Wakabaloola, Petrb, Frietjes, Luizpuodzius, OCCullens, Polytope24 and Anonymous: 30

- **M-theory** *Source:* https://en.wikipedia.org/wiki/M-theory?oldid=688093667 *Contributors:* AxelBoldt, CYD, Eloquence, BF, Bryan Derksen, Zundark, The Anome, Ap, Tim Chambers, Hari, Maury Markowitz, Stevertigo, Michael Hardy, Tim Starling, Gabbe, Tompagenet, Ixfd64, CesarB, Looxix~enwiki, JWSchmidt, Darkwind, Marco Krohn, Jeandré du Toit, Evercat, Schneelocke, Charles Matthews, Timwi, Reddi, Malcohol, Bevo, Jusjih, Slawojarek, Sander123, Fredrik, R3m0t, RedWolf, Blainster, DHN, Hadal, HaeB, Tobias Bergemann, David Gerard, Giftlite, DocWatson42, Jmnbpt, Barbara Shack, Fropuff, Moyogo, Sigfpe, Daen, Antandrus, Lumidek, ChrisCostello, Mike Rosoft, Spiffy sperry, Urvabara, Noisy, Discospinster, H0riz0n, Vsmith, Loren36, El C, Momotaro, Shanes, RoyBoy, Triona, Constantine, Smalljim, I9Q79oL78KiL0QTFHgyc, Giraffedata, Wolfrider~enwiki, Physicistjedi, MPerel, Gsklee, ShardPhoenix, Axl, Mac Davis, Kocio, Burn, Hu, Wtmitchell, SidP, DV8 2XL, Ringbang, Kazvorpal, Omnist, Sharkie, Joelpt, Angr, Firsfron, FeanorStar7, Pol098, WadeSimMiser, Mpatel, GregorB, Jugger90, Paxsimius, Mandarax, Chun-hian, Grammarbot, Rjwilmsi, Nightscream, Koavf, Zbxgscqf, Oblivious, Yug, Lionelbrits, Ruidlopes, The ARK, Latka, Mathbot, Diza, Phoenix2~enwiki, DVdm, Eric B, Bomb319, Loom91, Zafiroblue05, Bhny, Stephenb, KSchutte, Bovineone, Salsb, Erielhonan, Bobak, Asarelah, Dna-webmaster, Sandstein, Superdude99, Zzuuzz, Imaninjapirate, Arthur Rubin, Ilmari Karonen, Caco de vidro, DVD R W, Hide&Reason, Jmeden2000, Teo64x, Sardanaphalus, MartinGugino, RupertMillard, SmackBot, Kurochka, K-UNIT, Rwp, Rlbates99, Ajt, Ian Rose, Gilliam, Wlmg, DividedByNegativeZero, Mirokado, Bluebot, Cush, SMP, Ben.c.roberts, MalafayaBot, Nbarth, DHN-bot~enwiki, Colonies Chris, Joemah, N.MacInnes, Xiner, Nunocordeiro, Mbertsch, Addshore, EPM, Nakon, Kiplantt, Bigmantonyd, Martijn Hoekstra, Kabain52, Brdforallseasons, Sayden, Doug Bell, Jaganath, Shadowlynk, IronGargoyle, Jochietoch, Hu12, Jxh2154, Tawkerbot2, Valoem, Gebrah, Albertod4, Kurtan~enwiki, Harold f, Devourer09, Cyrusc, CRGreathouse, Olaf Davis, Lambertian, Friendlystar, Rowellcf, Bmk, Myasuda, DepartedUser2, Ekajati, Cydebot, Fluence, Mike Christie, Meno25, Gagueci, Kahananite, Michael C Price, Alexnye, IComputerSaysNo, Lord Satorious, Krowe, Mrockman, Thijs!bot, Epbr123, Daniel, Headbomb, NeilHalfway, James086, KrakatoaKatie, AntiVandalBot, Blue Tie, Alphachimpbot, J rowley, Shambolic Entity, SuperLuigi31, Buchhemi, Fetchcomms, 100110100, WolfmanSF, VoABot II, Madevin314, SHCarter, Rami R, Jqshenker, Just H, War wizard90, Rickard Vogelberg, Stephen Shenker, Theoretic, MartinBot, Kostisl, R'n'B, Euku, Numbo3, Maurice Carbonaro, Nly8nchz, Thucydides411, LordAnubisBOT, Janus Shadowsong, Peskydan, Isoko, Belovedfreak, Antony-22, Wesino, WJBscribe, Thomas795135, Blood Oath Bot, Idioma-bot, Sheliak, Gogobera, Jeff G., Rei-bot, Ask123, Pennstatephil, JhsBot, Mazarin07, Peace keeper II, Lamro, Antixt, Why Not A Duck, PhysPhD, Rknasc, Guystout, Drschawrz, YohanN7, SieBot, Robdunst, Paradoctor, Wing gundam, Holt27, Astroboyretro, Caidh, OKBot, Divinestuff, Wpac5, Ayleuss, Beofluff, Loren.wilton, ClueBot, Master Shake 9, The Thing That Should Not Be, Haemorrhage, Arakunem, Drmies, IMNTU, Yupjohnny, Huntthetroll, Patrik Andersson, Dank, Gardv, DumZiBoT, Jfosc, Maky, Truthnlove, Autocoast~enwiki, Albambot, Addbot, Uruk2008, Cuaxdon, CanadianLinuxUser, WikiUserPedia, Barak Sh, Tassedethe, Carapheonix, Togekiss101, Tide rolls, OlEnglish, Snaily, Legobot, Luckas-bot, Yobot, Fraggle81, Pcap, Foolo~enwiki, CinchBug, Tempodivalse, AnomieBOT, KDS4444, Götz, Charlesvi, Dalton h, Marcka, Alexzabbey, Jim1138, IRP, AdjustShift, Materialscientist, Citation bot, Quebec99, Ruike, TinucherianBot II, Ekwos, Techwiz2000, Omnipaedista, Peanuts4life, Pinethicket, Vicenarian, Tom.Reding, EDG161, Jusses2, Serols, ActivExpression, SkyMachine, Gerda Arendt, Tkachyk, 122589423KM, தகீர்வி சிவகுமார், Reaper Eternal, Apb91781, 786 zikhar, LcawteHuggle, Adam1217, EmausBot, GoingBatty, Pyschobbens, StringTheory11, Smiwi, Suslindisambiguator, SporkBot, PoisonGM, Besneatte, Maschen, SBaker43, Denholm Reynholm, RockMagnetist, ClueBot NG, Blueshift333, Rgwkenyon, Helpful Pixie Bot, Bibcode Bot, BG19bot, SharkinthePool, Msaunier, MusikAnimal, Copernicus01, Elginfball10, Qed3, Blaspie55, Zujua, Kooky2, Mediran, Chris5631, FEYKATD, Ecila3, Ubed junejo, Lugia2453, Frosty, AHusain314, Armanschwarz, Among Men, Faizan, Epicgenius, Diekilldie, EddieHugh, Dustin V. S., RaphaelQS, Beakr, DavidLeighEllis, Tedsanders, TFA Protector Bot, Vampre1122, Polytope24, Evandas, Oneidiotsavant, Pretickle, TheRealTheKoi, Shantsforeverandalways, QuantumMatt101, AKS.9955, FACBot, Kh3368, Sizeofint, Jyhtgqwqsdfghjydwq, Mberkson12, KasparBot, Cmealo, Yadav.aakash.500 and Anonymous: 450

- **Ekpyrotic universe** *Source:* https://en.wikipedia.org/wiki/Ekpyrotic_universe?oldid=681247385 *Contributors:* Bryan Derksen, Roadrunner, Bth, Wapcaplet, Sannse, Reddi, Lord Kenneth, JerryFriedman, Barbara Shack, Herbee, Lockeownzj00, Julianonions, Rich Farmbrough, Pjacobi, El C, Art LaPella, I9Q79oL78KiL0QTFHgyc, Jeodesic, MCiura, Monado, Snowolf, Jordan14, Joke137, YurikBot, Lexicon, 2over0, Closedmouth, SmackBot, Hammiesink, Pwjb, JorisvS, Dreftymac, Skapur, Buddy13, SJGSM, Lavateraguy, Nishidani, Thijs!bot, The Captain Returns, Headbomb, Peter Gulutzan, R'n'B, Joshua Issac, VolkovBot, NDUTU~enwiki, UnitedStatesian, Kennison, Hrafn, Singinglemon~enwiki, Kwork2, Addbot, DOI bot, OlEnglish, MuZemike, Luckas-bot, Omnipaedista, Bgagaga, SpaceFlight89, Reaper Eternal, Theoar, Georalex1,

28.5. TEXT AND IMAGE SOURCES, CONTRIBUTORS, AND LICENSES

Flying Fische, Bibcode Bot, Davidiad, ChrisGualtieri, Jp4gs, Steven Rogers, Monkbot, Tetra quark, Sleepy Geek, TheHecster and Anonymous: 23

- **String theory landscape** *Source:* https://en.wikipedia.org/wiki/String_theory_landscape?oldid=680316596 *Contributors:* Maury Markowitz, Sho Uemura, Fropuff, Eequor, I9Q79oL78KiL0QTFHgyc, Kusma, DV8 2XL, Ringbang, GregorB, BlaiseFEgan, Joke137, YurikBot, Bhny, Archelon, IAMTHEEGGMAN, Gadget850, SmackBot, Kurochka, Colonies Chris, CmdrObot, Mbell, Headbomb, Peter Gulutzan, Landscape~enwiki, Tim Shuba, Glen, Dr. Morbius, IgorSF, Jrcla2, Gbawden, Martarius, Niceguyedc, Addbot, DOI bot, Cuaxdon, SeniorInt, Macquaire, Omnipaedista, Citation bot 2, Citation bot 1, Tom.Reding, Xaviertan, Ale And Quail, RA0808, ThePowerofX, Isocliff, Bibcode Bot, Polytope24, Tetra quark and Anonymous: 28

- **Holographic principle** *Source:* https://en.wikipedia.org/wiki/Holographic_principle?oldid=689834128 *Contributors:* Derek Ross, Vicki Rosenzweig, Bryan Derksen, Timo Honkasalo, The Anome, William Avery, Stevertigo, Bcrowell, Mcarling, KAMiKAZOW, Emperor, AugPi, Reddi, Xaven, Phil Boswell, Sverdrup, Tobias Bergemann, Graeme Bartlett, Nat Krause, LeYaYa, Tromer, Jason Quinn, HorsePunchKid, Togo~enwiki, Tdent, Brianhe, Leibniz, Pjacobi, Dbachmann, Bender235, Pink18, Smalljim, I9Q79oL78KiL0QTFHgyc, Scentoni, PWilkinson, Guy Harris, Sligocki, Gpvos, Mindmatrix, StradivariusTV, Mpatel, Joke137, Christopher Thomas, Prothonotar, Rjwilmsi, Koavf, Jfraatz, Utopos, Chobot, Roboto de Ajvol, Wavelength, Bhny, Chris Capoccia, JocK, Bmdavll, Addps4cat, Gzabers, Kermit2, 2over0, Closedmouth, Sardanaphalus, SmackBot, ZerodEgo, Armeria, Mirokado, Bluebot, TimBentley, Nick Levine, Wen D House, Pwjb, LoveEncounterFlow, Byelf2007, Lambiam, JorisvS, Ckatz, PEiP, Hypnosifl, Stephen B Streater, Jynus, UncleDouggie, Lahiru k, Vyznev Xnebara, Foresee, Mhs5392, Peterdjones, Michael C Price, Alexnye, Doug Weller, Johnfn, Qwyrxian, Al Lemos, Headbomb, Davidhorman, Mmortal03, Escarbot, Knotwork, Len Raymond, JAnDbot, Kungfoofairy, Mikhailfranco, Kentucho, Mange01, Maurice Carbonaro, NerdyNSK, Cipral, McSly, Potatoswatter, Borat fan, Sheliak, Satani, Philip Trueman, TXiKiBoT, PhysPhD, AlleborgoBot, Cowlinator, Richard1968, Fcady2007, Mark Germine, M.D., Emfetz, Likebox, Robertcurrey, MrWikiMiki, Cmcelwain, Ln2069, Hamiltondaniel, Kallog, Quinling, ClueBot, Drmies, F-j123, HenrikErlandsson, M4gnum0n, Wndl42, SchreiberBike, DumZiBoT, TimothyRias, Andrw, DOI bot, Simonm223, Crus4d3, LaaknorBot, Tassedethe, Verbal, Legobot, Flash.starwalker, Luckas-bot, Yobot, Systemizer, Pcap, AnomieBOT, Materialscientist, Citation bot, ArthurBot, Coretheapple, RibotBOT, Charvest, Anthropodeus, Giddeon Fox, Prari, FrescoBot, Citation bot 2, Citation bot 1, DrilBot, Tom.Reding, MastiBot, Cjrcl, Lightlowemon, Dr. Salvia, WikitanvirBot, Eekerz, GoingBatty, Netheril96, Wikipelli, Italia2006, ZéroBot, Cogiati, Kzl.zawlin, Isocliff, ClueBot NG, Raidr, Helpful Pixie Bot, Bibcode Bot, Gordonben, Bmusician, CitationCleanerBot, Caypartisbot, BattyBot, Jimw338, Epicgenius, Christophe1946, Polytope24, JaconaFrere, Almaionescu, Monkbot, Stringer63, Hardkhora, Atreus57, Animikhroy967, Toes11111111231111231111123, Nøkkenbuer, Christos Theopoulos, Npmats, The 1editr and Anonymous: 124

- **Simulated reality** *Source:* https://en.wikipedia.org/wiki/Simulated_reality?oldid=688914173 *Contributors:* TwoOneTwo, Eloquence, Bryan Derksen, The Anome, William Avery, SimonP, Apollia, Hephaestos, Twilsonb, Nixdorf, Ixfd64, Gdvorsky, Yann, Skysmith, Mmorabito67, ZoeB, Error, Evercat, Sethmahoney, Brigman, Charles Matthews, The Anomebot, Dtgm, Kevin de Vries~enwiki, Zoicon5, Abscissa, Jake Nelson, Morwen, Saltine, Djungelurban, SEWilco, Ledge, Buridan, Andychrist, Jeffq, Branddobbe, Peyotekun, Paranoid, Jotomicron, Rfc1394, LGagnon, Ungvichian, Miles, Hooloovoo, Johnjosephbachir, Jthiesen, Nat Krause, Philwelch, Tom harrison, Rrgmitchell, Zigger, Monedula, Everyking, Gracefool, Bobblewik, Gadfium, Gdm, LucasVB, Onco p53, Loremaster, Kaldari, Karol Langner, DNewhall, Nils~enwiki, CSTAR, Latitude0116, CesarFelipe, Erik Garrison, Histrion, Robin klein, Beef, Grstain, Kingal86, Silly Dan, AAAAA, Pyrop, Discospinster, Brianhe, Habbit, Vsmith, JimR, Aardark, Bender235, Konstantin~enwiki, Aranel, PhilHibbs, Lyght, TMC1982, Viriditas, Kickstart70, Pearle, Danski14, Chira, Keenan Pepper, Plumbago, Axl, Rebroad, Knowledge Seeker, Alfvaen, LFaraone, Coolsi, GabrielF, Kazvorpal, Podstawko, Bobrayner, JarlaxleArtemis, Mlorrey, Tabletop, Flamingspinach, GregorB, Waldir, Toussaint, Stefanomione, Marudubshinki, Josh Parris, Search4Lancer, Rjwilmsi, George Burgess, Eyu100, Nick R, LjL, Pygy, Platypus222, Flarn2006, Wragge, TiagoTiago, Mathrick, Codex Sinaiticus, Diza, Ahunt, Hibana, Benlisquare, E Pluribus Anthony, YurikBot, ~Viper~, RussBot, Kevs, Ksyrie, A314268, Joncolvin, Dialectric, Darker Dreams, OOZ662, MaxVeers, Cardsplayer4life, WAS 4.250, Emijrp, Johndburger, Closedmouth, E Wing, Levil, Shawnc, Ilmari Karonen, SmackBot, Nahald, Amcbride, Carnaptime, Dreamer.redeemer, Xkoalax, TestPilot, DXBari, Kev585, Jagged 85, Kittynboi, Agentbla, HalfShadow, Xaosflux, Hmains, Des1974, Chris the speller, Mael-Num, CharonM72, The Chef, Vivekk, A Max J, Tamfang, Rfwoolf, Txinviolet, Joeyo, Klimov, Jmlk17, Cybercobra, Astroview120mm, RossF18, Qmwne235, Gizzakk, Deadflagblues, Ckatz, Waggers, Romeu, E-Kartoffel, TPIRFanSteve, Drae, Zepheus, Cat's Tuxedo, Iridescent, Cls14, CapitalR, Poweron, Cheesechimp, Lahiru k, Slippyd, Castorquinn, Jpxt2000, Alexander Iwaschkin, CmdrObot, Dj samson, Vyznev Xnebara, Dgw, Ndru01, Halbared, Gregbard, Bonzaiimonkey, Kiptrev, Peterdjones, Dancter, Capedia, Christian75, Paddles, Inkington, Thijs!bot, Sebasbronzini, Coelacan, Xaurtmj, Jobber, Keraunos, Headbomb, Oubiwann, Bobo159, Second Quantization, Gordon-Ross, SteloKim, Navigatr85, Escarbot, WikiSlasher, Luna Santin, Tim Shuba, Oddity-, Wahwahpedal, Qwerty Binary, Kaini, Golgofrinchian, Dereckson, Skomorokh, Dmackeybog, Pheonixstar, Robina Fox, Martinkunev, Some thing, MegX, Jgb37, Felix116, Coffee2theorems, Andrewthomas10, Mbarbier, Mad Dick Bones, Wwmbes, Lyonscc, Thibbs, James Kell, Red Dawn, Urco, Oicumaybewright, SwedishPsycho, Gwern, Mmoneypenny, Anaxial, Threedots dead, R'n'B, Mange01, 5Q5, Morris729, Nsigniacorp, Jeepday, Touisiau, Tarotcards, SteveChervitzTrutane, Trilobitealive, Nova314, Juliancolton, Mircea85, Russell Freeman, Johntenley, Sodaplayer, Pdcook, EmperorFedor, Davidweiner23, Idioma-bot, GIBBOUS3, Wikieditor06, Philomathoholic, VolkovBot, Omegastar, Irresistance, Vincenttoups, Lordvolton, Macslacker, TonyFleet, AllGloryToTheHypnotoad, Ahkey, Jamelan, Yossarian223, Kilmer-san, Richwil, Enviroboy, Duke56, Sapphic, Nathanleeds, Lyinginbedmon, S8333631, HyuugasPWN, SieBot, Al-gabr, Human05, Moonriddengirl, Paradoctor, USAFAN123, DibbleMint, Hyugens~enwiki, James.Denholm, Oxymoron83, Zharradan.angelfire, Rrogjenks, Cyberdaemon~enwiki, CharlesGillingham, Gamall Wednesday Ida, Johnny Go-Time, Gatortaur, MichaelWattam, Artman772000, EyalBrill, Raneksi, Martarius, ClueBot, Hermine hesse, Gluetube, Zprogrammer, Cirt, Mr.Atoz, Ipnoaddressforya, Rion2032, Resoru, Cenarium, Change93, A plague of rainbows, Tired time, Danisnub, XLinkBot, D.M. from Ukraine, UhOhFeeling, Bookbrad, Addbot, Lewisfred78, DOI bot, Jaypc, OktoberSunset, Gizziusa, Cfrobertson, Woland1234, Tassedethe, Lightbot, SeniorInt, Zhitelew, Yobot, Krichter82, Vrinan, AnomieBOT, Piano non troppo, Citation bot, E2eamon, Teleprinter Sleuth, LilHelpa, I Feel Tired, Magroo444222, P99am, Gap9551, GrouchoBot, Market troll, Altdotme, Omai Gohd~enwiki, Pmoteles, FrescoBot, Paine Ellsworth, Machine Elf 1735, Citation bot 1, Rgeraci, I dream of horses, Kylin.Ma, RedBot, Tlhslobus, Full-date unlinking bot, AustralianMelodrama, Mapfn, Derild4921, DARTH SIDIOUS 2, Eekerz, Ajraddatz, GoingBatty, Scgtrp, TeleComNasSprVen, Ponydepression, Ό οίστρος, David J Johnson, Wingman417, Staszek Lem, Donner60, Eventualentropy, AndyTheGrump, EdoBot, Llightex, Anderthought, Helpful Pixie Bot, Legoless, Remisebastien, BG19bot, Bokachoda-mathamota, Couchjudgement, Norgizfox5041, Nathanielfirst, NeutrinoDo, Igedan, Melissa Bennett, Belnova, Coladar, Rolf h nelson, Jpendergraph, Stamptrader, Phleg1, Corfu455, Cphwb556, Lfpose, Rnjckhpr, Amortias, Shreeharioffice, BrettReaper and Anonymous: 434

- **Black-hole cosmology** *Source:* https://en.wikipedia.org/wiki/Black-hole_cosmology?oldid=651210086 *Contributors:* Rjwilmsi, Vegaswikian,

Lambiam, Peter Gulutzan, GorillaWarfare, Polyamorph, RjwilmsiBot, EmausBot, Alpha Quadrant, Crux007, ClueBot NG, Bibcode Bot, BG19bot, Seshavatharam.bhc, Mohamed-Ahmed-FG, Tetra quark and Anonymous: 19

- **Anthropic principle** Source: https://en.wikipedia.org/wiki/Anthropic_principle?oldid=686949977 Contributors: The Epopt, Derek Ross, Mav, Wesley, Bryan Derksen, The Anome, Malcolm Farmer, RK, XJaM, Roadrunner, SimonP, B4hand, DrRetard, Boud, Stormwriter, DopefishJustin, Menchi, Cyde, TakuyaMurata, GTBacchus, Alfio, Looxix~enwiki, Snoyes, Angela, Timwi, Pablo Mayrgundter, Reddi, Timc, Fairandbalanced, Samsara, AaronSw, Banno, Phil Boswell, Robbot, Fredrik, Goethean, Peak, Gandalf61, Tim Ivorson, Mirv, Tualha, Sverdrup, Academic Challenger, Desmay, Wikibot, Robinh, Johnstone, Xanzzibar, Paul Richter, Gene Ward Smith, Barbara Shack, Tom harrison, Snowdog, Alibaba, Highlander~enwiki, Gracefool, Golbez, Toby Woodwark, Andycjp, Pcarbonn, Karol Langner, Lumidek, Robin klein, Klemen Kocjancic, D6, Rfl, Rich Farmbrough, Rhobite, FT2, Vsmith, Lulu of the Lotus-Eaters, Edgarde, RJHall, Carlon, TheMile, Rbj, I9Q79oL78KiL0QTFHgyc, Timl, Tritium6, KarlHallowell, QuantumEleven, Orangemarlin, Lycanthrope, Nurban, Plumbago, Mc6809e, Swift, Deacon of Pndapetzim, Deathphoenix, DV8 2XL, Mattbrundage, Ringbang, Euphrosyne, Tomato~enwiki, Japanese Searobin, Pseudovector, Siafu, WilliamKF, Dandv, JFG, BlaiseFEgan, Joke137, Btyner, DaveApter, Marudubshinki, Aarghdvaark, Graham87, Drbogdan, Rjwilmsi, Zbxgscqf, Staecker, A ghost, Bubba73, Reinis, Cassowary, Billjefferys, Fragglet, Sderose, Diza, YurikBot, RussBot, Gaius Cornelius, Joncolvin, Ptcamn, Thiseye, JulesH, Number 57, Mattgrommes, Crumley, Georgewilliamherbert, Closedmouth, Teply, Asterion, Nekura, Robertd, SmackBot, Island1, Ashenai, Mitteldorf, 1dragon, InverseHypercube, McGeddon, Huhnra, Edgar181, Portillo, Rmosler2100, Jjalexand, Concerned cynic, Thumperward, Goldfinger820, Can't sleep, clown will eat me, Cybercobra, Infovoria, Localzuk, Richard001, Lpgeffen, Monoape, Luís Felipe Braga, Ligulembot, Vina-iwbot~enwiki, Bejnar, Byelf2007, John, Loodog, Jaganath, Wickethewok, Danburke, Ocatecir, RomanSpa, Hypnosifl, Ryulong, BranStark, Jlrobertson, DedalusJMMR~enwiki, Az1568, Albertod4, Friendly Neighbour, CRGreathouse, CmdrObot, Olaf Davis, JohnCD, Gregbard, Brianroemen, Cydebot, A876, Peterdjones, Joeseither, Fcn, Wexcan, Mbell, CSvBibra, Headbomb, Second Quantization, Peter Gulutzan, Kosmocentric, Mdriver1981, Mentifisto, WikiSlasher, EdgarCarpenter, Bm gub, FForeclosers, Dr. Submillimeter, Res2216firestar, MER-C, Andonic, Bpmullins, Magioladitis, VoABot II, Andrewthomas10, Swpb, Quark7, Caroldermoid, KConWiki, Epstewart, Dirac66, A3nm, David Eppstein, Glen, JaGa, WLU, Info D, Gwern, Tirral, Sm8900, Richard Tierney, AstroHurricane001, Claus L. Rasmussen, Acalamari, SpigotMap, Mikael Häggström, LittleHow, Richard D. LeCour, OriEri, Kenneth M Burke, Jarry1250, Michaelpremsrirat, Fences and windows, Mrbrownn, Rei-bot, Ask123, Charlesdrakew, Arioch7, Telecineguy, Insanity Incarnate, Monty845, PaddyLeahy, GirasoleDE, Paradoctor, METIfan, Crash Underride, Soler97, Lightmouse, Jwjdiamond, Miguel.mateo, OKBot, Iknowyourider, Firefly322, Cheesefondue, Myrvin, Epistemion, Martarius, ClueBot, Justin W Smith, J8079s, SuperHamster, Niceguyedc, ChandlerMapBot, Excirial, Alexbot, Dmyersturnbull, Tnxman307, Hans Adler, XLinkBot, Pgallert, WikHead, Aunt Entropy, Silylene, Addbot, DOI bot, Guoguo12, Discrepancy, TutterMouse, Cst17, ChenzwBot, West.andrew.g, Numbo3-bot, Jarble, Legobot, Yobot, Ht686rg90, AnomieBOT, ^musaz, IRP, Citation bot, ArthurBot, Xqbot, Nickkid5, Gap9551, GrouchoBot, Lukebarnesy, Dukejansen, Omnipaedista, RibotBOT, Shadowjams, WebCiteBOT, Mr Owl1234, Joso98, Nagualdesign, Wikianiki, Schnufflus, Machine Elf 1735, Citation bot 1, Anthony on Stilts, Momergil, Marsiancba, MarcelB612, Jordgette, Chico889, Cyanophycean314, Joehubris, Ammodramus, Ti-30X, RjwilmsiBot, Tesseract2, John of Reading, Bludsucker, Qrsdogg, Bettymnz4, Wikipelli, TeleComNasSprVen, GlacierSupremacy, Solomonfromfinland, Hhhippo, Fæ, StringTheory11, Dondervogel 2, Ppw0, G-13114, ClueBot NG, Wikiphysicsgr, Justlettersandnumbers, Plusorminuszero, Бертран, Jack Ponting, Helpful Pixie Bot, Bibcode Bot, Horn.imh, Island001, Joydeep, Trevayne08, CitationCleanerBot, MrBill3, Rho21111, Queen4thewin, Jgates104, Cyberbot II, Shaarang tenneti, YFdyh-bot, Ersober, Josophie, Shrikarsan, Andrey.a.mitin, Aubreybardo, NormDrez, Trackteur, Cirksena, Velvel2, I'm your Grandma., Tetra quark, AlanSkeptic, Isambard Kingdom, KasparBot and Anonymous: 353

- **Wilkinson Microwave Anisotropy Probe** Source: https://en.wikipedia.org/wiki/Wilkinson_Microwave_Anisotropy_Probe?oldid=686116852 Contributors: Bryan Derksen, Andre Engels, Roadrunner, Zoe, Patrick, Boud, Modster, Looxix~enwiki, Bogdangiusca, Stone, Tempshill, Omegatron, Lumos3, Mark Krueger, Delpino, Harp, Just Another Dan, Bobblewik, Keith Edkins, ConradPino, JoJan, Bbbl67, Infradig, Spiffy sperry, Noisy, Rich Farmbrough, Guanabot, JimR, Foolip, Brian0918, RJHall, Kwamikagami, Susvolans, Mtruch, Viriditas, I9Q79oL78KiL0QTFHgyc, A2Kafir, Matthewcieplak, Gary, JHG, Burn, MBlackstone, Killing Vector, FeanorStar7, Georgia guy, StradivariusTV, Mpatel, Joke137, Emerson7, Rnt20, Drbogdan, Rjwilmsi, Mike Peel, Mordecai, FlaBot, Zotel, Whosasking, YurikBot, Hairy Dude, JabberWok, Van der Hoorn, Gaius Cornelius, Grafen, Psi-kat, DarthVader, Chrisbrl88, Reyk, Geoffrey.landis, Philip Stevens, Sardanaphalus, SmackBot, Ashill, Unyoyega, Nickst, CMD Beaker, Lainagier, Onsly, Jprg1966, Silly rabbit, Nbarth, Colonies Chris, KieferSkunk, WDGraham, Chlewbot, Pissant, Vina-iwbot~enwiki, Andrei Stroe, Ohconfucius, CFLeon, Rijkbenik, Perfectblue97, Beetstra, Hypnosifl, Xiaphias, Igfm2, Spebudmak, Newone, Dave Runger, George100, Kurtan~enwiki, Friendly Neighbour, CRGreathouse, CmdrObot, Laplacian, Olaf Davis, Vyznev Xnebara, N2e, Mlsmith10, Phædrus, AndrewHowse, Cydebot, Dchristle, Arb, Int3gr4te, Meaglin, Thijs!bot, Bobtheowl2, Headbomb, Darklilac, MER-C, CosineKitty, .anacondabot, Magioladitis, RogierBrussee, VoABot II, Swpb, Jatkins, Vssun, Archolman, Alro, R'n'B, CommonsDelinker, 739ajal22, Ohms law, Sheliak, Funandtrvl, Xenonice, Sdsds, TXiKiBoT, Ng.j, BotKung, Vchimpanzee, Djmckee1, AlleborgoBot, SieBot, NetKismet, MrWikiMiki, Lightmouse, OKBot, Agdewijn, Mr. Stradivarius, WikiLaurent, ClueBot, Anonymous799, Starkiller88, Jdgilbey, General Epitaph, Wwheaton, F-j123, Agge1000, Sun Creator, Brews ohare, Coinmanj, Scog, Chaosdruid, Panos84, Jonverve, LieAfterLie, Kerichill, Kbdankbot, Addbot, DOI bot, Toyokuni3, Quantumobserver, Aldebaran66, AnomieBOT, Hunnjazal, Citation bot, ArthurBot, GrouchoBot, Fotaun, Prari, FrescoBot, Originalwana, Argumzio, 117Avenue, Citation bot 1, AstaBOTh15, Gus the mouse, Jonesey95, Tom.Reding, Emqueuekay, Full-date unlinking bot, MathEconMajor, Meier99, Puzl bustr, Dinamik-bot, Lizinvt, RjwilmsiBot, Tesseract2, John of Reading, Docjudith, GoingBatty, 0Zero0, ZéroBot, Pippo skaio, CrimsonBot, H3llBot, Ewa5050, Mathewsian, Hoeksas, Mhvk, ClueBot NG, Mariguld, Frietjes, Mlhalpern, Bibcode Bot, BG19bot, Ninney, Glacialfox, Cosmoedit, BattyBot, Cyberbot II, Khazar2, Dexbot, PHert, CuriousMind01, Wjs64, Frosty, Ihjdekeijzer, Kogge, Krotera, WPGA2345, DavidLLarson, Monkbot, Unatnas1986, Tetra quark and Anonymous: 94

- **Kolmogorov complexity** Source: https://en.wikipedia.org/wiki/Kolmogorov_complexity?oldid=685623539 Contributors: Damian Yerrick, AxelBoldt, The Anome, Gareth Owen, Andre Engels, Arvindn, RTC, Michael Hardy, Dominus, Kku, Bcrowell, Axlrosen, TakuyaMurata, Minesweeper, Snoyes, Tim Retout, Charles Matthews, Dcoetzee, Doradus, Populus, Fredrik, Altenmann, TittoAssini, Tobias Bergemann, Giftlite, Muness, DavidCary, Fastfission, Mcapdevila, Rpyle731, Behnam, Gracefool, Eequor, Bobblewik, Neilc, CSTAR, APH, Sam Hocevar, Shiftchange, D6, Xrchz, NathanHurst, Guanabot, ArnoldReinhold, Lulu of the Lotus-Eaters, Mani1, Dylanr, Euyyn, Army1987, Jfusion, Tromp, Flammifer, Tresoldi, Andrewbadr, Wrs1864, Haham hanuka, Crust, Sligocki, Pion, Mononoke~enwiki, Jheald, Xiaoyanggu, Macterra, OwenX, Pol098, GregorB, Graham87, Rjwilmsi, Michal.burda, Vegaswikian, StefSybo, Mathbot, Maxal, Gramschmidt, Enon, Vonkje, CiaPan, YurikBot, Hairy Dude, Alpt, Rodasmith, Grubber, Gaius Cornelius, Pseudomonas, Kasajian, Trovatore, R.e.s., Twin Bird, Gareth Jones, Mccready, Jessemerriman, Ott2, 2over0, Arthur Rubin, GrEp, SmackBot, RDBury, Selfworm, BiT, Mcld, Bluebot, Nbarth, Kostmo, Jmax-, HLwiKi, UU, Calbaer, Radagast83, B jonas, LoveMonkey, Unomano, Lambiam, Wvbailey, IDSIAupdate, CBM, Gregbard, Cydebot, Stasinos,

Headbomb, Oddity-, Hermel, Erxnmedia, Ichaer, The Transhumanist, Magioladitis, Althai, A3nm, David Eppstein, Robin S, Alexei Kopylov, Leyo, Maurice Carbonaro, Touisiau, Ale2006, Aelkiss, LokiClock, Mgalle, Pleroma, Serprex, Gopher292, XLinkBot, GDibyendu, Multipundit, Se'taan, Ender2101, Legobot, Luckas-bot, Yobot, Electron cloud, PMLawrence, 8ung3st, AnomieBOT, Hiihammuk, Citation bot, Vivohobson, Xqbot, Jeffwang, Nippashish, Sophus Bie, FrescoBot, Louperibot, Citation bot 1, J. Sketter, Kiefer.Wolfowitz, RobinK, Surement, Rjwilmsi-Bot, EmausBot, Crglab, Bdijkstra, Tommy2010, Uploadvirus, Bethnim, ZéroBot, Traxs7, Prvák, OnePt618, RockMagnetist, Helpful Pixie Bot, BG19bot, Gedis, BattyBot, Dexbot, Jochen Burghardt, Etothei, Abrudno, Monkbot, JMP EAX and Anonymous: 166

- **Possible world** *Source:* https://en.wikipedia.org/wiki/Possible_world?oldid=674631799 *Contributors:* Michael Hardy, Owl, Angela, Poor Yorick, Peter Damian (original account), Radgeek, J D, Banno, Cobra libre, Wiglaf, Jason Quinn, Gdm, Piotrus, Cjewell, CSTAR, Kesac, JimWae, Sam Hocevar, Rich Farmbrough, Wclark, El C, Chalst, Swalesy2, Posiduck, Obradovic Goran, SlimVirgin, DV8 2XL, Mel Etitis, Noetica, Rjwilmsi, Pitan, Julescubtree, Saswann, YurikBot, RussBot, Greycat, KSchutte, Tomisti, BorgQueen, Paul Erik, SmackBot, McGeddon, Jagged 85, Mhss, Chris the speller, Dbtfz, Physis, Santa Sangre, Kripkenstein, Gco, Gregbard, Peterdjones, Pjwerner, Doug Weller, JamesAM, Letranova, Thijs!bot, Mbell, Tito-, Frotz, B9 hummingbird hovering, Pomte, RickardV, Heyitspeter, GAdam, Harfarhs, Paulherrick, Ontoraul, Broadbot, Classicalecon, PixelBot, Addbot, Hsansom, Yobot, AnomieBOT, XZeroBot, Omnipaedista, BrideOfKripkenstein, Machine Elf 1735, Alltat, The Mysterious El Willstro, Maroit, RUBEN TESOLIN, ClueBot NG, Helpful Pixie Bot, BG19bot, Wikikrax, Shusane, Sonora Fabian, Wolfgang42, Susumu Maeda, Arash1971, Chrismorey, GreyWinterOwl, PinkBalloonCat, Riyukx and Anonymous: 53

- **Modal realism** *Source:* https://en.wikipedia.org/wiki/Modal_realism?oldid=688846553 *Contributors:* William Avery, Populus, Dodger~enwiki, Karol Langner, Atemperman, Rich Farmbrough, Treborbassett, DV8 2XL, Velho, Mel Etitis, Noetica, Mandarax, BD2412, Tomisti, SmackBot, Dnavarro, Yajinden, DabMachine, Switchercat, Jibal, Gregbard, Peterdjones, Mbell, Marek69, NorwegianBlue, Johnabdl, Swpb, RickardV, Heyitspeter, Sapphic, Bagels13, Mild Bill Hiccup, PixelBot, Pfhorrest, Addbot, JEN9841, AnomieBOT, Rockypedia, BrideOfKripkenstein, Machine Elf 1735, Mamerto puebla, Dream of Nyx, Solomon7968, Modalizer, Luot, VisitorQ80, Dough34, ZeppoShemp and Anonymous: 26

- **Counterpart theory** *Source:* https://en.wikipedia.org/wiki/Counterpart_theory?oldid=675250800 *Contributors:* Zundark, Edward, Topbanana, Jerzy, Barbara Shack, Rich Farmbrough, RJFJR, Joriki, BD2412, Koavf, Lockley, Intgr, Open2universe, SmackBot, Iridescent, CmdrObot, Gregbard, Cydebot, Doug Weller, JamesAM, Wylfing, Skomorokh, RickardV, Squids and Chips, Belastro, Ethidium, Leushenko, SamuelTheGhost, Spfanstiel, Tassedethe, Omnipaedista, BrideOfKripkenstein, Machine Elf 1735, GoingBatty, Dough34, Johnsoniensis and Anonymous: 25

- **The Fabric of Reality** *Source:* https://en.wikipedia.org/wiki/The_Fabric_of_Reality?oldid=679342269 *Contributors:* Stevertigo, Edward, Zumbo, Mschlindwein, Rich Farmbrough, Jamiemichelle, GregorB, BD2412, Bubba73, Rats, Diza, Avalon, SmackBot, TestPilot, Fuhghettaboutit, TenPoundHammer, ArglebargleIV, JzG, Mets501, CmdrObot, Geniustwin, Ankit jn, Peterdjones, Ttiotsw, Drpixie, Headbomb, Nick Number, Charles01, Albany NY, Kraxler, R'n'B, Persephone19, J.delanoy, Tokyogirl79, Wikiisawesome, Lamro, Hrafn, TJRC, VideoRanger2525, Frank Romein~enwiki, Mr. Granger, Martarius, SchreiberBike, XLinkBot, Addbot, Download, Luckas-bot, Yobot, ElkeK, Rudimae, HRoestBot, Grok42, Hessamnia, UnderHigh, John of Reading, Slightsmile, TeleComNasSprVen, TatiG81278, Prussia81278, Jlloyd81278, Kartasto, Llightex, ClueBot NG, Helpful Pixie Bot, BG19bot, BattyBot, Mamlekat and Anonymous: 31

28.5.2 Images

- **File:AdS3.svg** *Source:* https://upload.wikimedia.org/wikipedia/commons/4/47/AdS3.svg *License:* CC BY-SA 3.0 *Contributors:* This file was derived from: AdS3 (new).png
Original artist:

- derivative work: Alex Dunkel (Maky)

- **File:Ambox_current_red.svg** *Source:* https://upload.wikimedia.org/wikipedia/commons/9/98/Ambox_current_red.svg *License:* CC0 *Contributors:* self-made, inspired by Gnome globe current event.svg, using Information icon3.svg and Earth clip art.svg *Original artist:* Vipersnake151, penubag, Tkgd2007 (clock)

- **File:Ambox_important.svg** *Source:* https://upload.wikimedia.org/wikipedia/commons/b/b4/Ambox_important.svg *License:* Public domain *Contributors:* Own work, based off of Image:Ambox scales.svg *Original artist:* Dsmurat (talk · contribs)

- **File:Ambox_question.svg** *Source:* https://upload.wikimedia.org/wikipedia/commons/1/1b/Ambox_question.svg *License:* Public domain *Contributors:* Based on Image:Ambox important.svg *Original artist:* Mysid, Dsmurat, penubag

- **File:Ambox_rewrite.svg** *Source:* https://upload.wikimedia.org/wikipedia/commons/1/1c/Ambox_rewrite.svg *License:* Public domain *Contributors:* self-made in Inkscape *Original artist:* penubag

- **File:Aristarchus_working.jpg** *Source:* https://upload.wikimedia.org/wikipedia/commons/2/2b/Aristarchus_working.jpg *License:* Public domain *Contributors:* ? *Original artist:* ?

- **File:BH_LMC.png** *Source:* https://upload.wikimedia.org/wikipedia/commons/5/5e/BH_LMC.png *License:* CC BY-SA 2.5 *Contributors:* Own work *Original artist:* User:Alain r

- **File:Baby_Universe.jpg** *Source:* https://upload.wikimedia.org/wikipedia/commons/5/56/Baby_Universe.jpg *License:* Public domain *Contributors:* ? *Original artist:* ?

- **File:BigBangNoise.jpg** *Source:* https://upload.wikimedia.org/wikipedia/commons/6/62/BigBangNoise.jpg *License:* Public domain *Contributors:* map.gsfc.nasa.gov/m ig/030644/030644.html *Original artist:* NASA

- **File:BrainGate.jpg** *Source:* https://upload.wikimedia.org/wikipedia/commons/f/fd/BrainGate.jpg *License:* Public domain *Contributors:* Transferred from en.wikipedia to Commons. *Original artist:* PaulWicks at English Wikipedia

- **File:CMB_Timeline300_no_WMAP.jpg** *Source:* https://upload.wikimedia.org/wikipedia/commons/6/6f/CMB_Timeline300_no_WMAP.jpg *License:* Public domain *Contributors:* Original version: NASA; modified by Ryan Kaldari *Original artist:* NASA/WMAP Science Team

- **File:CMB_Timeline75.jpg** *Source:* https://upload.wikimedia.org/wikipedia/commons/6/60/CMB_Timeline75.jpg *License:* Public domain *Contributors:* http://map.gsfc.nasa.gov/media/060915/index.html (direct link) *Original artist:* NASA/WMAP Science Team
- **File:Calabi_yau.jpg** *Source:* https://upload.wikimedia.org/wikipedia/commons/f/f3/Calabi_yau.jpg *License:* Public domain *Contributors:* Mathematica output, created by author *Original artist:* Jbourjai
- **File:Calabi_yau_formatted.svg** *Source:* https://upload.wikimedia.org/wikipedia/commons/8/8c/Calabi_yau_formatted.svg *License:* Public domain *Contributors:* This file was derived from: Calabi yau.jpg
  Original artist:
- derivative work: Polytope24
- **File:Commons-logo.svg** *Source:* https://upload.wikimedia.org/wikipedia/en/4/4a/Commons-logo.svg *License:* ? *Contributors:* ? *Original artist:* ?
- **File:Compactification_example.svg** *Source:* https://upload.wikimedia.org/wikipedia/commons/f/f5/Compactification_example.svg *License:* CC BY-SA 4.0 *Contributors:* Brian Greene (2004). The Elegant Universe (DVD). Part II (String's the thing): WGBH Boston Video. Event occurs at 43:55. OCLC 54019786 *Original artist:* Alex Dunkel (Maky)
- **File:Crab_Nebula.jpg** *Source:* https://upload.wikimedia.org/wikipedia/commons/0/00/Crab_Nebula.jpg *License:* Public domain *Contributors:* HubbleSite: gallery, release. *Original artist:* NASA, ESA, J. Hester and A. Loll (Arizona State University)
- **File:Crystal_Clear_app_kedit.svg** *Source:* https://upload.wikimedia.org/wikipedia/commons/e/e8/Crystal_Clear_app_kedit.svg *License:* LGPL *Contributors:* Sabine MINICONI *Original artist:* Sabine MINICONI
- **File:D3-brane_et_D2-brane.PNG** *Source:* https://upload.wikimedia.org/wikipedia/commons/8/88/D3-brane_et_D2-brane.PNG *License:* Public domain *Contributors:* Image:D-brane.PNG, oeuvre personnelle. *Original artist:* <a href='//commons.wikimedia.org/w/index.php?title=User:Rogilbert,,&,,action=edit,,&,,redlink=1' class='new' title='User:Rogilbert (page does not exist)'>Rogilbert</a>
- **File:Earth'{}s_Location_in_the_Universe_SMALLER_(JPEG).jpg** *Source:* https://upload.wikimedia.org/wikipedia/commons/0/0f/Earth%27s_Location_in_the_Universe_SMALLER_%28JPEG%29.jpg *License:* CC BY-SA 3.0 *Contributors:* Own work *Original artist:* Andrew Z. Colvin
- **File:Earth-moon.jpg** *Source:* https://upload.wikimedia.org/wikipedia/commons/5/5c/Earth-moon.jpg *License:* Public domain *Contributors:* NASA [1] *Original artist:* Apollo 8 crewmember Bill Anders
- **File:Edit-clear.svg** *Source:* https://upload.wikimedia.org/wikipedia/en/f/f2/Edit-clear.svg *License:* Public domain *Contributors:* The *Tango! Desktop Project*. *Original artist:*
  The people from the Tango! project. And according to the meta-data in the file, specifically: "Andreas Nilsson, and Jakub Steiner (although minimally)."
- **File:Edward_Witten.jpg** *Source:* https://upload.wikimedia.org/wikipedia/commons/9/97/Edward_Witten.jpg *License:* Public domain *Contributors:* Own work *Original artist:* Ojan
- **File:End_of_universe.jpg** *Source:* https://upload.wikimedia.org/wikipedia/commons/9/98/End_of_universe.jpg *License:* Public domain *Contributors:* ? *Original artist:* ?
- **File:Flammarion.jpg** *Source:* https://upload.wikimedia.org/wikipedia/commons/8/87/Flammarion.jpg *License:* Public domain *Contributors:* Camille Flammarion, *L'Atmosphère: Météorologie Populaire* (Paris, 1888), *pp.* 163 *Original artist:* Anonymous
- **File:Folder_Hexagonal_Icon.svg** *Source:* https://upload.wikimedia.org/wikipedia/en/4/48/Folder_Hexagonal_Icon.svg *License:* Cc-by-sa-3.0 *Contributors:* ? *Original artist:* ?
- **File:Formation_of_galactic_clusters_and_filaments.jpg** *Source:* https://upload.wikimedia.org/wikipedia/commons/7/7d/Formation_of_galactic_clusters_and_filaments.jpg *License:* CC BY 3.0 us *Contributors:* http://cosmicweb.uchicago.edu/filaments.html *Original artist:* Andrey Kravtsov (the University of Chicago) and Anatoly Klypin (New Mexico State University)
- **File:History_of_the_Universe.svg** *Source:* https://upload.wikimedia.org/wikipedia/commons/d/db/History_of_the_Universe.svg *License:* CC BY-SA 3.0 *Contributors:* Own work *Original artist:* Yinweichen
- **File:Horizonte_inflacionario.svg** *Source:* https://upload.wikimedia.org/wikipedia/commons/b/b4/Horizonte_inflacionario.svg *License:* CC-BY-SA-3.0 *Contributors:* Transferred from en.wikipedia to Commons.; original: *I created this work in Adobe Illustrator. Original artist:* Joke137 at English Wikipedia
- **File:Hugh-Everett.jpg** *Source:* https://upload.wikimedia.org/wikipedia/en/c/cf/Hugh-Everett.jpg *License:* Fair use *Contributors:*
  http://ucispace.lib.uci.edu/handle/10575/1060
  http://sites.uci.edu/ucisca/2011/09/15/hugh-everett-iii-and-quantum-physics/ *Original artist:* ?
- **File:Ilc_9yr_moll4096.png** *Source:* https://upload.wikimedia.org/wikipedia/commons/3/3c/Ilc_9yr_moll4096.png *License:* Public domain *Contributors:* http://map.gsfc.nasa.gov/media/121238/ilc_9yr_moll4096.png *Original artist:* NASA / WMAP Science Team
- **File:Knot_table-blank_unknot.svg** *Source:* https://upload.wikimedia.org/wikipedia/commons/f/f0/Knot_table-blank_unknot.svg *License:* Public domain *Contributors:*
  Knot table.svg
  Original artist:
- derivative work: 84user (removed Unknot label)
- **File:Kolmogorov_complexity_and_computable_lower_bounds.gif** *Source:* https://upload.wikimedia.org/wikipedia/commons/e/e6/Kolmogorov_complexity_and_computable_lower_bounds.gif *License:* CC BY-SA 3.0 *Contributors:* Own work *Original artist:* Jochen Burghardt

28.5. TEXT AND IMAGE SOURCES, CONTRIBUTORS, AND LICENSES

- **File:LeeSmolinAtHarvard.JPG** *Source:* https://upload.wikimedia.org/wikipedia/commons/5/54/LeeSmolinAtHarvard.JPG *License:* CC BY 3.0 *Contributors:* Own work by the original uploader *Original artist:* Lumidek at English Wikipedia
- **File:Limits_of_M-theory.svg** *Source:* https://upload.wikimedia.org/wikipedia/commons/b/b8/Limits_of_M-theory.svg *License:* CC BY-SA 3.0 *Contributors:*
Limits of M-theory.png

Original artist:

- derivative work: Alex Dunkel (Maky)
- **File:Mandelpart2_red.png** *Source:* https://upload.wikimedia.org/wikipedia/commons/d/d4/Mandelpart2_red.png *License:* Public domain *Contributors:*
- Color adjusted from en:Image:Mandelpart2.png to provide a likeness to Image:Mandelpart2.jpg. Created with Fractal Forge 2.8.2; modified by Adobe Photoshop 7.0 *Original artist:* Reguiieee at English Wikipedia
- **File:Many-worlds-bloch.png** *Source:* https://upload.wikimedia.org/wikipedia/en/d/d6/Many-worlds-bloch.png *License:* Cc-by-sa-3.0 *Contributors:* ? *Original artist:* ?
- **File:Many-worlds.svg** *Source:* https://upload.wikimedia.org/wikipedia/commons/8/82/Many-worlds.svg *License:* CC-BY-SA-3.0 *Contributors:* en:Image:Many-worlds.png *Original artist:* Traced by User:Stannered
- **File:Merge-arrow.svg** *Source:* https://upload.wikimedia.org/wikipedia/commons/a/aa/Merge-arrow.svg *License:* Public domain *Contributors:* ? *Original artist:* ?
- **File:Mergefrom.svg** *Source:* https://upload.wikimedia.org/wikipedia/commons/0/0f/Mergefrom.svg *License:* Public domain *Contributors:* ? *Original artist:* ?
- **File:Mexican_hat_potential_polar.svg** *Source:* https://upload.wikimedia.org/wikipedia/commons/7/7b/Mexican_hat_potential_polar.svg *License:* Public domain *Contributors:* Own work by uploader, with gnuplot *Original artist:* RupertMillard
- **File:Microwave_Sky_polarization.png** *Source:* https://upload.wikimedia.org/wikipedia/commons/6/64/Microwave_Sky_polarization.png *License:* Public domain *Contributors:* http://map.gsfc.nasa.gov/media/060917/ *Original artist:* NASA / WMAP Science Team
- **File:Multiverse_-_level_II.svg** *Source:* https://upload.wikimedia.org/wikipedia/commons/3/34/Multiverse_-_level_II.svg *License:* Public domain *Contributors:* Vectorisation of Multiverse - level II.GIF (by K1234567890y), by Lokal_Profil *Original artist:*
- Original by K1234567890y
- **File:NASA-HS201427a-HubbleUltraDeepField2014-20140603.jpg** *Source:* https://upload.wikimedia.org/wikipedia/commons/6/69/NASA-HS201427a-HubbleUltraDeepField2014-20140603.jpg *License:* Public domain *Contributors:* http://hubblesite.org/newscenter/archive/releases/2014/27/image/a/ (image link) *Original artist:* NASA, ESA, H. Teplitz and M. Rafelski (IPAC/Caltech), A. Koekemoer (STScI), R. Windhorst (Arizona State University), and Z. Levay (STScI)
- **File:NASA_logo.svg** *Source:* https://upload.wikimedia.org/wikipedia/commons/e/e5/NASA_logo.svg *License:* Public domain *Contributors:* Converted from Encapsulated PostScript at http://grcpublishing.grc.nasa.gov/IMAGES/Insig-cl.eps *Original artist:* National Aeronautics and Space Administration
- **File:Nearsc.gif** *Source:* https://upload.wikimedia.org/wikipedia/commons/d/d8/Nearsc.gif *License:* CC BY-SA 2.5 *Contributors:* http://www.atlasoftheuniverse.com/nearsc.html *Original artist:* Richard Powell
- **File:Open_and_closed_strings.svg** *Source:* https://upload.wikimedia.org/wikipedia/commons/5/56/Open_and_closed_strings.svg *License:* Public domain *Contributors:* Own work *Original artist:* Xoneca
- **File:PIA16874-CobeWmapPlanckComparison-20130321.jpg** *Source:* https://upload.wikimedia.org/wikipedia/commons/6/64/PIA16874-CobeWmapPlanckComparison-20130321.jpg *License:* Public domain *Contributors:* http://photojournal.jpl.nasa.gov/catalog/PIA16874 (direct link) *Original artist:* NASA/JPL-Caltech/ESA
- **File:Paths-many-worlds.svg** *Source:* https://upload.wikimedia.org/wikipedia/commons/3/3b/Paths-many-worlds.svg *License:* CC-BY-SA-3.0 *Contributors:* en:Image:Paths-many-worlds.png *Original artist:* Traced by User:Stannered
- **File:People_icon.svg** *Source:* https://upload.wikimedia.org/wikipedia/commons/3/37/People_icon.svg *License:* CC0 *Contributors:* OpenClipart *Original artist:* OpenClipart
- **File:Portal-puzzle.svg** *Source:* https://upload.wikimedia.org/wikipedia/en/f/fd/Portal-puzzle.svg *License:* Public domain *Contributors:* ? *Original artist:* ?
- **File:Question_book-new.svg** *Source:* https://upload.wikimedia.org/wikipedia/en/9/99/Question_book-new.svg *License:* Cc-by-sa-3.0 *Contributors:*
Created from scratch in Adobe Illustrator. Based on Image:Question book.png created by User:Equazcion *Original artist:* Tkgd2007
- **File:Schroedingers_cat_film.svg** *Source:* https://upload.wikimedia.org/wikipedia/commons/c/c8/Schroedingers_cat_film.svg *License:* CC0 *Contributors:* Own work *Original artist:* Christian Schirm
- **File:Sobel_North_America.gif** *Source:* https://upload.wikimedia.org/wikipedia/commons/f/f4/Sobel_North_America.gif *License:* CC-BY-SA-3.0 *Contributors:* No machine-readable source provided. Own work assumed (based on copyright claims). *Original artist:* No machine-readable author provided. Roke~commonswiki assumed (based on copyright claims).
- **File:Sound-icon.svg** *Source:* https://upload.wikimedia.org/wikipedia/commons/4/47/Sound-icon.svg *License:* LGPL *Contributors:* Derivative work from Silsor's versio *Original artist:* Crystal SVG icon set

- **File:Spacetime_dimensionality.svg** *Source:* https://upload.wikimedia.org/wikipedia/commons/5/56/Spacetime_dimensionality.svg *License:* CC BY-SA 3.0 *Contributors:* On the dimensionality of spacetime, by Max Tegmark. This version extracted directly from the PostScript file. *Original artist:* Max Tegmark
- **File:Splittings-1.png** *Source:* https://upload.wikimedia.org/wikipedia/commons/2/2b/Splittings-1.png *License:* CC-BY-SA-3.0 *Contributors:* enwiki *Original artist:* en:User:Jecowa
- **File:Spontaneous_symmetry_breaking_(explanatory_diagram).png** *Source:* https://upload.wikimedia.org/wikipedia/commons/a/a5/Spontaneous_symmetry_breaking_%28explanatory_diagram%29.png *License:* CC BY-SA 3.0 *Contributors:* Own work *Original artist:* FT2
- **File:Standard_Model_of_Elementary_Particles.svg** *Source:* https://upload.wikimedia.org/wikipedia/commons/0/00/Standard_Model_of_Elementary_Particles.svg *License:* CC BY 3.0 *Contributors:* Own work by uploader, PBS NOVA [1], Fermilab, Office of Science, United States Department of Energy, Particle Data Group *Original artist:* MissMJ
- **File:StringTheoryDualities.svg** *Source:* https://upload.wikimedia.org/wikipedia/commons/8/8a/StringTheoryDualities.svg *License:* CC BY-SA 3.0 *Contributors:*
- StringTheoryDualities.jpg *Original artist:*
- derivative work: Alex Dunkel (Maky)
- **File:Stylised_Lithium_Atom.svg** *Source:* https://upload.wikimedia.org/wikipedia/commons/e/e1/Stylised_Lithium_Atom.svg *License:* CC-BY-SA-3.0 *Contributors:* based off of Image:Stylised Lithium Atom.png by Halfdan. *Original artist:* SVG by Indolences. Recoloring and ironing out some glitches done by Rainer Klute.
- **File:Symbol_template_class.svg** *Source:* https://upload.wikimedia.org/wikipedia/en/5/5c/Symbol_template_class.svg *License:* Public domain *Contributors:* ? *Original artist:* ?
- **File:ThomasDiggesmap.JPG** *Source:* https://upload.wikimedia.org/wikipedia/commons/3/3e/ThomasDiggesmap.JPG *License:* Public domain *Contributors:* Thomas Digges map: public domain, copied from <a data-x-rel='nofollow' class='external text' href='http://www.astrosociety.org/pubs/mercury/30_05/copernicus.html'>here</a> *Original artist:* w:Thomas Digges (1546?–1595)
- **File:Uniform_tiling_433-t0_(formatted).svg** *Source:* https://upload.wikimedia.org/wikipedia/commons/2/21/Uniform_tiling_433-t0_%28formatted%29.svg *License:* CC BY-SA 4.0 *Contributors:* This file was derived from: Uniform tiling 433-t0.png *Original artist:*
- derivative work: Polytope24
- **File:Universe_content_pie_chart.jpg** *Source:* https://upload.wikimedia.org/wikipedia/commons/3/30/Universe_content_pie_chart.jpg *License:* Public domain *Contributors:* http://map.gsfc.nasa.gov/media/080998/index.html *Original artist:* Credit: NASA / WMAP Science Team
- **File:WMAP.ogv** *Source:* https://upload.wikimedia.org/wikipedia/commons/1/10/WMAP.ogv *License:* Public domain *Contributors:* Goddard Multimedia *Original artist:* NASA/Goddard Space Flight Center
- **File:WMAP_2008.png** *Source:* https://upload.wikimedia.org/wikipedia/commons/2/28/WMAP_2008.png *License:* Public domain *Contributors:* NASA / WMAP Science Team *Original artist:* NASA / WMAP Science Team
- **File:WMAP_2008_23GHz.png** *Source:* https://upload.wikimedia.org/wikipedia/commons/3/33/WMAP_2008_23GHz.png *License:* Public domain *Contributors:* LAMBDA WMAP Images; [1] *Original artist:* NASA / WMAP Science Team
- **File:WMAP_2008_23GHz_foregrounds.png** *Source:* https://upload.wikimedia.org/wikipedia/commons/a/ad/WMAP_2008_23GHz_foregrounds.png *License:* Public domain *Contributors:* LAMBDA WMAP Images; [1] *Original artist:* NASA / WMAP Science Team
- **File:WMAP_2008_33GHz.png** *Source:* https://upload.wikimedia.org/wikipedia/commons/4/48/WMAP_2008_33GHz.png *License:* Public domain *Contributors:* LAMBDA WMAP Images; [1] *Original artist:* NASA / WMAP Science Team
- **File:WMAP_2008_33GHz_foregrounds.png** *Source:* https://upload.wikimedia.org/wikipedia/commons/d/d1/WMAP_2008_33GHz_foregrounds.png *License:* Public domain *Contributors:* LAMBDA WMAP Images; [1] *Original artist:* NASA / WMAP Science Team
- **File:WMAP_2008_41GHz.png** *Source:* https://upload.wikimedia.org/wikipedia/commons/4/4c/WMAP_2008_41GHz.png *License:* Public domain *Contributors:* LAMBDA WMAP Images; [1] *Original artist:* NASA / WMAP Science Team
- **File:WMAP_2008_41GHz_foregrounds.png** *Source:* https://upload.wikimedia.org/wikipedia/commons/f/fc/WMAP_2008_41GHz_foregrounds.png *License:* Public domain *Contributors:* LAMBDA WMAP Images; [1] *Original artist:* NASA / WMAP Science Team
- **File:WMAP_2008_61GHz.png** *Source:* https://upload.wikimedia.org/wikipedia/commons/2/28/WMAP_2008_61GHz.png *License:* Public domain *Contributors:* LAMBDA WMAP Images; [1] *Original artist:* NASA / WMAP Science Team
- **File:WMAP_2008_61GHz_foregrounds.png** *Source:* https://upload.wikimedia.org/wikipedia/commons/f/f3/WMAP_2008_61GHz_foregrounds.png *License:* Public domain *Contributors:* LAMBDA WMAP Images; [1] *Original artist:* NASA / WMAP Science Team
- **File:WMAP_2008_94GHz.png** *Source:* https://upload.wikimedia.org/wikipedia/commons/e/e2/WMAP_2008_94GHz.png *License:* Public domain *Contributors:* LAMBDA WMAP Images; [1] *Original artist:* NASA / WMAP Science Team
- **File:WMAP_2008_94GHz_foregrounds.png** *Source:* https://upload.wikimedia.org/wikipedia/commons/b/b2/WMAP_2008_94GHz_foregrounds.png *License:* Public domain *Contributors:* LAMBDA WMAP Images; [1] *Original artist:* NASA / WMAP Science Team
- **File:WMAP_2008_TT_and_TE_spectra.png** *Source:* https://upload.wikimedia.org/wikipedia/commons/e/ea/WMAP_2008_TT_and_TE_spectra.png *License:* Public domain *Contributors:* LAMBDA WMAP Images; [1] *Original artist:* NASA / WMAP Science Team
- **File:WMAP_2008_universe_content.png** *Source:* https://upload.wikimedia.org/wikipedia/commons/2/21/WMAP_2008_universe_content.png *License:* Public domain *Contributors:* [1] *Original artist:* NASA / WMAP Science Team
- **File:WMAP_2010.png** *Source:* https://upload.wikimedia.org/wikipedia/commons/2/2d/WMAP_2010.png *License:* Public domain *Contributors:* http://wmap.gsfc.nasa.gov/media/101080 *Original artist:* ?

28.5. TEXT AND IMAGE SOURCES, CONTRIBUTORS, AND LICENSES

- **File:WMAP_2010_23GHz.png** *Source:* https://upload.wikimedia.org/wikipedia/commons/1/1d/WMAP_2010_23GHz.png *License:* Public domain *Contributors:* http://lambda.gsfc.nasa.gov/product/map/current/m_images.cfm *Original artist:* NASA / WMAP Science Team
- **File:WMAP_2010_33GHz.png** *Source:* https://upload.wikimedia.org/wikipedia/commons/a/a4/WMAP_2010_33GHz.png *License:* Public domain *Contributors:* http://lambda.gsfc.nasa.gov/product/map/current/m_images.cfm *Original artist:* NASA / WMAP Science Team
- **File:WMAP_2010_41GHz.png** *Source:* https://upload.wikimedia.org/wikipedia/commons/8/8d/WMAP_2010_41GHz.png *License:* Public domain *Contributors:* http://lambda.gsfc.nasa.gov/product/map/current/m_images.cfm *Original artist:* NASA / WMAP Science Team
- **File:WMAP_2010_61GHz.png** *Source:* https://upload.wikimedia.org/wikipedia/commons/6/66/WMAP_2010_61GHz.png *License:* Public domain *Contributors:* http://lambda.gsfc.nasa.gov/product/map/current/m_images.cfm *Original artist:* NASA / WMAP Science Team
- **File:WMAP_2010_94GHz.png** *Source:* https://upload.wikimedia.org/wikipedia/commons/3/33/WMAP_2010_94GHz.png *License:* Public domain *Contributors:* http://lambda.gsfc.nasa.gov/product/map/current/m_images.cfm *Original artist:* NASA / WMAP Science Team
- **File:WMAP_collage.jpg** *Source:* https://upload.wikimedia.org/wikipedia/commons/2/2f/WMAP_collage.jpg *License:* Public domain *Contributors:* [1]; [2] *Original artist:* NASA / WMAP Science Team
- **File:WMAP_launch.jpg** *Source:* https://upload.wikimedia.org/wikipedia/commons/5/52/WMAP_launch.jpg *License:* Public domain *Contributors:* http://map.gsfc.nasa.gov/media/990458/index.html (direct link) *Original artist:* NASA/KSC
- **File:WMAP_orbit.jpg** *Source:* https://upload.wikimedia.org/wikipedia/commons/f/f5/WMAP_orbit.jpg *License:* Public domain *Contributors:* description, image *Original artist:* NASA / WMAP Science Team
- **File:WMAP_receivers.png** *Source:* https://upload.wikimedia.org/wikipedia/commons/d/df/WMAP_receivers.png *License:* Public domain *Contributors:* [1] *Original artist:* NASA / WMAP Science Team
- **File:WMAP_spacecraft.jpg** *Source:* https://upload.wikimedia.org/wikipedia/commons/4/4e/WMAP_spacecraft.jpg *License:* Public domain *Contributors:* [1]; converted from the high-resolution TIFF version to a JPEG *Original artist:* NASA / WMAP Science Team
- **File:WMAP_spacecraft_diagram.jpg** *Source:* https://upload.wikimedia.org/wikipedia/commons/b/be/WMAP_spacecraft_diagram.jpg *License:* Public domain *Contributors:* ? *Original artist:* ?
- **File:WMAP_trajectory_and_orbit.jpg** *Source:* https://upload.wikimedia.org/wikipedia/commons/8/8a/WMAP_trajectory_and_orbit.jpg *License:* Public domain *Contributors:* description, image *Original artist:* NASA / WMAP Science Team
- **File:Wiki_letter_w_cropped.svg** *Source:* https://upload.wikimedia.org/wikipedia/commons/1/1c/Wiki_letter_w_cropped.svg *License:* CC-BY-SA-3.0 *Contributors:*
- Wiki_letter_w.svg *Original artist:* Wiki_letter_w.svg: Jarkko Piiroinen
- **File:Wikibooks-logo-en-noslogan.svg** *Source:* https://upload.wikimedia.org/wikipedia/commons/d/df/Wikibooks-logo-en-noslogan.svg *License:* CC BY-SA 3.0 *Contributors:* Own work *Original artist:* User:Bastique, User:Ramac et al.
- **File:Wikibooks-logo.svg** *Source:* https://upload.wikimedia.org/wikipedia/commons/f/fa/Wikibooks-logo.svg *License:* CC BY-SA 3.0 *Contributors:* Own work *Original artist:* User:Bastique, User:Ramac et al.
- **File:Wikinews-logo.svg** *Source:* https://upload.wikimedia.org/wikipedia/commons/2/24/Wikinews-logo.svg *License:* CC BY-SA 3.0 *Contributors:* This is a cropped version of Image:Wikinews-logo-en.png. *Original artist:* Vectorized by Simon 01:05, 2 August 2006 (UTC) Updated by Time3000 17 April 2007 to use official Wikinews colours and appear correctly on dark backgrounds. Originally uploaded by Simon.
- **File:Wikiquote-logo.svg** *Source:* https://upload.wikimedia.org/wikipedia/commons/f/fa/Wikiquote-logo.svg *License:* Public domain *Contributors:* ? *Original artist:* ?
- **File:Wikisource-logo.svg** *Source:* https://upload.wikimedia.org/wikipedia/commons/4/4c/Wikisource-logo.svg *License:* CC BY-SA 3.0 *Contributors:* Rei-artur *Original artist:* Nicholas Moreau
- **File:Wikiversity-logo-Snorky.svg** *Source:* https://upload.wikimedia.org/wikipedia/commons/1/1b/Wikiversity-logo-en.svg *License:* CC BY-SA 3.0 *Contributors:* Own work *Original artist:* Snorky
- **File:Wikiversity-logo.svg** *Source:* https://upload.wikimedia.org/wikipedia/commons/9/91/Wikiversity-logo.svg *License:* CC BY-SA 3.0 *Contributors:* Snorky (optimized and cleaned up by verdy_p) *Original artist:* Snorky (optimized and cleaned up by verdy_p)
- **File:Wiktionary-logo-en.svg** *Source:* https://upload.wikimedia.org/wikipedia/commons/f/f8/Wiktionary-logo-en.svg *License:* Public domain *Contributors:* Vector version of Image:Wiktionary-logo-en.png. *Original artist:* Vectorized by Fvasconcellos (talk · contribs), based on original logo tossed together by Brion Vibber

28.5.3 Content license

- Creative Commons Attribution-Share Alike 3.0

Made in the USA
San Bernardino, CA
07 January 2017